Applied Probability
Control
Economics
Information and Communication
Modeling and Identification
Numerical Techniques
Optimization

Applications of Mathematics

9

I.A. Ibragimov
Y.A. Rozanov

Gaussian Random Processes

Translated by A.B. Aries

Springer-Verlag
New York Heidelberg Berlin

I.A. Ibragimov
Lomi
Fontanka 25
Leningrad 191011
U.S.S.R.

Y.A. Rozanov
V.A. Steklov Mathematics Institute
Zazilov St. 42
Moscow 3-333
U.S.S.R.

AMS Subject Classifications: 60G10, 60G15, 60G35

Library of Congress Cataloging in Publication Data
Ibragimov, Il'dar Abdulovich.
Gaussian random processes.
(Applications of mathematics ; 9)
Translation of Gaussovskie sluchainye protsessy.
Bibliography: p.
1. Stochastic processes. I. Rozanov, Iurii
Antol'evich, joint author. II. Title.
QA274.4.I2613 519.2 78-6705

The original Russian edition GAUSSOVSKIE SLUCHAINYE PROTSESSY
was published in 1970 by Nauka.

Printed in the United States of America.

9 8 7 6 5 4 3 2 1

ISBN 0-387-90302-x Springer-Verlag New York
ISBN 3-540-90302-x Springer-Verlag Berlin Heidelberg

To Andrei Nickolajevich Kolmogorov

Preface

The book deals mainly with three problems involving Gaussian stationary processes. The first problem consists of clarifying the conditions for mutual absolute continuity (equivalence) of probability distributions of a "random process segment" and of finding effective formulas for densities of the equivalent distributions. Our second problem is to describe the classes of spectral measures corresponding in some sense to regular stationary processes (in particular, satisfying the well-known "strong mixing condition") as well as to describe the subclasses associated with "mixing rate". The third problem involves estimation of an unknown mean value of a random process, this random process being stationary except for its mean, i.e., it is the problem of "distinguishing a signal from stationary noise". Furthermore, we give here auxiliary information (on distributions in Hilbert spaces, properties of sample functions, theorems on functions of a complex variable, etc.).

Since 1958 many mathematicians have studied the problem of equivalence of various infinite-dimensional Gaussian distributions (detailed and systematic presentation of the basic results can be found, for instance, in [23]). In this book we have considered Gaussian stationary processes and arrived, we believe, at rather definite solutions.

The second problem mentioned above is closely related with problems involving ergodic theory of Gaussian dynamic systems as well as prediction theory of stationary processes. From a probabilistic point of view, this problem involves the conditions for weak dependence of the "future" of the process on its "past". The employment of these conditions has resulted in a fruitful theory of limit theorems for weakly dependent variables (see, for instance, [14], [22]); the best known condition of this kind is obviously the so-called condition of "strong mixing". The problems arising in considering regularity conditions reduce in the case of Gaussian processes to a peculiar approximation problem related to linear spectral theory. The book contains the results of investigations of this problem which helped solve it almost completely.

The problem of estimating the mean is perhaps the oldest and most widely known in mathematical statistics. There are two approaches to the solution of this problem: first, the best unbiased estimates can be constructed on the basis of the spectral density of stationary noise; otherwise the least squares method can be applied.

We suggest one common class of "pseudobest" estimates to include best unbiased estimates as well as classical least squares estimates. For these "pseudobest" estimates explicit expressions are given, consistency conditions are found, asymptotic formulas are derived for the error correlation matrix, and conditions for asymptotic effectiveness are defined. It should be mentioned that the results relevant to regularity conditions and the mean estimation are formulated in spectral terms and can automatically be carried over (within the "linear theory") to arbitrary wide-sense stationary processes.

Each chapter has its own numbering of formulas, theorems, etc. For example, formula (4.21) means formula 21 of Section 4 of the same chapter where the reference is made. For the convenience of the reader we provide references to textbooks or reference books. The references are listed at the end of the book.

Contents

CHAPTER I

Preliminaries

I.1 Gaussian Probability Distributions in a Euclidean Space

A probability distribution P in an n-dimensional vector space $\mathbb{R}^n$ is said to be *Gaussian* if the characteristic function

$$\varphi(u) = \int_{\mathbb{R}^n} e^{i(u, x)} \mathsf{P}(dx), \qquad u \in \mathbb{R}^n$$

(here $(u, x) = \sum u_k x_k$ denotes the scalar product of vectors $u = (u_1, \ldots, u_n)$ and $x = (x_1, \ldots, x_n)$) has the form

$$\varphi(u) = \exp\left\{ i(u, a) - \frac{1}{2}(Bu, u) \right\}, \qquad u \in \mathbb{R}^n, \tag{1.1}$$

where $a = (a_1, \ldots, a_n) \in \mathbb{R}^n$ is the *mean* and B is a linear self-adjoint non-negative definite operator called a *correlation operator*; the matrix $\{B_{kj}\}$ defining B is said to be a *correlation matrix*. In this case

$$(u, a) = \int_{\mathbb{R}^n} (u, x) \mathsf{P}(dx),$$

$$(Bu, v) = \int_{\mathbb{R}^n} [(u, x) - (u, a)][(v, x) - (v, a)] \mathsf{P}(dx), \tag{1.2}$$

$$u, v \in \mathbb{R}^n.$$

The distribution P with mean value a and correlation operator B is concentrated in an m-dimensional hyperplane L of $\mathbb{R}^n$ (m being the rank of the correlation matrix), which can be expressed as

$$L = a + B\mathbb{R}^n$$

(L being the totality of all vectors $y \in \mathbb{R}^n$ of the form $y = a + Bx$, $x \in \mathbb{R}^n$).

In fact,

$$\mathsf{P}(\mathbb{R}^n \backslash L) = 0,$$

the distribution P being absolutely continuous with respect to Lebesgue measure dy in the hyperplane L, so that

$$\mathsf{P}(\Gamma) = \int_{\Gamma \cap L} p(y)\,dy, \tag{1.3}$$

where the distribution density $p(y)$, $y \in L$, has the form

$$p(y) = \frac{1}{(2\pi)^{m/2} \det B} \exp\left\{ -\frac{1}{2}(B^{-1}(y-a), (y-a)) \right\}. \tag{1.4}$$

Here $\det B$ denotes the determinant of the matrix that prescribes the operator B in the subspace $\mathbb{R}^m = B\mathbb{R}^n$, and B^{-1} is the inverse on this subspace.

I.2 Gaussian Random Functions with Prescribed Probability Measure

Let $(\Omega, \mathfrak{A}, \mathsf{P})$ be a probability space, i.e., a measurable space of elements $\omega \in \Omega$ with probability measure P on a σ-algebra $\mathfrak{A}$ of sets $A \subseteq \Omega$.

Any real measurable function $\xi = \xi(\omega)$ on a space Ω is said to be a random variable. The totality of random variables $\xi(t) = \xi(\omega, t)$ (parameter t runs through a set T) is said to be a random function of parameter $t \in T$. The random variables $\xi(t)$ themselves are said to be values of this random function $\xi = \xi(t)$; for fixed $\omega \in \Omega$ the real function $\xi(\omega, \cdot) = \xi(\omega, t)$ of $t \in T$ is said to be a *sample function* or a *trajectory* of the random function $\xi = \xi(t)$.

We shall consider another space X of real functions $x = x(t)$ of $t \in T$, which includes all trajectories $\xi = \xi(\omega, t)$, $t \in T$, of the random function $\xi = \xi(t)$. (For instance, the space $X = \mathbb{R}^T$ of all real functions $x = x(t)$, $t \in T$, possesses this property.) Denote by $\mathfrak{B}$ the minimal σ-algebra of sets of X containing all cylinder sets of this space, i.e., sets of the form

$$[x(t_1), \ldots, x(t_n)] \in \Gamma \tag{2.1}$$

(the set indicated by (2.1) consists of the functions $x = x(t)$ for which the values $[x(t_1), \ldots, x(t_n)]$ at the points $t_1, \ldots, t_n \in T$ prescribe a vector that belongs to a Borel set Γ in an n-dimensional vector space $\mathbb{R}^n$). The mapping $\xi = \xi(\omega)$ under which each $\omega \in \Omega$ corresponds to a pertinent sample function $\xi(\omega, \cdot) = \xi(\omega, t)$ of $t \in T$—an element of the space X—is a measurable mapping in a probability space $(\Omega, \mathfrak{A}, \mathsf{P})$ onto a measurable space $(X, \mathfrak{B})$. The sets $A \in \mathfrak{A}$ of the form $A = \{\xi \in B\}$—the preimages of sets $B \in \mathfrak{B}$ under the mapping $\xi = \xi(\omega, \cdot)$ indicated—form (in the aggregate) a σ-algebra. This σ-algebra $\mathfrak{A}_\xi$ is minimal among σ-algebras of the sets containing all sets of the form

$$[\xi(t_1), \ldots, \xi(t_n)] \in \Gamma \tag{2.2}$$

(the set indicated consists of the elements $\omega \in \Omega$ for which the values $[\xi(\omega, t_1), \ldots, \xi(\omega, t_n)]$ prescribe a vector belonging to a Borel set Γ of an n-dimensional vector space $\mathbb{R}^n$), or, in other words, the σ-algebra $\mathfrak{A}_\xi$ is to be generated by values $\xi(t)$, $t \in T$. Probability measure P_ξ defined on a σ-algebra $\mathfrak{B}$ by the relation

$$\mathsf{P}_\xi(B) = \mathsf{P}\{\xi \in B\}, \qquad B \in \mathfrak{B}, \tag{2.3}$$

is said to be a *probability distribution of the random function* $\xi = \xi(t)$ (on the pertinent function space X).

We shall discuss next the question: When is the family of real variables $\xi(t) = \xi(\omega, t)$ given on a space Ω (parameter t runs through a set T) a random function with the given probability distribution P_ξ? More precisely, when does there exist probability measure P in the space Ω related with the given distribution P_ξ by means of (2.3)? We assume in this case that the set $\xi(\Omega)$ of all sample functions $\xi(\omega, \cdot) = \xi(\omega, t)$ of $t \in T$ belongs to the space X.

It is readily seen that *such a probability measure* P *exists if and only if the set* $\xi(\Omega)$ *has a complete exterior measure*, i.e.,

$$\mathsf{P}_\xi(B) = 1 \quad \text{for } B \supseteq \xi(\Omega) \tag{2.4}$$

for any measurable set $B \in X$.

In fact, if P_ξ is the probability distribution of the random function $\xi = \xi(t)$, for any set $\bar{B} \in X$ in the complement of the set $\xi(\Omega)$ the set $\{\xi \in \bar{B}\}$ is empty and

$$\mathsf{P}_\xi(\bar{B}) = \mathsf{P}\{\xi \in \bar{B}\} = 0.$$

On the other hand, for any sets $B_1, B_2 \in \mathfrak{B}$, such that $\{\xi \in B_1\} = \{\xi \in B_2\}$, the symmetric difference $B_1 \circ B_2 = (B_1 \backslash B_2) \cup (B_2 \backslash B_1)$ is contained in the complement of the set $\xi(\Omega)$, and, under the condition (2.4), $\mathsf{P}_\xi(B_1 \circ B_2) = 0$, $\mathsf{P}_\xi(B_1) = \mathsf{P}_\xi(B_2)$. Hence the relation

$$\mathsf{P}\{\xi \in B\} = \mathsf{P}_\xi(B), \qquad B \in \mathfrak{B}, \tag{2.5}$$

defines the single-valued function $\mathsf{P} = \mathsf{P}(A)$ on the σ-algebra $\mathfrak{A}_\xi$ of all sets of the form $A = \{\xi \in B\}$, $B \in \mathfrak{B}$, generated by $\xi(t)$, $t \in T$. Obviously, P is a probability measure and $\xi = \xi(t)$ is a random function on a probability space $(\Omega, \mathfrak{A}, \mathsf{P})$ with the given probability distribution P_ξ.

The measure P on the σ-algebra $\mathfrak{A}_\xi$ generated by the variables $\xi(t)$, $t \in T$, can be defined uniquely by finite-dimensional distributions $\mathsf{P}_{t_1, \ldots, t_n}$ of which each is a Borel measure on $\mathbb{R}^n$ defined by

$$\mathsf{P}_{t_1, \ldots, t_n}(\Gamma) = \mathsf{P}\{[\xi(t_1), \ldots, \xi(t_n)] \in \Gamma\}, \tag{2.6}$$

$\mathsf{P}_{t_1, \ldots, t_n}$ being the probability distribution of the random vector $[\xi(t_1), \ldots, \xi(t_n)]$. In fact,

$$\mathsf{P}(A) = \inf \sum_k \mathsf{P}(A_k), \tag{2.7}$$

where the lower bound is taken over all sets A_k of the form (2.2), whose union covers the set $A \in \mathfrak{A}$. In particular, this fact refers to the probability distribution P_ξ on the corresponding function space X—it is the probability measure on the σ-algebra $\mathfrak{B}$ generated by the given variables $\xi(t) = \xi(x, t)$ on the space X, i.e., the variables of the form

$$\xi(t, x) = x(t), \qquad x \in X \tag{2.8}$$

(where the parameter t, fixed for each functional $\xi(x, t) = x(t)$ of $x \in X$, runs through the set T).

Denote by $\Gamma \times \mathbb{R}^{n-m}$ the Borel set in an n-dimensional space of vectors $[x(t_1), \ldots, x(t_n)]$ such that $[x(t_{i_1}), \ldots, x(t_{i_m})] \in \Gamma$ (Γ is a Borel set in an m-dimensional subspace $\mathbb{R}^m \subseteq \mathbb{R}^n$) with the remaining coordinates $x(t_i)$ arbitrary. Finite-dimensional distributions are "compatible" in the sense that

$$\mathsf{P}_{t_1, \ldots, t_n}(\Gamma \times \mathbb{R}^{n-m}) = \mathsf{P}_{t_{i_1}, \ldots, t_{i_m}}(\Gamma) \tag{2.9}$$

for all sets of the above type.

Let $X = \mathbb{R}^T$ be the space of all real functions $x = x(t)$, $t \in T$. According to a well-known theorem due to Kolmogorov,* any given family of distributions $\mathsf{P}_{t_1, \ldots, t_n}$ prescribes a continuous additive function P (defined by (2.6), where the variables have the explicit form (2.8)), on the algebra of all cylinder sets (2.1). This function extends uniquely to a probability measure P on the σ-algebra $\mathfrak{B}$. The random function $\xi = \xi(t)$ with values $\xi(t) = \xi(x, t)$ in the probability space $(X, \mathfrak{B}, \mathsf{P})$ has finite-dimensional distributions coinciding with the initial compatible distributions $\mathsf{P}_{t_1, \ldots, t_n}$.

Starting from probability distribution $\mathsf{P} = \mathsf{P}_\xi$ on the function space X, under the condition (2.4) we can define (see (2.5)) a probability measure on the corresponding space Ω.

The random functions $\xi = \xi(t)$ and $\tilde{\xi} = \tilde{\xi}(t)$ with values in the same space are said to be *equivalent* if with probability one (for almost all $\omega \in \Omega$)

$$\xi(\omega, t) = \tilde{\xi}(\omega, t)$$

for each fixed $t \in T$. Obviously, the finite-dimensional distributions of equivalent random functions coincide. Taking an equivalent random function $\xi = \xi(t)$ with the trajectories in any function space X, we can define (see (2.3)) a probability measure in this space as well.

Random variables are said to be *Gaussian* if their finite-dimensional distributions are Gaussian. More precisely, (when we deal with a random function $\xi = \xi(t)$ with parameter $t \in T$ under some parametrization), the values $\xi(t) = \xi(\omega, t)$ and the function $\xi = \xi(t)$ itself are said to be *Gaussian* if all finite-dimensional distributions $\mathsf{P}_{t_1, \ldots, t_n}$ are Gaussian. Probability measure P on a σ-algebra $\mathfrak{A}_\xi$ generated by all $\xi(t)$ is also said to be *Gaussian.*

Each of the finite-dimensional distributions $\mathsf{P}_{t_1, \ldots, t_n}$ of the Gaussian random function $\xi = \xi(t)$ has mean value $[a(t_1), \ldots, a(t_n)]$ and correlation

* See [10], p. 150.

matrix $\{B(t_k, t_j)\}$ where $a(t)$, $t \in T$, is *the mean value of* the function $\xi = \xi(t)$, and $B(s, t)$, $s, t \in T$, is *its correlation function*:*

$$a(t) = \mathsf{M}\xi(t),$$
$$B(s, t) = \mathsf{M}[\xi(s) - a(s)][\xi(t) - a(t)], \qquad s, t \in T. \tag{2.10}$$

Therefore, the Gaussian measure P on a σ-algebra $\mathfrak{A}_\xi$ can be defined uniquely by means of its mean value $a(t)$, $t \in T$, and its correlation function $B(s, t)$, $s, t \in T$.

The mean value $a(t)$, $t \in T$, can be arbitrary, and the correlation function $B(s, t)$, $s, t \in T$, need only satisfy the positive definiteness condition

$$\sum_{k,j=1}^{n} c_k c_j B(t_k, t_j) \geqslant 0 \tag{2.11}$$

for any $t_1, \ldots, t_n \in T$ and real $c_1, \ldots, c_n$.

For any function $a(t)$, $t \in T$, and a positive definite correlation function $B(s, t)$, $s, t \in T$, there exists a Gaussian random function with the mean $a(t)$, $t \in T$, and a correlation function $B(s, t)$, $s, t \in T$. Actually, Gaussian distributions $\mathsf{P}_{t_1, \ldots, t_n}$ with the mean $[a(t_1), \ldots, a(t_n)]$ and correlation matrices $\{B(t_k, t_j)\}$ are compatible distributions, and define a Gaussian measure P in the space $X = \mathbb{R}^T$ of all real functions $x = x(t)$ of $t \in T$, on the σ-algebra $\mathfrak{B} = \mathfrak{A}_\xi$, which can be generated by the given values $\xi(t) = \xi(x, t)$ on X of the form (2.8) (parameter t runs through the set T).

I.3 Lemmas on the Convergence of Gaussian Variables

Let $\xi_n = \xi_n(\omega)$, $n = 1, 2, \ldots$, be a sequence of random variables on a probability space $(\Omega, \mathfrak{A}, \mathsf{P})$. The sequence ξ_n, $n = 1, 2, \ldots$, is said to be *convergent in probability on a set* $A \in \mathfrak{A}$ *to some variable* $\xi = \xi(\omega)$ if for any $\varepsilon > 0$

$$\lim_{n\to\infty} \mathsf{P}(\{|\xi_n - \xi| > \varepsilon\} \cap A) = 0. \tag{3.1}$$

Let us recall that a sequence ξ_n, $n = 1, 2, \ldots$, converges in probability if and only if this sequence is Cauchy,† i.e., on the same set A the sequence $\Delta_{nm} = \xi_n - \xi_m$, $n, m = 1, 2, \ldots$, converges to zero in probability.

Lemma 1. *If a sequence of Gaussian variables* ξ_n, $n = 1, 2, \ldots$, *converges in probability on a set* $A \in \mathfrak{A}$ *of positive measure* ($\mathsf{P}(A) > 0$), *it is convergent in the mean*:

$$\lim_{n\to\infty} \mathsf{M}[\xi_n - \xi]^2 = 0. \tag{3.2}$$

* $\mathsf{M}\xi$ denotes the expectation of a random variable $\xi = \xi(\omega)$ on a probability space $(\Omega, \mathfrak{A}, P)$: $\mathsf{M}\xi = \int_\Omega \xi(\omega)\mathsf{P}(d\omega)$.

† See, for example, [10], p. 90.

Proof. We shall consider Gaussian variables $\Delta_{nm} = \xi_n - \xi_m$. For any $\varepsilon > 0$

$$\mathsf{P}\{|\Delta_{nm}| > \varepsilon\} = 2\int_\varepsilon^\infty \frac{1}{\sqrt{2\pi}\sigma_{nm}} \exp\left\{-\frac{(x - a_{nm})^2}{2\sigma_{nm}^2}\right\} dx,$$

where $a_{nm} = \mathsf{M}\Delta_{nm}$, $\sigma_{nm}^2 = \mathsf{P}(\Delta_{nm} - a_{nm})^2$. Suppose that the sequence ξ_n, $n = 1, 2, \ldots$, is not convergent in the mean; this is equivalent to

$$\overline{\lim_{n, m \to \infty}} (a_{nm}^2 + \sigma_{nm}^2) > 0.$$

It can be easily seen that under this condition for a positive ε we have

$$\overline{\lim_{n, m \to \infty}} \mathsf{P}\{|\Delta_{nm}| > \varepsilon\} \geqslant 1 - p/2,$$

where $p = \mathsf{P}(A) > 0$. But then

$$\overline{\lim_{n, m \to \infty}} \mathsf{P}(\{|\Delta_{nm}| > \varepsilon\} \cap A) \geqslant p/2,$$

which fact contradicts (3.1). Hence

$$\lim_{n, m \to \infty} \mathsf{M}\Delta_{nm}^2 = \lim_{n, m \to \infty} (a_{nm}^2 + \sigma_{nm}^2) = 0,$$

i.e., the sequence ξ_n, $n = 1, 2, \ldots$, is Cauchy (a fundamental) in the mean and, therefore, is convergent in the mean.

In particular, if a sequence of Gaussian variables ξ_n, $n = 1, 2, \ldots$, is convergent with positive probability (i.e., convergent for all ω from a set $A \in \mathfrak{A}$ of positive measure), it is convergent in the mean. □

Let us consider a sequence of independent Gaussian variables ξ_n, $n = 1, 2, \ldots$.

Lemma 2. *The series* $\sum_{n=1}^\infty \xi_n^2$ *is convergent with positive probability if and only if the series* $\sum_{n=1}^\infty \mathsf{M}\xi_n^2$ *is convergent.*

Proof. Obviously,

$$\sum_{n=1}^\infty \mathsf{M}\xi_n^2 = \mathsf{M} \sum_{n=1}^\infty \xi_n^2,$$

and hence the convergence of the series $\sum_{n=1}^\infty \mathsf{M}\xi_n^2$ implies that the variable $\xi^2 = \sum_{n=1}^\infty \xi_n^2(\omega)$ is finite for almost all $\omega \in \Omega$, i.e., the series $\sum_{n=1}^\infty \xi_n^2$ is convergent with probability one. Let the series $\sum_{n=1}^\infty \xi_n^2$ be convergent with positive probability (by the well-known zero–one law* this series is convergent with probability one as well). Then the sequence ξ_n, $n = 1, 2, \ldots$, converges to 0 in the mean: $\mathsf{M}\xi_n^2 \to 0$ for $n \to \infty$ (see Lemma 1). Let $a_n =$

* See, for example, [10], p. 157.

$M\xi_n$, $\sigma_n^2 = M(\xi_n - a_n)^2$. Then

$$a_n^2 + \sigma_n^2 = M(\xi_n')^2 + \int_{|x|>1} x^2 \frac{1}{\sqrt{2\pi}\sigma_n} \exp\left\{-\frac{(x-a_n)^2}{2\sigma_n^2}\right\} dx,$$

where the random variables $\xi_n' = \xi_n'(\omega)$ are defined as

$$\xi_n'(\omega) = \begin{cases} \xi_n(\omega) & \text{for } |\xi_n| \leqslant 1 \\ 0 & \text{for } |\xi_n| > 1. \end{cases}$$

For $a_n^2 + \sigma_n^2 \to 0$ we have

$$\int_{|x|>1} x^2 \frac{1}{\sqrt{2\pi}\sigma_n} \exp\left\{-\frac{(x-a_n)^2}{2\sigma_n^2}\right\} dx = o(a_n^2 + \sigma_n^2),$$

so that

$$M(\xi_n')^2 \sim a_n^2 + \sigma_n^2.$$

By the well-known three-series theorem* a necessary condition for the series $\sum_{n=1}^{\infty} \xi_n^2$ of independent variables ξ_n^2, $= 1, 2, \ldots$, to be convergent is that $\sum_{n=1}^{\infty} M(\xi_n')^2 < \infty$. But $M(\xi_n')^2 \sim a_n^2 + \sigma_n^2$, and, consequently, it follows from the convergence of the series $\sum_{n=1}^{\infty} \xi_n^2$ that $\sum_{n=1}^{\infty} (a_n^2 + \sigma_n^2) < \infty$. □

I.4 Gaussian Variables in a Hilbert Space

A random variable ξ in a Euclidean n-dimensional space $\mathbb{R}^n$ is said to be *Gaussian* if its probability distribution is Gaussian.

The random variable $\xi \in \mathbb{R}^n$ is Gaussian if and only if the real variable $\xi(u) = (u, \xi)$ (equal to the scalar product of the elements $u, \xi \in \mathbb{R}^n$) is Gaussian for each $u \in \mathbb{R}^n$.

In fact, the value at a point $u \in \mathbb{R}^n$ of the characteristic function $\varphi(u)$ of the random variable $\xi \in \mathbb{R}^n$ coincides with the value of the characteristic function of the real random variable $\xi(u) = (u, \xi)$ at the point 1 and has the form

$$\varphi(u) = Me^{i(u, \xi)} = \exp\left\{i(u, a) - \frac{1}{2}(Bu, u)\right\}, \qquad u \in \mathbb{R}^n$$

(see (1.1), where (u, a) is the mean and (Bu, u) is the variance of the Gaussian variable $\xi(u) = (u, \xi)$).

It is clear that the random variable $\xi \in \mathbb{R}^n$ is Gaussian if and only if the random function of the form $\xi(u) = (u, \xi)$ of $u \in \mathbb{R}^n$ is Gaussian.

Let U be a complete separable Hilbert space and let $\xi = \xi(\omega)$ be a function on a probability space $(\Omega, \mathfrak{A}, P)$ with the values in U. The random element ξ of a Hilbert space U is said to be a random variable in U if the scalar product (u, ξ) for each $u \in U$ is a real random variable, i.e., it is a measurable function on the probability space $(\Omega, \mathfrak{A}, P)$.

* See, for example, [10], p. 166.

The random variable ξ in a Hilbert space U is said to be *Gaussian* if the real random variable $\xi(u) = (u, \xi)$ is Gaussian for each $u \in U$. This fact is equivalent to the fact that the random function $\xi(u) = (u, \xi)$ of $u \in U$ is Gaussian since values $\xi(u) = (u, \xi)$ as well as any vector values $[\xi(u_1), \ldots, \xi(u_n)]$ are Gaussian.

In fact, for any vector $\lambda = [\lambda_1, \ldots, \lambda_n]$ in $\mathbb{R}^n$ the scalar product $\sum_{k=1}^n \lambda_k \xi(u_k)$ is equal to

$$\sum_{k=1}^{n} \lambda_k \xi(u_k) = \left(\sum_{k=1}^{n} \lambda_k u_k, \xi \right) = (u, \xi),$$

where $u = \sum_{k=1}^n \lambda_k u_k \in U$; by hypothesis, the variable $\xi(u) = (u, \xi)$ is Gaussian.

Obviously, the mean

$$a(u) = \mathsf{M}(u, \xi), \qquad u \in U,$$

of the random function $\xi(u) = (u, \xi)$, $u \in U$, is a linear functional, and the correlation function

$$B(u, v) = \mathsf{M}[(u, \xi) - a(u)][(v, \xi) - a(v)], \qquad u, v \in U,$$

is a bilinear positive functional on the Hilbert space U. In this case, since the scalar product (u, ξ) is a continuous function of $u \in U$ for each fixed $\omega \in \Omega$, a Gaussian function $\xi(u) = (u, \xi)$ of $u \in U$ must be continuous in the mean (see Lemma 1):

$$\lim_{\|u - v\| \to 0} \mathsf{M}[(u, \xi) - (v, \xi)]^2 = 0 \tag{4.1}$$

($\|u\|$ denotes the norm of the element $u \in U$). But

$$\mathsf{M}[(u, \xi) - (v, \xi)]^2 = a(u - v)^2 + B(u - v, u - v)$$

and (4.1) implies that the functionals $a(u)$ and $B(u, v)$ are continuous.

Being a linear continuous functional, the mean $a(u)$ can be expressed as

$$a(u) = (u, a), \qquad u \in U, \tag{4.2}$$

for some element a in U. Any element $a \in U$ having the property that

$$(u, a) = \int_\Omega (u, \xi(\omega))\mathsf{P}(d\omega) \tag{4.3}$$

for all $u \in U$ is said to be *the mean** of a random variable $\xi \in U$. Being a continuous positive bilinear functional, the correlation function $B(u, v)$ can be expressed

$$B(u, v) = (Bu, v), \qquad u, v \in U, \tag{4.4}$$

where B is a linear positive (i.e., nonnegative self-adjoint) operator in a Hilbert space U called a *correlation operator*.

Let us show that the correlation operator B is completely continuous.

* For the integrability of functions with values in a Hilbert space, see, for example, [12], p. 59.

In fact, any orthonormalized sequence $v_1, v_2, \ldots$ goes to zero weakly, so that the Gaussian variables $\xi_n = (v_n, \xi)$, $n = 1, 2, \ldots$, where $\xi = \xi(\omega) \in U$, goes to zero as $n \to \infty$ for all $\omega \in \Omega$. Therefore (see Lemma 1), they are convergent in the mean, i.e.,

$$\mathsf{M}\xi_n^2 = (Bv_n, v_n) \to 0$$

(here and further on we assume for simplicity of notation that the mean $a \in U$ is 0). If the operator B was not assumed completely continuous, outside some ε-neighborhood of zero there would be an infinite number of spectral points (taking into account the multiplicity), and, therefore, an infinite number of invariant orthogonal subspaces for each element of which

$$(Bu, u) = \int_{|\lambda| > \varepsilon} \lambda d(E_\lambda u, u) \geqslant \varepsilon \|u\|^2,$$

where $B = \int \lambda \, dE_\lambda$ is the spectral representation of the continuous self-adjoint operator B.

Further, we shall choose a complete orthonormalized basis of eigen-elements $v_1, v_2, \ldots$ of this completely continuous symmetric positive operator B corresponding to eigenvalues $\sigma_1^2, \sigma_2^2, \ldots$. The corresponding variables $\xi_k = (v_k, \xi)$, $k = 1, 2, \ldots$, are uncorrelated:

$$\mathsf{M}\xi_k\xi_j = (Bv_k, v_j) = \begin{cases} \sigma_k^2 & \text{for } j = k, \\ 0 & \text{for } j \neq k. \end{cases}$$

In this case

$$\sum_1^\infty \xi_k^2(\omega) = \sum_1^\infty (v_k, \xi(\omega))^2 = \|\xi(\omega)\|^2.$$

As is well known, uncorrelated Gaussian variables are independent and the convergence of the series $\sum_1^\infty \xi_k^2(\omega)$ (for all ω) implies the convergence of the series $\sum_1^\infty \mathsf{M}\xi_k^2$ (see Lemma 2). Consequently,

$$\sum_1^\infty (Bv_k, v_k) = \sum_1^\infty \mathsf{M}\xi_k^2 = \sum_1^\infty \sigma_k^2 < \infty,$$

i.e., the correlation operator B is a nuclear operator:* for any orthonormal system $u_1, u_2, \ldots, \in U$,

$$\sum_1^\infty (Bu_k, u_k) < \infty. \tag{4.5}$$

Therefore, if we have a Gaussian random variable $\xi \in U$, the random function $\xi(u) = (u, \xi)$ of parameter $u \in U$ has a mean of the form (4.2) and a correlation function of the form (4.4) where the correlation operator B is a nuclear operator on the Hilbert space U.

Next, let $\xi(u)$, $u \in U$, be an arbitrary Gaussian random function with a mean of the form (4.2) and a correlation function of the form (4.4), where B

* See, for example, [7], p. 55.

is a nuclear operator on the Hilbert space U. Then there exists an equivalent random function $\xi(u)$, $u \in U$, and a Gaussian random variable $\xi = \xi(\omega)$ in U such that

$$\xi(u) = (u, \xi), \qquad u \in U. \tag{4.6}$$

The variable $\xi \in U$ indicated can be defined for almost all elementary outcomes ω by the formula

$$\xi(\omega) = \sum_{k=1}^{\infty} \xi(v_k)v_k, \tag{4.7}$$

where $v_1, v_2, \ldots$ is the complete orthonormal system of eigenelements of the nuclear operator B, and, by virtue of the relation

$$\mathsf{M} \sum_{k=1}^{\infty} \xi(v_k)^2 = \sum_{k=1}^{\infty} B(v_k, v_k) < \infty$$

for independent Gaussian variables $\xi(v_1), \xi(v_2), \ldots$, the series $\sum_{k=1}^{\infty} \xi(v_k)^2$ is convergent with probability one. In fact, $\xi(u)$, $u \in U$, is a random linear functional in the sense that with probability one

$$\xi(\lambda_1 u_1 + \lambda_2 u_2) = \lambda_1 \xi(u_1) + \lambda_2 \xi(u_2)$$

for any real λ_1, λ_2 and any elements $u_1, u_2 \in U$ since, as we can easily verify,

$$\mathsf{M}[\xi(\lambda_1 u_1 + \lambda_2 u_2) - \lambda_1 \xi(u_1) - \lambda_2 \xi(u_2)]^2 = 0.$$

Furthermore, the random functional $\xi(u)$ is continuous in the mean (see (4.1) and below), and, since

$$u = \lim_{n \to \infty} \sum_{k=1}^{n} (u, v_k)v_k,$$

we have

$$\xi(u) = \lim_{n \to \infty} \xi\left(\sum_{k=1}^{n} (u, v)v_k\right) = \lim_{n \to \infty} \sum_{k=1}^{n} (u, v_k)\xi(v_k)$$

(in the sense of convergence in the mean); at the same time with probability one

$$(u, \xi) = \lim_{n \to \infty} \sum_{k=1}^{n} \left(u, \sum_{k=1}^{n} \xi(v_k)v_k\right) = \lim_{n \to \infty} \sum_{k=1}^{n} (u, v_k)\xi(v_k),$$

so that we have the equality (4.6) with probability one for each value $\xi(u)$ of the primary random function $\xi(u)$, $u \in U$.

Thus, we have arrived at the following result.*

Theorem 1. *The Gaussian functional $\xi(u)$, $u \in U$, on a Hilbert space U can be represented by* (4.6) *if and only if the mean $a(u)$, $u \in U$, is a continuous*

* A survey of results related to distributions in linear spaces can be found, for example, in Yu. V. Prokhorov, "The method of characteristic functionals," *Proceedings of the 4th Berkeley Symposium, Vol. 2*, 1961, pp. 403–419.

linear functional, and the correlation function $B(u, v)$, u, $v \in U$, is a continuous bilinear functional, the corresponding operator B being a nuclear operator in the representation (4.4) *of the bilinear functional $B(u, v)$.*

As is well known, a Hilbert space U can be identified with the adjoint space X of all linear continuous functionals on U. In fact, each functional $x \in X$ can be defined uniquely by the formula $x = (u, x)$, $u \in U$, where $x \in U$ is a fixed element of the space U. Following the exposition in Section 2, any Gaussian random function $\xi(u)$, $u \in U$, of the form (4.6) corresponds to a probability distribution P_ξ in a Hilbert space $X = U$ (*sample functionals* $\xi(u) = (u, \xi(\omega))$, $u \in U$, belong to X). The Gaussian measure P_ξ is defined on the σ-algebra $\mathfrak{B}$ generated by cylinder sets of the space $X = U$ of the form

$$[(u_1, x), \ldots, (u_n, x)] \in \Gamma, \tag{4.8}$$

where $u_1, \ldots, u_n \in U$ and Γ are Borel sets in an n-dimensional Euclidean space; the σ-algebra $\mathfrak{B}$ is generated, obviously, by variables of the form

$$\xi(x, u) = (u, x), \qquad x \in X, \tag{4.9}$$

where the parameter u runs through the space $U = X$.

For any element $a \in U$ and a positive nuclear operator B there exists a Gaussian random function $\xi(u)$, $u \in U$, of the form (4.6) with a mean $a \in U$ and a correlation operator B. There also exists a Gaussian function of this type with values defined by (4.9) on the probability space $(X, \mathfrak{B}, \mathsf{P}_\xi)$.

Example. Gaussian Variables in the Function Space $\mathscr{L}^2(T)$. Let $\xi = \xi(t)$ be a Gaussian random process on an interval T of the real line with mean $a(t)$, $t \in T$, and let correlation function $B(s, t)$, s, $t \in T$, satisfy the following condition: for all $s, t \in T$,

$$\begin{aligned} \lim_{s \to t} [a(s) - a(t)] &= 0, \\ \lim_{s \to t} [B(s, s) - 2B(s, t) + B(t, t)] &= 0. \end{aligned} \tag{4.10}$$

This condition implies that the random process $\xi = \xi(t)$ is continuous in the mean:

$$\lim_{s \to t} \mathsf{M}[\xi(s) - \xi(t)]^2 = 0.$$

As is well known,* in this case there exists an equivalent measurable process (with values $\tilde{\xi}(t) = \tilde{\xi}(\omega, t)$) such that the function $\tilde{\xi} = \tilde{\xi}(\omega, t)$ with respect to a pair of variables (ω, t) on the product of spaces $\Omega \times T$ is measurable. We shall consider that the initial Gaussian process $\xi = \xi(t)$ itself is Gaussian. Assume that the condition

$$\int_T a^2(t)\, dt < \infty, \qquad \int_T B(t, t)\, dt < \infty, \tag{4.11}$$

* See, for example, [10], p. 209.

is satisfied which implies that

$$\int_T \mathsf{M}\xi^2(t)\,dt < \infty.$$

By Fubini's theorem on iterated integration

$$\int_T \mathsf{M}\xi^2(t)\,dt = \int_\Omega \int_T \xi^2(\omega, t)\,dt\,\mathsf{P}(d\omega) < \infty$$

and almost all sample functions $\xi(\omega, \cdot) = \xi(\omega, t)$ of $t \in T$ belong to a Hilbert space $\mathscr{L}^2(T)$ of real square-integrable functions $u = u(t)$ of $t \in T$ with the scalar product

$$(u, v) = \int_T u(t)v(t)\,dt.$$

We redefine the values $\xi(\omega, t)$ for those $\omega \in \Omega$ for which the sample functions $\xi(\omega, \cdot) = \xi(\omega, t)$, $t \in T$, do not appear in $\mathscr{L}^2$ (the set of such $\omega \in \Omega$ has measure 0) and go over to a measurable Gaussian process $\xi = \xi(t)$ whose sample functions belong to the Hilbert space $\mathscr{L}^2$. This random function $\xi = \xi(\omega, \cdot)$ can be treated as a random element in the Hilbert space $\mathscr{L}^2$.

Consider the scalar products

$$(u, \xi) = \int_T u(t)\xi(\omega, t)\,dt, \qquad u \in U.$$

Since $\xi = \xi(\omega, t)$ is a measurable function of the variables (ω, t), for each fixed $u \in U$ the real function (u, ξ) of $\omega \in \Omega$ is also measurable, i.e., is a random variable. Therefore, $\xi = \xi(\omega, \cdot)$ is a random variable in a Hilbert space $\mathscr{L}^2(T)$. By Fubini's theorem the random function $\xi(u) = (u, \xi)$ of $u \in U$ has the mean

$$a(u) = \int_T u(t)a(t)\,dt = (u, a), \qquad u \in U, \tag{4.12}$$

and the correlation function

$$B(u, v) = \int_T \int_T u(s)v(t)B(s, t)\,ds\,dt = (Bu, v), u, \qquad v \in U, \tag{4.13}$$

where the correlation operator B is given by a kernel $B(s, t)$:

$$Bu(t) = \int_T B(s, t)u(s)\,ds.$$

The random variable $\xi \in \mathscr{L}^2$ is Gaussian. In fact, as can be easily seen, for a continuous function $u = u(t)$ of $t \in T$ the random variable $(u, \xi) = \int_T u(t)\xi(t)\,dt$ is the limit in the mean of Gaussian variables of the form

$$\sum_{k=1}^{n} u(t_k)\xi(t_k)(t_k - t_{k-1}),$$

where $t_0 \leqslant t_1 \leqslant \cdots \leqslant t_n$ are subdivision points of an interval T and for an arbitrary function $u \in \mathscr{L}^2$ the variable (u, ξ) is the limit in the mean of Gaussian variables (u_n, ξ) (here $u_n = u_n(t)$, $n = 1, 2, \ldots$, is a sequence of continuous functions convergent in the mean to the function $u = u(t)$). The limiting value for a sequence of Gaussian variables is, as known, also Gaussian.*

I.5 Conditional Probability Distributions and Conditional Expectations

Let $\xi(u) = \xi(\omega, u)$, $u \in U$, be a family of Gaussian random variables on a probability space Ω and let $\mathfrak{A}(U)$ be the σ-algebra of sets in Ω generated by all variables $\xi(u) = \xi(\omega, u)$ on Ω (parameter u runs through U). We shall assume for simplicity that $\mathsf{M}\xi(u) = 0$, $u \in U$. Denote by $\boldsymbol{H}(U)$ the Hilbert space of random variables η (measurable with respect to the σ-algebra $\mathfrak{A}(U)$) with the scalar product

$$\langle \eta_1, \eta_2 \rangle = \mathsf{M}(\eta_1 \cdot \eta_2). \tag{5.1}$$

Denote by $H(U)$ the closed linear hull of all variables $\xi(u)$, $u \in U$. Let S and T be some subsets in U.

We shall consider an arbitrary variable $\eta \in H(S)$ and the projection $\hat{\eta}$ of it onto a subspace $H(T)$. Since $H(U)$ is the aggregate of Gaussian variables, the difference $\Delta = \eta - \hat{\eta}$ as a Gaussian variable orthogonal to all variables $\xi(t)$, $t \in T$, is independent of all these variables. Therefore,

$$\eta = \hat{\eta} + \Delta, \tag{5.2}$$

where $\hat{\eta}$ is the variable described above, measurable with respect to the σ-algebra $\mathfrak{A}(T)$, and Δ is a Gaussian variable with zero mean and variance $\sigma^2 = \mathsf{M}(\eta - \hat{\eta})^2$; this Gaussian variable is independent of $\xi(t)$, $t \in T$. It can be easily seen† that the conditional distribution of the variable η with respect to the σ-algebra $\mathfrak{A}(T)$ always exists and that for almost all ω it is a Gaussian distribution with the mean

$$\mathsf{M}\{\eta | \mathfrak{A}(T)\} = \hat{\eta}(\omega) \tag{5.3}$$

and the constant variance

$$\sigma^2 = \mathsf{M}\{(\eta - \hat{\eta})^2 | \mathfrak{A}(T)\} = \mathsf{M}(\eta - \hat{\eta})^2. \tag{5.4}$$

Let us recall that the conditional expectation $\mathsf{M}\{\eta | \mathfrak{A}(T)\}$ is geometrically the projection of the variable $\eta \in \boldsymbol{H}(U)$ onto the subspace $\boldsymbol{H}(T)$, and in our case this is the variable $\hat{\eta}$, i.e., the projection onto the subspace $\boldsymbol{H}(T)$.

It is well known that Gaussian variables have finite moments of all orders. Denote by $H^n(U)$ the closure of the subspace of all variables of the form

$$\eta = \varphi[\xi(u_1), \ldots, \xi(u_k)], \tag{5.5}$$

where $\varphi(x_1, \ldots, x_k)$ is a polynomial of degree not higher than n in the arbitrary number of variables $x_1, \ldots, x_k$, and $u_1, \ldots, u_k \in U$.

* See, for example, [10], p. 33.

† See, for example, [6], p. 75.

Theorem 2. *For any variable $\eta \in H^n(S)$ its conditional expectation $\hat{\eta} = \mathsf{M}\{\eta|\mathfrak{A}(T)\}$ belongs to the subspace $H^n(T)$ (with the same index n):*

$$\mathsf{M}\{\eta|\mathfrak{A}(T)\} \in H^n(T). \tag{5.6}$$

Proof. We can consider without loss of generality that the set S and the set T are finite (say, $S = \{s_1, \ldots, s_k\}$ and $T = \{t_1, \ldots, t_k\}$). In fact, we can go over to the general case by passage to the limit,* since

$$\eta = \lim_{m \to \infty} \eta_m, \qquad \mathsf{M}\{\eta|\mathfrak{A}(T)\} = \lim_{m \to \infty} \mathsf{M}\{\eta_m|\mathfrak{A}(T)\},$$

where the convergence is in the mean, and

$$\eta_m = \varphi_m[\xi(s_{k_1}), \ldots, \xi(s_{k_m})],$$

and also

$$\mathsf{M}[\eta|\mathfrak{A}(T)] = \lim_{m \to \infty} \mathsf{M}[\eta|\xi(t_{k_1}), \ldots, \xi(t_{k_m})].$$

Let $\eta = \varphi[\xi(s_1), \ldots, \xi(s_k)]$, where $\varphi(x_1, \ldots, x_k)$ is a polynomial of degree not higher than n. As we have seen, the theorem is certainly true for $n = 1$. Assume that it is true for all indices not exceeding $n - 1$. Denote by $\hat{\xi}(s_j)$ the projections of variables $\xi(s_j)$, $j = 1, \ldots, k$, onto the subspace $H(T)$. The differences $\xi(s_j) - \hat{\xi}(s_j)$, $j = 1, \ldots, k$, are independent of $\xi(t)$, $t \in T$. Set

$$\zeta = \varphi[\xi(s_1) - \hat{\xi}(s_1), \ldots, \xi(s_k) - \hat{\xi}(s_k)].$$

The variable ζ is independent of $\xi(t)$, $t \in T$, and in the expansion

$$\eta - \zeta = \sum_{j=1}^{k} \frac{\partial}{\partial x_j} \varphi[\xi(s_1), \ldots, \xi(s_k)]\hat{\xi}(s_j) + \ldots$$

the right-hand side is a linear combination of expressions of the form

$$\varphi_m[\xi(s_1), \ldots, \xi(s_k)] \cdot \psi_{n-m}[\hat{\xi}(s_1), \ldots, \hat{\xi}(s_k)],$$

where $\varphi_m(x_1, \ldots, x_k)$ and $\psi_{n-m}(x_1, \ldots, x_k)$ are polynomials of degree not higher than m and $n - m$, respectively, with $m \leqslant n - 1$. By hypothesis the conditional expectation $\hat{\eta}_m = \mathsf{M}[\eta_m|\mathfrak{A}(T)]$ of the variable $\eta_m = \varphi_m[\xi(s_1), \ldots, \xi(s_k)]$ belongs to the subspace $H^m(T)$, $m \leqslant n - 1$. Obviously, the product $\hat{\eta}_m \cdot \psi_{n-m}[\hat{\xi}(s_1), \ldots, \hat{\xi}(s_k)]$ belongs to $H^n(T)$ and hence also the linear combination of these products, the conditional expectation of the difference $\eta - \zeta$ equal to $\hat{\eta} - M\zeta$ being such a linear combination. Therefore, $\hat{\eta}_m = \mathsf{M}[\eta|\mathfrak{A}(T)] \in H^n(T)$, as was to be proved. □

We shall define next Hermite polynomials in several variables.

Let $\mathsf{P}(dx)$ be Gaussian measure in a k-dimensional space $\mathbb{R}^k$ of vectors $x = [x_1, \ldots, x_k]$ and let $\boldsymbol{H}$ be a Hilbert space of all real square integrable

* See, for example [6], pp. 29, 287.

functions $\varphi = \varphi(x)$ of $x \in \mathbb{R}^k$ with the scalar product

$$\langle \varphi, \psi \rangle = \int_{\mathbb{R}^k} \varphi(x)\psi(x)P(dx).$$

As is well known* a Gaussian distribution has finite moments of all orders, and the aggregate of all polynomials $\varphi = \varphi(x_1, \ldots, x_k)$ of the variables $x_1, \ldots, x_k$ is a set everywhere dense in $\boldsymbol{H}$. Any polynomial $\varphi = \varphi(x_1, \ldots, x_k)$ of degree p orthogonal to all polynomials of degree lower than p will be said to be *a Hermite polynomial.*

Denote by H_p the aggregate of all Hermite polynomials of the same degree p. Obviously, H_p is a finite-dimensional subspace, and the Hilbert space $\boldsymbol{H}$ is the sum of orthogonal subspaces H_p, $p = 0, 1, \ldots$:

$$\boldsymbol{H} = \sum_{p=0}^{\infty} \oplus H_p.$$

Let us consider a Gaussian vector variable $[\xi(u_1), \ldots, \xi(u_k)]$. Denote by $H_p(u_1, \ldots, u_k)$ the aggregate of all variables of the form

$$\eta = \varphi[\xi(u_1), \ldots, \xi(u_k)],$$

where $\varphi = \varphi(x_1, \ldots, x_k)$ is the Hermite polynomial of degree p with respect to the distribution P of the vector variable $[\xi(u_1), \ldots, \xi(u_k)]$. Obviously

$$\boldsymbol{H}(u_1, \ldots, u_k) = \sum_{p=0}^{\infty} \oplus H_p(u_1, \ldots, u_k). \tag{5.7}$$

Lemma 3. *For any $s_1, \ldots, s_k$ and $t_1, \ldots, t_l$, and for any p and q (with $p \neq q$), the subspaces $H_p(s_1, \ldots, s_k)$ and $H_q(t_1, \ldots, t_l)$ are orthogonal:*

$$H_p(s_1, \ldots, s_k) \perp H_q(t_1, \ldots, t_l). \tag{5.8}$$

Proof. Let $p < q$ for definiteness. We shall consider an arbitrary variable $\eta \in H_p(s_1, \ldots, s_k)$ and the conditional expectation $\hat{\eta} = \mathsf{M}\{\eta | \mathfrak{U}(t_1, \ldots, t_l)\}$ of this variable. By Theorem 2 the variable $\hat{\eta}$ belongs to the subspace $H^p(t_1, \ldots, t_l) = \sum_{r=0}^{p} \oplus H_r(t_1, \ldots, t_l)$. But $\hat{\eta}$ is the projection of the variable η onto the whole space $\boldsymbol{H}(t_1, \ldots, t_l)$ so that the difference $\eta - \hat{\eta}$ is orthogonal to $\boldsymbol{H}(t_1, \ldots, t_l)$ and, in particular, $\eta - \hat{\eta} \perp H_q(t_1, \ldots, t_l)$. At the same time $\hat{\eta} \perp H_q(t_1, \ldots, t_l)$, since $\hat{\eta}$ belongs to $H^p(t_1, \ldots, t_l)$, orthogonal to $H_q(t_1, \ldots, t_l)$ for $p < q$. Consequently,

$$\eta = [(\eta - \hat{\eta}) + \hat{\eta}] \perp H_q(t_1, \ldots, t_l),$$

which was to be proved. □

* This fact can be explained as follows. Obviously, the system of functions of the form $e^{i(t, x)}$, $(t, x) = \sum t_\lambda x_\lambda$, is complete in a complex space $\boldsymbol{H}$, and

$$\left\| e^{i(t, x)} - \sum_{n=1}^{N} \frac{[i(t, x)]^n}{n!} \right\|^2 \leqslant C \frac{\sigma^{2(n+1)}}{(n+1)!} \to 0, \quad \text{where } \sigma^2 = \int (t, x)^2 P(dx).$$

Furthermore, we shall define the subspace $H_p(U)$ as the closed linear hull of all subspaces $H_p(u_1, \ldots, u_k)$ where $u_1, \ldots, u_k \in U$ (index p is the same for all $u_1, \ldots, u_k$). Obviously, by virtue of (5.7) and (5.8) we have

$$H^n(U) = \sum_{p=0}^{n} \oplus H_p(U), \qquad \boldsymbol{H}(U) = \sum_{p=0}^{\infty} \oplus H_p(U), \tag{5.9}$$

where $H_0(U)$ contains only constant variables; $H_1(U) = H(U)$ is the closed linear hull of variables $\xi(u) \in U$; $H_2(U)$ is the closed linear hull of variables $[\xi(u_1)\xi(u_2) - B(u_1, u_2)]$, where $u_1, u_2 \in U$; $H_3(U)$ is the closed linear hull of variables $[\xi(u_1)\xi(u_2)\xi(u_3) - \xi(u_1)B(u_2, u_3) - \xi(u_2)B(u_1, u_3) - \xi(u_3)B(u_1, u_2)]$ where $u_1, u_2, u_3 \in U$; etc.

As seen from (5.9), the conditional expectation $\mathsf{M}[\eta|\mathfrak{A}(T)]$ of any variable $\eta \in H_p(S)$ with respect to the σ-algebra $\mathfrak{A}(T)$ belongs to the subspace $H_p(T)$ (with the same index p).

We recall the general formula for products of Gaussians:

$$\mathsf{M}\xi(u_1) \cdots \xi(u_n) = \sum \prod B(u_k, u_j), \tag{5.10}$$

where the sum is to be taken over all subdivisions of the set $(u_1, \ldots, u_n)$ into pairs (u_k, u_j), and the product is to be taken over all pairs (u_k, u_j) of the corresponding subdivision.

We obtain this formula from the relation

$$\mathsf{M}\xi(u_1) \ldots \xi(u_n) = \frac{\partial^n}{\partial\lambda_1 \ldots \partial\lambda_n} \varphi(0),$$

where $\varphi(\lambda) = \mathsf{M} \exp\{i(\lambda, \xi)\}$, the characteristic function of the Gaussian vector $\xi = [\xi(u_1), \ldots, \xi(u_n)]$, is of the form (see (1.1))

$$\varphi(\lambda) = \exp\left\{-\frac{1}{2}\sum_{k,j=1}^{n} \lambda_k\lambda_j B(u_k, u_j)\right\}, \qquad \lambda = [\lambda_1, \ldots, \lambda_n].$$

I.6 Gaussian Stationary Processes and the Spectral Representation

A Gaussian random process $\xi(t) = \xi(\omega, t)$ with values in a probability space Ω, where the parameter t takes integer (discrete) or real values ($-\infty < t < \infty$), is said to be *stationary* if its mean is constant

$$a(t) = \mathsf{M}\xi(t) \equiv a$$

and the correlation function $B(s, t)$ depends on the difference $(s - t)$ only:

$$B(s, t) = \mathsf{M}[\xi(s) - a][\xi(t) - a] = B(s - t) \tag{6.1}$$

(in what follows we shall take $a = 0$).

The function $B(t)$ in (6.1) is said to be *a correlation function* of the stationary process $\xi(t)$; it can be expressed as

$$B(t) = \int e^{i\lambda t} F(d\lambda), \tag{6.2}$$

where $F(d\lambda)$ is called the *spectral measure* of the stationary process $\xi(t)$ (positive bounded measure). In (6.2) the integration is over $-\pi \leqslant \lambda \leqslant \pi$ in the case of discrete "time" t and over $-\infty < \lambda < \infty$ in the case of continuous time t.

The stationary process $\xi(t)$ permits a spectral representation of the form

$$\xi(t) = \int e^{i\lambda t}\Phi(d\lambda), \tag{6.3}$$

where $\Phi(d\lambda)$ is called the *stochastic spectral measure* such that

$$\mathsf{M}\Phi(\Delta_1)\overline{\Phi(\Delta_2)} = F(\Delta_1 \cap \Delta_2).$$

Each variable η from the closed linear hull $H(T)$ of the values $\xi(t)$, $t \in T$, permits a spectral representation of the form

$$\eta = \int \varphi(\lambda)\Phi(d\lambda), \tag{6.4}$$

where $\varphi(\lambda)$ is the function from the space $L_T(F)$, the real linear hull of the functions $e^{i\lambda t}$ of λ, $t \in T$, closed with respect to the scalar product

$$\langle \varphi_1, \varphi_2 \rangle_F = \int \varphi_1(\lambda)\overline{\varphi_2(\lambda)}F(d\lambda). \tag{6.5}$$

The stochastic integral given by (6.4) is defined for any function $\varphi \in L_T(F)$ and yields $\eta \in H(T)$. The correspondence $\eta \leftrightarrow \varphi(\lambda)$ is a unitary isomorphism* of the Hilbert spaces $H(T)$ and $L_T(F)$:

$$\langle \eta_1, \eta_2 \rangle = \langle \varphi_1, \varphi_2 \rangle_F. \tag{6.6}$$

In the case where the parameter t is continuous and the set T is a finite interval, we can define the space $L_T(F)$ as the closure of the subspace L^0 (by the scalar product given by (6.5)) of all functions of the form

$$\varphi(\lambda) = \int_T e^{i\lambda t}u(t)\,dt, \tag{6.7}$$

where the $u = u(t)$ are infinitely differentiable functions vanishing outside of the interval T. Since the functions $\varphi(\lambda)$ decrease faster than $|\lambda|^{-n}$ as $\lambda \to \infty$, the scalar product (6.5) can be defined on the subspace L^0 with the help of a finite spectral measure as well as any σ-finite measure $G(d\lambda)$ satisfying the condition

$$\int (1 + \lambda^2)^{-n}G(d\lambda) < \infty$$

for some integer n.

Let us set

$$\langle \varphi_1, \varphi_2 \rangle_G = \int \varphi_1(\lambda)\overline{\varphi_2(\lambda)}G(d\lambda) \tag{6.8}$$

and define the complete Hilbert space $L_T(G)$ to be the closure of all functions of the form (6.7) by the scalar product given by (6.8).

Let $L_T(G)$ be a Hilbert space of the type indicated. (6.4) prescribes the random functional $\eta = \eta(\varphi)$ defined on the everywhere dense subspace of

* For this see, for example, [24].

functions $L_T(G) \cap L_T(F)$. We want to know under what conditions $\eta = \eta(\varphi)$ is (to within an equivalence) a random element from the conjugate space of $L_T(G)$, i.e., a Gaussian linear continuous functional on the Hilbert space $L_T(G)$.

Suppose $\eta = \eta(\varphi)$ is a random element from the conjugate space of $L_T(G)$, i.e.,

$$\eta(\varphi) = \langle \varphi, \eta \rangle_G, \tag{6.9}$$

where $\eta = \eta(\lambda)$ is a Gaussian function with trajectories in the Hilbert space $L_T(G)$. The correlation operator B can then be found from the relations

$$\langle B\varphi_1, \varphi_2 \rangle_G = \mathsf{M}\eta(\varphi_1)\eta(\varphi_2) = \langle \varphi_1, \varphi_2 \rangle_F = \langle A\varphi_1, A\varphi_2 \rangle_F = (A^*A\varphi_1, \varphi_2\rangle_G,$$

where A is the operator on the Hilbert space $L_T(G)$ into the Hilbert space $L_T(F)$ determined by the equality

$$A\varphi(\lambda) = \varphi(\lambda), \qquad \varphi \in L_T(G) \cap L_T(F), \tag{6.10}$$

and where A^* is its adjoint on $L_T(F)$ into $L_T(G)$. B will be a nuclear operator like any correlation operator.

On the other hand, if the operator $B = A^*A$ is nuclear by Theorem 1 the Gaussian linear functional $\eta = \eta(\varphi)$ with correlation functional $\langle B\varphi_1, \varphi_2 \rangle_G$ is equivalent to a Gaussian element in the space adjoint to $L_T(G)$ determined by (6.9).

Let us note that we have the inclusion

$$L_T(G) \subseteq L_T(F)$$

not only for a nuclear operator but also for any bounded operator $B = A^*A$, since

$$\|\varphi\|_F^2 = \langle A\varphi, A\varphi \rangle_F = \langle B\varphi, \varphi \rangle_G \leqslant \|B\| \cdot \|\varphi\|_G^2.$$

We note that for the finite measure $G(d\lambda)$, (6.9) is equivalent to a spectral representation of the initial stationary process $\xi(t)$, $t \in T$:

$$\xi(t) = \int e^{-i\lambda t}\eta(\lambda)G(d\lambda), \qquad t \in T. \tag{6.11}$$

In fact, the functions $\varphi(\lambda) = e^{i\lambda t}$ are complete in the Hilbert space $L_T(G)$ and, from (6.11), $\eta(e^{i\lambda t}) = \langle e^{i\lambda t}, \eta \rangle_G$, $t \in T$, extends to the whole space $L_T(G)$, the closed linear hull of functions of the type $\varphi(\lambda) = e^{i\lambda t}$.

If we want to deal with the initial random process $\xi(t)$, $t \in T$, rather than its functional $\eta(\varphi)$, $\varphi \in L_T(G)$, we need to introduce the space X of all real functions $x = x(t)$, $t \in T$, permitting a *spectral representation* as

$$x(t) = \int e^{-i\lambda t}\psi(\lambda)G(d\lambda), \qquad t \in T, \tag{6.12}$$

where $\psi(\lambda) \in L_T(G)$ and the values of $x(t)$ coincide with values of the linear continuous functional $\langle \varphi, \psi \rangle_G$ with $\varphi(\lambda) = e^{i\lambda t}$. It is seen that (6.12) provides the one-to-one correspondence between $x \in X$ and $\psi \in L_T(G)$.

If we introduce a scalar product so that

$$\langle x_1, x_2 \rangle = \langle \psi_1, \psi_2 \rangle_G, \tag{6.13}$$

where ψ_1 and ψ_2 correspond to x_1 and x_2, respectively, X will become a complete Hilbert space.

If we consider the special case where $G(d\lambda) = (1/2\pi)\,d\lambda$ and the Hilbert space $L_T(G)$ consists of functions of the form

$$\psi(\lambda) = \int_T e^{i\lambda t} x(t)\,dt, \tag{6.14}$$

where $x = x(t)$ belongs to the usual $L^2(T)$ space of real square-integrable functions with the scalar product

$$\langle x_1, x_2 \rangle = \int_T x_1(t)x_2(t)\,dt,$$

(6.14) yields the usual Fourier transform.

By the well-known Plancherel formula,

$$\langle x_1, x_2 \rangle = \langle \psi_1, \psi_2 \rangle_G = \int \psi_1(\lambda)\overline{\psi_2(\lambda)}\,\frac{d\lambda}{2\pi},$$

which implies that if (6.11) and (6.12) are understood in the sense of a Fourier transform, the Hilbert space X of all square-integrable functions $x = x(t)$, $t \in T$, is seen formally as a special case of the general scheme of construction of a Hilbert space with scalar product given by (6.13).

We have actually proved the following:

Theorem 3. *A random process $\xi(t)$, $t \in T$, is (to within an equivalence) a random element of a Hilbert space X if and only if the product $B = A^*A$ is a nuclear operator on a Hilbert space $L_T(G)$ where the operator A is defined by* (6.10).

We shall prove below that for a rather wide class of absolutely continuous measures $F(d\lambda)$ and $G(d\lambda)$ with densities $f(\lambda) = F(d\lambda)/d\lambda$ and $g(\lambda) = G(d\lambda)/d\lambda$, *the operator A^*A is nuclear if*

$$\int \frac{f(\lambda)}{g(\lambda)}\,d\lambda < \infty \tag{6.15}$$

(there exist many cases where the operator A^*A is not nuclear if the condition given by (6.15) is not satisfied).

We shall examine in Chapter III an operator of the form $\Delta = A_1^*A_1 - E$ where A_1^* is an operator of the same type as A but which maps $L_T(G)$ onto a space $L_1(G_1)$ constructed with respect to the measure $G_1(d\lambda)$ having density $g_1(\lambda) = g(\lambda) + f(\lambda)$.

Since $f(\lambda) = g_1(\lambda) - g(\lambda)$, we have

$$\begin{aligned}\langle \Delta\varphi, \psi \rangle_G &= \langle A_1^*A_1\varphi, \psi \rangle - \langle \varphi, \psi \rangle_G \\ &= \langle \varphi, \psi \rangle_{G_1} - \langle \varphi, \psi \rangle_G \\ &= \langle \varphi, \psi \rangle_F = \langle A^*A\varphi, \psi \rangle_G\end{aligned}$$

for any ϕ, $\psi \in L_T(G)$; therefore, the operator A^*A coincides with Δ. We prove (see Chapter III, Theorem 17) that the condition

$$\int \left[\frac{f(\lambda)}{g(\lambda)}\right]^2 d\lambda < \infty$$

is sufficient for the operator $\Delta = A^*A$ to be a Hilbert–Schmidt operator. Hence, if we introduce the space $L_T(F_1)$ with $f_1(\lambda) = \sqrt{f(\lambda)g(\lambda)}$, the operators of the type indicated,

$$L_T(G) \overset{B_1}{\to} L_T(F_1) \quad \text{and} \quad L_T(F_1) \overset{B_2}{\to} L_T(F),$$

will be such that $B_1^*B_1$ and $B_2^*B_2$ are Hilbert–Schmidt operators under the condition given by (6.15). It can be easily seen that $A^*A = B_1^*B_2^*B_2B_1$ is nuclear (see, for example, [7], p. 57).

I.7 Properties of the Sample Functions*

I.7.1 Differentiability in the Mean: Some Asymptotic Relationships

Let $\xi(t)$, $-\infty < t < \infty$, be a Gaussian stationary process with continuous time t.

A process $\xi(t)$ is said to be *differentiable* (in the mean) if there exists a limit (in the mean)

$$\lim_{h\to 0} \frac{\xi(t+h) - \xi(t)}{h} = \xi'(t).$$

This limit exists if and only if in the Hilbert space $L_T(F)$, where $T = (-\infty, \infty)$, the limit

$$\lim_{h\to 0} \frac{e^{i\lambda h} - 1}{h} e^{i\lambda t} = i\lambda e^{i\lambda t}$$

exists; obviously, this is equivalent to the condition

$$\int_{-\infty}^{\infty} \lambda^2 F(d\lambda) < \infty \tag{7.1}$$

(here $F(d\lambda)$ is the spectral measure of the stationary process $\xi(t)$).

If

$$\xi(t) = \int_{-\infty}^{\infty} e^{i\lambda t}\Phi(d\lambda)$$

is the spectral representation of the Gaussian stationary process $\xi(t)$, its derivative $\xi'(t)$, $-\infty < t < \infty$, being also a Gaussian stationary process will be

$$\xi'(t) = \int_{-\infty}^{\infty} e^{i\lambda t}(i\lambda)\Phi(d\lambda).$$

It can easily be seen that the condition given by (7.1) is equivalent to

$$\mathsf{M}[\Delta_h\xi(t)]^2 = \Delta_{-h}\Delta_h B(0) = O\{h^2\} \tag{7.2}$$

* For more detail, see [5].

as $h \to 0$, where $B(t)$ is the correlation function and Δ_n denotes the operator of taking the difference ($\Delta_h B(t) = B(t + h) - B(t)$). In fact, by virtue of (7.2) we have for any Λ that

$$\int_{-\Lambda}^{\Lambda} \lambda^2 F(d\lambda) \leqslant C \int_{-\Lambda}^{\Lambda} \frac{1 - \cos \lambda h}{h^2} F(d\lambda) \leqslant C \frac{\Delta_{-h}\Delta_h B(0)}{h^2}$$

for sufficiently small h (no matter how large the prescribed Λ is) where C is a constant; hence (7.1) follows from (7.2).

Let us consider a nondifferentiable stationary process $\xi(t)$ and examine the conditions on the spectral measure $F(d\lambda)$ for which we have the relation

$$\Delta_{-h}\Delta_h B(0) = O\{|h|^{2\alpha}\}, \tag{7.3}$$

where $0 < \alpha < 1$.

We have

$$\frac{\Delta_{-h}\Delta_h B(0)}{h^{2\alpha}} = \int_{-\infty}^{\infty} \left(\frac{1 - \cos \lambda h}{h^2}\right)^{\alpha} (1 - \cos \lambda h)^{1-\alpha} F(d\lambda)$$
$$\leqslant C \int_{-\infty}^{\infty} \lambda^{2\alpha} F(d\lambda);$$

therefore, (7.3) is satisfied if

$$\int_{-\infty}^{\infty} \lambda^{2\alpha} F(d\lambda) < \infty. \tag{7.4}$$

We discuss in more detail the case where there exists a spectral density $f(\lambda) = F(d\lambda)/d\lambda$. For sufficiently large λ, $|\lambda| \geqslant \Lambda$, let

$$f(\lambda) = |\lambda|^{-\beta}$$

(where $\beta > 1$ since the spectral density $f(\lambda)$ must be an integrable function). For a nondifferentiable stationary process we must have $\beta \leqslant 3$. Let $\beta < 3$. Then

$$\frac{\Delta_{-h}\Delta_h B(0)}{h^{2\alpha}} = 2 \int_{\Lambda}^{\infty} \frac{1 - \cos \lambda h}{\lambda^{\beta} h^{2\alpha}} d\lambda + O\{h^{2(1-\alpha)}\}.$$

Substituting the variable $\lambda h = \mu$ we have*

$$\int_{\Lambda}^{\infty} \frac{1 - \cos \lambda h}{\lambda^{\beta} h^{2\alpha}} d\lambda = h^{\beta - 2\alpha - 1} \int_{\Lambda h}^{\infty} \frac{1 - \cos u}{u^{\beta}} du$$
$$\sim h^{\beta - 2\alpha - 1} \int_{0}^{\infty} \frac{1 - \cos h}{u^{\beta}} du$$

and, therefore,

$$\frac{\Delta_{-h}\Delta_h B(0)}{h^{2\alpha}} \sim Ch^{\beta - 2\alpha - 1},$$

where

$$C = 2 \int_{0}^{\infty} \frac{1 - \cos u}{u^{\beta}} du.$$

* $\alpha \sim \beta$ for the variables α and β implies that $\lim \alpha/\beta = 1$.

It can easily be deduced from the relations thus obtained that if

$$f(\lambda) = O\{|\lambda|^{-\beta}\}, \tag{7.5}$$

(7.3) will be satisfied for $2\alpha = \beta - 1$; if

$$\overline{\lim_{\lambda \to \infty}} f(\lambda)|\lambda|^{\beta} = \infty, \tag{7.6}$$

we shall have for $2\alpha = \beta - 1$ that

$$\overline{\lim_{h \to 0}} \frac{\Delta_{-h}\Delta_h B(0)}{h^{2\alpha}} = \infty.$$

Similarly, for $\beta = 3$ we obtain for a spectral density $f(\lambda)$ of the type given by (7.5)

$$\frac{\Delta_{-h}\Delta_h B(0)}{h^2} \leqslant C \int_{Ah}^{\infty} \frac{1 - \cos u}{u^3} \, du + O(1) = O\{|\ln|h||\},$$

and, therefore,

$$\Delta_{-h}\Delta_h B(0) = O\{h^2|\ln|h||\}. \tag{7.7}$$

I.7.2 Continuity

Let us consider the nondifferentiable Gaussian stationary process $\xi(t)$, $-\infty < t < \infty$, with correlation function $B(t)$ satisfying (7.3).

Theorem 4. *Under the condition given by (7.3) there exists an equivalent Gaussian process $\xi(t)$ for each trajectory of which, for sufficiently small h,*

$$|\Delta_h \xi(t)| \leqslant C|h|^{\alpha}|\ln|h||^{1/2} \tag{7.8}$$

uniformly with respect to t in each finite interval, where C is a constant.

Proof. For sufficiently small h we have

$$\begin{aligned}
&\mathsf{P}\{|\Delta_h \xi(t)| > c'|h|^{\alpha}|\ln|h||^{1/2}\} \\
&\leqslant \mathsf{P}\left\{\frac{|\Delta_h \xi(t)|}{|\Delta_{-h}\Delta_h B(0)|^{1/2}} > \frac{c'}{c''}|\ln|h||^{1/2}\right\} \\
&= \frac{2}{\sqrt{2\pi}} \int_{(c'/c'')|\ln|h||^{1/2}} e^{-x^2/2} \, dx \\
&\leqslant \frac{2}{\sqrt{2\pi}} \int_{(c'/c'')|\ln|h||^{1/2}} x e^{-x^2/2} \, dx \\
&= \frac{2}{\sqrt{2\pi}} e^{1/2(c'/c'')^2 \ln|h|} = \frac{2}{\sqrt{2\pi}} |h|^{\beta}, \qquad \beta = \frac{1}{2}\left(\frac{c'}{c''}\right)^2,
\end{aligned}$$

where c'' is the constant in the relation $\Delta_{-h}\Delta_h B(0) \leqslant c''|h|^{2\alpha}$ and c' is chosen so that $\beta > 1$.

Consider the initial process $\xi(t)$ at binary-rational points of the form $t = k/2^n$, assuming for simplicity that $0 \leqslant t \leqslant 1$. For $h = 2^{-n}$ we have

$$\mathsf{P}\left\{\max_k \left|\Delta_h\xi\left(\frac{k}{2^n}\right)\right| > c'|h|^\alpha|\ln|h||^{1/2}\right\} \leqslant \sum_{k=0}^{2^n-1}\left(\frac{2}{\sqrt{2\pi}}|h|^\beta\right) = \frac{2}{\sqrt{2\pi}}2^{-(\beta-1)n}.$$

Since $\beta > 1$ and the series $\sum_{n=1}^{\infty} 2^{-(\beta-1)n}$ is convergent, it follows from the Borel–Cantelli lemma that for sufficiently small h

$$\left|\Delta_h\xi\left(\frac{k}{2^n}\right)\right| \leqslant c'|h|^\alpha|\ln|h||^{1/2}$$

uniformly for $k = 0, \ldots, 2^n - 1$.

It can easily be seen that any interval $[k/2^n, k_1/2^{n_1}]$ can be taken as the sum of intervals $[r/2^m, (r+1)/2^m]$ where r and m are integers, but not more than two intervals of the type indicated for any m. Therefore, for any h (with appropriate n within $2^{-n} \leqslant h \leqslant 2^{-n+1}$)

$$h = \frac{k_1}{2^{n_1}} - \frac{k}{2^n} = \sum_m{}^* 2^{-m},$$

where $\sum^*$ denotes the summation over corresponding m. Taking this fact into account, we get for sufficiently small h

$$\begin{aligned}
\left|\Delta_h\xi\left(\frac{k}{2^n}\right)\right| &\leqslant \sum_m{}^* \left|\Delta_{2^{-m}}\xi\left(\frac{r}{2^m}\right)\right| \\
&\leqslant \sum_m{}^* c'2^{-\alpha m}|\ln 2^m|^{1/2} \\
&\leqslant 2^{-\alpha n}(\ln 2^n)^{1/2}2c' \sum_{m=n}^{\infty} 2^{-\alpha(m-n)}\left(\frac{\ln 2^m}{\ln 2^n}\right)^{1/2} \\
&\leqslant |h|^\alpha|\ln|h||^{1/2}2c' \sum_{k=0}^{\infty} 2^{-\alpha k}\left(1 + \frac{k}{n}\right)^{1/2} \\
&\leqslant C|h|^\alpha|\ln|h||^{1/2}.
\end{aligned}$$

Therefore, the trajectories of the process $\xi(t)$ considered satisfy with probability 1 the condition given by (7.8) on the set of all binary-rational points; in particular, for almost all $\omega \in \Omega$ the trajectories $\xi(\omega, \cdot) = \xi(\omega, t)$ are uniformly continuous functions on the set dense everywhere of binary-rational points t_{kn}. Obviously, for an arbitrary point t the limit $\lim_{t_{kn}\to t} \xi(\omega, t_{kn})$ exists and coincides with the initial value of $\xi(t) = \xi(\omega, t)$ for almost all $\omega \in \Omega$. It is seen that the trajectories of the equivalent process with values defined as

$$\xi(\omega, t) = \lim_{t_{kn}\to t} \xi(\omega, t_{kn})$$

satisfy the condition given by (7.8) for almost all ω. Defining the variables $\xi(\omega, t)$ for the remaining ω (assuming, for example, $\xi(\omega, t) \equiv 0$), we obtain an equivalent process $\xi(t)$ whose trajectories satisfy the condition given by (7.8). The theorem is proved. □

I.7.3 Limit Theorems

Let us consider a nondifferentiable Gaussian stationary process $\xi(t)$ with correlation function $B(t)$. Assume that on an interval $(0, \tau)$ the second derivative $B''(t)$ exists (with the exception of a finite number of points) with discontinuities of the first kind only, i.e., the finite limits $B''(t-0) = \lim_{h \to 0} B(t-h)$ and $B''(t+0) = \lim_{h \to 0} B(t+h)$ exist within the interval $(0, \tau)$ for any point t. We recall (see Section I.1) that for the nondifferentiable process $\xi(t)$

$$\lim_{h \to 0} \frac{\Delta_{-h}\Delta_h B(0)}{h^2} = \infty; \tag{7.9}$$

therefore, the derivative $B''(t)$ has discontinuities of the second kind at the point $t = 0$. More precisely,

$$\lim_{h \to 0} B''(h) = -\infty.$$

Theorem 5. *We have the following limiting relation:*

$$\lim_{h \to 0} \frac{1}{N} \sum_{k=0}^{N-1} \frac{[\Delta_h \xi(kh)]^2}{\Delta_{-h}\Delta_h B(0)} = 1, \tag{7.10}$$

where $h = \tau/N$ and the limit is understood in the sense of the mean convergence.

Proof. Let n be a number of points of discontinuity of $B''(t)$. Let us take an arbitrarily small $\varepsilon > 0$.

Each point of discontinuity t, $0 \leqslant t \leqslant \tau$, can be enclosed in some interval so that the total length of these points does not exceed ε. Let us denote by I_ε the complement of the union of these intervals (I_ε is the union of a finite number of intervals). It is clear that the function $B''(t)$ is uniformly continuous on the set I_ε and

$$\Delta_{-h}\Delta_h B(t) = O\{B''(t)h^2\} \tag{7.11}$$

uniformly with respect to $t \in I_\varepsilon$. Taking into account (7.7) we have

$$\Delta_{-h}\Delta_h B(t) = o\{\Delta_{-h}\Delta_h B(0)\} \tag{7.12}$$

uniformly with respect to $t \in I_\varepsilon$. Furthermore, since the function $\Delta_{-h}\Delta_h B(t)$ as well as the correlation function $B(t)$ are positive definite, it follows that for all t

$$|\Delta_{-h}\Delta_h B(t)| \leqslant |\Delta_{-h}\Delta_h B(0)|. \tag{7.13}$$

We have

$$M\Delta_h\xi(s)\Delta_h\xi(t) = \Delta_{-h}\Delta_h B(s-t)$$

and, by (5.10),

$$\begin{aligned} M[\Delta_h\xi(s)\cdot\Delta_h\xi(t)\cdot\Delta_h\xi(u)\cdot\Delta_h\xi(v)] &= \Delta_{-h}\Delta_h B(s-t)\cdot\Delta_{-h}\Delta_h B(u-v) \\ &\quad + \Delta_{-h}\Delta_h B(s-u)\cdot\Delta_{-h}\Delta_h B(t-v) \\ &\quad + \Delta_{-h}\Delta_h B(s-v)\cdot\Delta_{-h}\Delta_h B(t-u). \end{aligned}$$

It is easy to calculate that

$$\begin{aligned} \sigma^2(h) &= \mathsf{M}\left[\frac{1}{N}\sum_{k=0}^{N-1}\frac{\Delta_h\xi(kh)^2}{\Delta_{-h}\Delta_h B(0)} - 1\right]^2 \\ &= \frac{2}{N^2}\sum_{k,\,j=0}^{N-1}\left[\frac{\Delta_{-h}\Delta_h B((k-j)h)}{\Delta_{-h}\Delta_h B(0)}\right]^2. \end{aligned} \tag{7.14}$$

For fixed j no more than $1 + [\delta h^{-1}]$ points of the form $(k-j)h$ can belong to each interval of the length δ; the total number of such points does not exceed $N(1 + [\delta h^{-1}])$. Therefore, the number of points of the form $(k-j)h$ in the complement of the set I_ε does not exceed the number $(Nn + N^2\varepsilon/\tau)$. It follows from (7.11)–(7.14) that for any $\varepsilon > 0$, for sufficiently small h

$$\sigma^2(h) \leqslant \frac{2}{N^2}\left(Nn + N^2\frac{\varepsilon}{\tau}\right) + 2\left\{\max_{t\in I_\varepsilon}\left[\frac{\Delta_{-h}\Delta_h B(t)}{\Delta_{-h}\Delta_h B(0)}\right]^2\right\} \leqslant C\varepsilon.$$

where C is a constant. The theorem is proved. □

Let us note that there exists a subsequence $h_1, h_2, \ldots,$ for which the limiting relation given by (7.10) is satisfied with probability 1. Furthermore, we can take as the subsequence $h_1, h_2, \ldots$ any sequence for which

$$\sum_{n=1}^{\infty}\sigma^2(h_n) < \infty \tag{7.15}$$

since under the condition given by (7.15) there exists a sequence $\varepsilon_n \to 0$ such that

$$\sum_{h=h_n}\mathsf{P}\left\{\left|\frac{1}{N}\sum_{k=0}^{N-1}\frac{[\Delta_h\xi(kh)]^2}{\Delta_{-h}\Delta_h B(0)} - 1\right| \geqslant \varepsilon_n\right\} < \infty,$$

and it follows from this by the Borel–Cantelli lemma that we have convergence with probability 1 in (7.10) for $h = h_n$. It is interesting to examine the question of the rate (defined in (7.14)) at which the function $\sigma^2(h)$ decreases as $h \to 0$:

$$\sigma^2(h) = \frac{2}{N^2}\frac{h^2}{[\Delta_{-h}\Delta_h B(0)]^2}\sum_{k,\,j=0}^{N-1}[\Delta_{-h}\Delta_h B((k-j)h)]^2.$$

Lemma 4. *Under the assumptions made before on the function $B''(t)$ we have the following estimate*:

$$
\begin{aligned}
d^2(h) &= \sum_{k,\, j=0}^{N-1} [\Delta_{-h}\Delta_h B((k-j)h)]^2 \\
&= O\{\max(|\Delta_{-h}\Delta_h B(0)|, |\Delta_{-h}\Delta_h B(0)|^2 |h|^{-1})\}.
\end{aligned}
\tag{7.16}
$$

Proof. It is seen that

$$
d^2(h) = O\left\{h^{-1}[\Delta_{-h}\Delta_h B(0)]^2 + h^2 \iint_{|s-t|>2h} [B''(s-t)]^2\, ds\, dt\right\}.
$$

Further, for any fixed $\delta > 0$

$$
\iint_{|s-t|>2h} [B''(s-t)]^2\, ds\, dt = O\left\{\int_{2h}^{\delta} B''(t)^2\, dt\right\},
$$

and if the function $B''(t)$ is monotone in a certain neighborhood $(0, \delta)$ then

$$
\int_{2h}^{\delta} B''(t)^2\, dt = B''(2h+\theta)[B'(\delta) - B'(2h)],
$$

where $2h \leqslant \theta \leqslant \delta$ and

$$
B''(2h+\theta) = O\{h^2 \Delta_{-h}\Delta_h B(0)\}.
$$

If the function $B''(t)$ is monotone, it retains the sign in some neighborhood $(0, \delta)$; therefore, $B'(t)$ is also monotone. Obviously, the function $\Delta_h B(t) = \int_t^{t+h} B'(s)\, ds$ as well will be monotone. Hence, when $|B'(2h)| \to \infty$ as $h \to 0$, then

$$
|B'(2h)| = O\{h^{-1}[\Delta_h B(h)]\},
$$

where $|\Delta_h B(h)| \leqslant |\Delta_h B(0)| = \frac{1}{2}|\Delta_{-h}\Delta_h B(0)|$ because the function $\Delta_h B(t)$ is monotone.

Therefore,

$$
\iint_{|s-t|>2h} [B''(s-t)]^2\, ds\, dt = O\begin{cases} h^{-2}|\Delta_{-h}\Delta_h B(0)|, & \text{if } B'(2h) \text{ is bounded} \\ h^{-3}|\Delta_{-h}\Delta_h B(0)|^2, & \text{if } B'(2h) \text{ is unbounded.} \end{cases}
$$

As a result we have

$$
d^2(h) = O\{\max(|h|^{-1}|\Delta_{-h}\Delta_h B(0)|^2, |\Delta_{-h}\Delta_h B(0)|)\}.
$$

The relation given by (7.16) yields the following estimate for the function $\sigma^2(h)$:

$$
\sigma^2(h) = O\left\{\max\left(|h|, \frac{h^2}{|\Delta_{-h}\Delta_h B(0)|}\right)\right\}.
\tag{7.17}
$$

In particular, under the condition given by (7.3) the function $\sigma^2(h)$ decreases as $h \to 0$ at least as $|h|^{\beta}$:

$$
\sigma^2(h) = O\{|h|^{\beta}\}, \qquad \beta = \max\{1, 2(1-\alpha)\}.
$$

Therefore the condition given by (7.15) will be satisfied, for example, for any sequence of the form $h_n = 2^{-n}$, $n = 1, 2, \ldots$.

For the type of stationary Gaussian processes considered, we have finally the following result, completing Theorem 5.

Theorem 6. *Under the condition that*

$$\Delta_{-h}\Delta_h B = o\{|h|^{1/2}\} \tag{7.18}$$

we have the limiting relation

$$\lim_{h\to 0} h^{-1}\frac{1}{N}\sum_{k=0}^{N-1} \Delta_h\xi(kh)\Delta_h\xi(t+kh) = B'(t-0) - B'(t+0), \tag{7.19}$$

where t is any fixed point of the interval $(0, \tau)$, $N = [h^{-1}(\tau - t)] - 1$, and convergence is in the mean.

Proof. Elementary calculations similar to those carried out before yield that the variable

$$\eta(h) = h^{-1}\frac{1}{N}\sum_{k=0}^{N-1} \Delta_h\xi(kh)\Delta_h\xi(t+kh)$$

has the mean

$$\mathsf{M}\eta(h) = \frac{\Delta_h\Delta_{-h}B(t)}{h} = \frac{B(t-h) - B(t)}{h} - \frac{B(t+h) - B(t)}{h},$$

its variance being such that

$$D\eta(h) \leqslant Cd^2(h).$$

It is seen from the estimate given by (7.16) for the variable $d^2(h)$ that we have (7.19) under the condition given by (7.18).

Of course, the relation in (7.19) holds with probability 1 for a sequence $h = h_n$, $n = 1, 2, \ldots$, decreasing sufficiently rapidly. □

CHAPTER II

The Structures of the Spaces $H(T)$ and $L_T(F)$

II.1 Preliminaries

II.1.1 Introduction

We have already seen (in Section I.6) that the Hilbert space of random variables $H(T)$ generated by a stationary process $\xi(t)$, $t \in T$ (with spectral measure $F(d\lambda)$), is isometric to a space of functions $L_T(F)$, which is the closed linear hull of the functions $e^{i\lambda t}$ for $\lambda \in [-\pi, \pi]$ in the case of discrete time t and for $\lambda \in [-\infty, \infty]$ in the case of continuous time t. This fact enables us to investigate stationary processes using analytic tools. To do this, it is useful first to study in detail the analytic structure of spaces $L_T(F)$, as we shall do in this chapter, restricting ourselves to the case where T is a finite interval or a half-line.

It is clear that we may limit ourselves to the interval $T = [-\tau, \tau]$ or $T = [0, \tau]$ and half-lines $T = (-\infty, 0]$ or $T = [0, \infty)$, since an arbitrary interval or the half-line T_1 can be obtained by a "shift" of T by some real t, and the corresponding space $L_{T_1}(F)$ is obtained by multiplying $L_T(F)$ by $e^{i\lambda t}$.

To make it easier for the reader to grasp the results to be obtained in this case, we shall assume that $\xi(t)$ is a stationary process with discrete time with spectral density $f(\lambda) = 1$. It is seen that here $L_T(F)$ consists of trigonometric polynomials $P(e^{i\lambda}) = \sum_{t \in T} c(t)e^{i\lambda t}$ if T is a finite interval, and in fact coincides with the Hardy space $\mathscr{H}^2$ within a circle (or outside of a circle) if T is a half-line; more precisely, $L_T(F)$ consists of square-integrable functions $\varphi(\lambda)$ that can be expanded in a single Fourier series:

$$\varphi(\lambda) = \sum_{-\infty}^{0} c(t)e^{i\lambda t} \quad \text{for } T = (-\infty, 0]$$

(or

$$\varphi(\lambda) = \sum_{0}^{\infty} c(t)e^{i\lambda t} \quad \text{for } T = [0, \infty)).$$

What is the space $L_T(F)$ if $f(\lambda) \neq 1$ in the case of continuous time processes?

It is useful to note initially that if the spectral measures $F(d\lambda)$ and $G(d\lambda)$ are related by the inequality

$$F(d\lambda) \geqslant G(d\lambda)$$

($F(d\lambda)$ majorizes $G(d\lambda)$), the corresponding Hilbert spaces $L_T(F)$ and $L_T(G)$ satisfy the inclusion

$$L_T(F) \subseteq L_T(G).$$

This is immediate from the fact that any fundamental sequence of functions of the form $\varphi_n(\lambda) = \sum_k c_{kn} e^{i\lambda t_{kn}}$, $n = 1, 2, \ldots$, in the space $L_T(F)$ convergent to a function $\varphi(\lambda) \in L_T(F)$, is, at the same time, fundamental in the space $L_T(G)$:

$$\|\varphi_m - \varphi_n\|_G \leqslant \|\varphi_m - \varphi_n\|_F \to 0$$

for $m, n \to \infty$. In this case the limiting function $\psi(\lambda) \in L_T(G)$ coincides almost everywhere relative to $G(d\lambda)$ with the limiting function $\varphi(\lambda) \in L_T(F)$ indicated above. Therefore, the functions $\varphi(\lambda)$ and $\psi(\lambda)$ coincide, being the elements of the Hilbert space $L_T(G)$, i.e., $\varphi(\lambda) \in L_T(G)$.

This fact immediately implies that if we have a spectral density $f(\lambda) = F(d\lambda)/d\lambda$ of the type*

$$f(\lambda) \asymp 1 \tag{1.1}$$

(in the case of discrete time t), as well as a spectral density $f(\lambda) = 1$, the space $L_{[0,\tau]}(F)$ consists of all the polynomials

$$P(e^{i\lambda}) = \sum_{t=0}^{\tau} c(t)e^{i\lambda t}$$

(with real coefficients $c(t)$, $0 \leqslant t \leqslant \tau$). Then the spaces $L_{(-\infty,0]}(F)$ (respectively, $L_{[0,\infty)}(F)$) consist of square-integrable functions that can be expanded as a Fourier series of the form

$$\varphi(e^{i\lambda}) = \sum_{-\infty}^{0} c(t)e^{i\lambda t} \qquad \left(\text{respectively, } \varphi(e^{i\lambda}) = \sum_{0}^{\infty} c(t)e^{i\lambda t}\right)$$

that coincide with boundary values (for $r \to 1$) of analytic functions $\varphi(z)$ in the circle $|z| < 1$ (outside of a circle) of the Hardy spaces $\mathscr{H}^2$ mentioned above:

$$\varphi(e^{i\lambda}) = \lim_{r \to 1} \varphi(z), \qquad z = re^{i\lambda}.$$

* We recall that for the variables α and β the relation $\alpha \asymp \beta$ implies that $0 \leqslant c_1 \leqslant \alpha/\beta \leqslant c_2 < \infty$.

In the case of continuous time t the condition

$$f(\lambda) \asymp (1 + \lambda^2)^{-n} \tag{1.2}$$

where n is an integer, is an analog of the condition given by (1.1).

For $n = 0$ the space $L_{[0,\tau]}(F)$ defined in Section I.6 consists obviously of square-integrable functions $\varphi(\lambda)$ that can be expressed as Fourier integrals of the form $\varphi(\lambda) = \int_0^\tau e^{i\lambda t}c(t)\,dt$. Similarly, the spaces $L_{(-\infty,0]}(F)$ (respectively, $L_{[0,\infty)}(F)$) consist of functions of the form $\varphi(\lambda) = \int_\infty^0 e^{i\lambda t}c(t)\,dt$ (respectively, $\varphi(\lambda) = \int_0^\infty e^{i\lambda t}c(t)d(t)$).

Making use of elementary knowledge only, we can show that *under the condition given by* (1.2) *the space* $L_{[0,\tau]}(F)$ *coincides with the class of functions that can be expressed as*

$$\varphi(\lambda) = P(i\lambda) + (1 + i\lambda)^n \int_0^\tau e^{i\lambda t}c(t)\,dt, \tag{1.3}$$

where $P(i\lambda)$ *is a polynomial of degree less than* $n - 1$ (*with real coefficients*) *and* $c(t)$ *is a square-integrable* (*real*) *function.*

In fact, all the functions $\varphi(\lambda) = (i\lambda)^k e^{i\lambda s}$, $k = 1, \ldots, n - 1$, being limits of the form

$$\varphi(\lambda) = \lim_{h \to 0} (i\lambda)^{k-1} \frac{e^{i\lambda(s+h)} - e^{i\lambda s}}{h},$$

belong to the space $L_{[0,\tau]}(F)$, so that each polynomial $P(i\lambda) = \sum_0^{n-1} c_k(i\lambda)^k$ is contained in $L_{[0,\tau]}(F)$. Furthermore, the functions $\varphi(\lambda) = (1 + i\lambda)^{n-1}(e^{(1+i\lambda)s} - 1)$, $0 \leqslant s \leqslant \tau$, contained in $L_{[0,\tau]}(F)$ can be expressed as

$$\varphi(\lambda) = (1 + i\lambda)^n \int_0^\tau e^{i\lambda t}c_s(t)\,dt,$$

where

$$c_s(t) = \begin{cases} e^t & \text{for } 0 \leqslant t \leqslant s, \\ 0 & \text{for } s \leqslant t \leqslant \tau. \end{cases}$$

It can be readily seen that the closed linear hull of functions $\varphi(\lambda) = (1 + i\lambda)^{n-1}(e^{(1+i\lambda)s} - 1)$, $0 \leqslant s \leqslant \tau$, and $\varphi(\lambda) = (i\lambda)^k$, $0 \leqslant k \leqslant n - 1$, forms the whole space $L_{[0,\tau]}(F)$ (being, by definition, the closed linear hull of functions $e^{i\lambda s}$, $0 \leqslant s \leqslant \tau$) since, starting from the functions $\varphi(\lambda)$, we can arrive at the functions $\varphi(\lambda) = e^{i\lambda s}$ by means of iterated integration:

$$\int_0^t (1 + i\lambda)^{n-1} e^{(1+i\lambda)s}\,ds = (1 + i\lambda)^{n-2}(e^{(1+i\lambda)t} - 1);$$

etc. It can also be seen that the linear hull of "step" functions $c_s(t)$ of the type indicated, in which the parameter s runs through the entire interval $[0, \tau]$, is everywhere dense in the Hilbert space $\mathscr{L}^2[0, \tau]$ of square-integrable functions $c(t)$, $0 \leqslant t \leqslant r$. Moreover, for functions $\varphi'(\lambda)$ and $\varphi''(\lambda)$ of the form

$$\varphi(\lambda) = (1 + i\lambda)^n \int_0^\tau e^{i\lambda t}c(t)\,dt,$$

where $c(t)$ is a linear combination of all step functions $c_s(t)$, we have by the well-known Parseval equality

$$\begin{aligned}\|\varphi'(\lambda) - \varphi''(\lambda)\|_F^2 &= \int_{-\infty}^{\infty} |\varphi'(\lambda) - \varphi''(\lambda)|^2 f(\lambda)\, d\lambda \\ &\asymp \int_{-\infty}^{\infty} |\varphi'(\lambda) - \varphi''(\lambda)|^2 |1 + i\lambda|^{2n}\, d\lambda \\ &= 2\pi \int_0^{\tau} |c'(t) - c''(t)|^2\, dt.\end{aligned}$$

It is clear that the closed linear hull of all functions $\varphi(\lambda)$ of the form indicated coincides with the class of functions $\varphi(\lambda)$ that can be expressed as

$$\varphi(\lambda) = (1 + i\lambda)^n \int_0^{\tau} e^{i\lambda t} c(t)\, dt,$$

where $c(t) \in \mathscr{L}^2[0, \tau]$. Adding to this all the functions $\varphi(\lambda) = (i\lambda)^k$, $0 \leqslant k \leqslant n - 1$, we obtain, obviously, the space $L_{[0, \tau]}(F)$.

(1.3) enables us to describe adequately the variables in the space $H(T)$ for $T = [0, \tau]$, the closed linear hull of variables $\xi(t)$, $0 \leqslant t \leqslant \tau$. In fact, if $\Phi(d\lambda)$ is the stochastic spectral measure of a stationary process $\xi(t)$, every variable $\eta \in H(T)$ can be represented as an integral $\eta = \int \varphi(\lambda)\Phi(d\lambda)$ (see Section I.6) in which $\varphi(\lambda) \in L_T(F)$. It is easily seen that

$$\eta = \sum_{k=0}^{n-1} \left[a_k \xi^{(k)}(0) + b_k \xi^{(k)}(\tau) + \int_0^{\tau} \xi^{(k)}(t) c_k(t)\, dt \right],$$

where a_k and b_k are real coefficients, $c_k(t)$ are square-integrable functions, and $\xi^{(k)}(t)$ are derivatives in the process, $k = 0, \ldots, n - 1$.

Let us note that if the spectral density $f(\lambda)$ satisfies the condition

$$f(\lambda) \geqslant c(1 + \lambda^2)^{-n}$$

only, the corresponding space $L_{[0, \tau]}(F)$ is contained in the space $L_T(G)$ associated with the spectral density $g(\lambda) = c(1 + \lambda^2)^{-n}$. Hence each function $\varphi(\lambda) \in L_{[0, \tau]}(F)$ can be represented by (1.3). It is also crucial to note that (1.3) determines an entire analytic function for all complex λ. Later on (see Section III.4) we shall show that the space $L_{[0, \tau]}(F)$ can be identified with the class of functions of the form (1.3) under the condition given by (1.2), as well as under the weaker condition

$$f(\lambda) \asymp (1 + \lambda^2)^{-n}, \qquad \lambda \to \infty$$

(i.e., $f(\lambda) \asymp (1 + \lambda^2)^{-n}$ for sufficiently large λ only). Under the condition given by (1.2) we can easily deduce from the representation (1.3) the general formula for functions $\varphi(\lambda)$ in the spaces $L_{(-\infty, 0]}(F)$ and $L_{[0, \infty)}(F)$. Every function $\varphi(\lambda) \in L_{[0, \infty)}(F)$ is, in fact, the limit of a sequence of functions $\varphi_k(\lambda) \in L_{[0, \tau_k]}(F)$, $\tau_k \to \infty$, that can be expressed as

$$\varphi_k(\lambda) = P_k(i\lambda) + (1 + i\lambda)^n \int_0^{\infty} e^{i\lambda t} c_k(t)\, dt, \qquad k = 1, 2, \ldots,$$

where the sequence $c_k(t)$, $k = 1, 2, \ldots$, is fundamental in a Hilbert space $\mathscr{L}^2(0, \infty)$ of square – integrable functions and convergent to a function $c(t)$, $0 \leqslant t \leqslant \infty$, in this space. It is seen that the limiting function $\varphi(\lambda) = \lim_{k\to\infty} \varphi_k(\lambda)$ can be expressed as

$$\varphi(\lambda) = P(i\lambda) + (1 + i\lambda)^n \int_0^\infty e^{i\lambda t} c(t)\, d(t),$$

where $P(i\lambda) = \lim P_k(i\lambda)$ is a polynomial of degree less than $n - 1$. It is also seen that this function (where $c(t) \in \mathscr{L}^2(0, \infty)$) belongs to the space $L_{[0, \infty)}(F)$.

Similarly, the space $L_{(-\infty, 0]}(F)$ coincides with the class of all functions that can be described by the formula

$$\varphi(\lambda) = P(i\lambda) + (1 + i\lambda)^n \int_{-\infty}^0 e^{i\lambda t} c(t)\, d(t).$$

We shall investigate further the structure of the corresponding spaces $L_T(F)$ when the spectral density $f(\lambda)$ does not necessarily satisfy the condition given by (1.2) but still vanishes at infinity although "not too rapidly," i.e.,

$$\int_{-\pi}^{\pi} \ln f(\lambda)\, d\lambda > -\infty$$

for discrete time t and

$$\int_{-\infty}^{\infty} \frac{\ln f(\lambda)}{1 + \lambda^2}\, d\lambda > -\infty$$

for continuous time t. In this case, retaining the earlier notation, we consider complex spaces $H(T)$ and $L_T(F)$.

II.1.2 Functions Analytic in a Circle

Let us denote by $\mathscr{H}^p$, $1 \leqslant p \leqslant \infty$, the class of analytic functions $\varphi(z)$ in a unit circle $|z| < 1$ for which

$$\lim_{r\to 1} \int_{-\pi}^{\pi} |\varphi(re^{i\lambda})|^p\, d\lambda < \infty, \qquad z = re^{i\lambda}.$$

If $\varphi \in \mathscr{H}^p$, then for almost all $\lambda \in [-\pi, \pi]$ the boundary values of $\varphi(e^{i\lambda}) = \lim_{r\to 1} \varphi(re^{i\lambda})$ exist and

$$\lim_{r\to 1} \int_{-\pi}^{\pi} |\varphi(re^{i\lambda})|^p\, d\lambda = \int_{-\pi}^{\pi} |\varphi(e^{i\lambda})|^p\, d\lambda.$$

The space $\mathscr{H}^p$ is a Banach space with a norm $\|\varphi\|^{(p)} = (\int_{-\pi}^{\pi} |\varphi(e^{i\lambda})|^p\, d\lambda)^{1/p}$. We can identify $\mathscr{H}^p$ with the closed subspace (in the known space* $\mathscr{L}^p(-\pi, \pi)$) of all functions $\varphi(e^{i\lambda}) \in \mathscr{L}^p(-\pi, \pi)$ for which

$$\int_{-\pi}^{\pi} \varphi(e^{i\lambda}) e^{in\lambda}\, d\lambda = 0, \qquad n = 1, 2, \ldots\ .$$

* The space $\mathscr{L}^p(a, b)$ consists of functions $\varphi(\lambda)$ on the interval $a \leqslant \lambda \leqslant b$ for which $\|\varphi\|^{(p)} = (\int_a^b |\varphi(\lambda)|^p\, d\lambda)^{1/p} < \infty$.

We shall denote by $\mathscr{H}^p$ the given subspace consisting of boundary values of the functions described analytic in a circle.

A function $\varphi(z)$ analytic inside the circle $|z| < 1$ is said to be *an outer function* if we can represent it as

$$\varphi(z) = a\exp\left\{\frac{1}{2\pi}\int_{-\pi}^{\pi}\frac{e^{i\lambda}+z}{e^{i\lambda}-z}\ln p(\lambda)\,d\lambda\right\}, \qquad |a| = 1,$$

where the real function $p(\lambda)$ is nonnegative and $\ln p \in \mathscr{L}^1(-\pi, \pi)$.

A function $\varphi(z)$ analytic inside a circle is said to be an *inner function* if $|\varphi(z)| \leqslant 1$ and $|\varphi(e^{i\lambda})| = 1$ for almost all $\lambda \in [-\pi, \pi]$.

By the *Blaschke product* we shall mean an analytic function $B(z)$ of the form

$$B(z) = az^p \prod \left[\frac{\bar{\alpha}_n}{|\alpha_n|}\frac{\alpha_n - z}{1 - \bar{\alpha}_n z}\right]^{p_n}, \qquad |a| = 1,$$

where $p, p_1, p_2, \ldots$ are nonnegative integers, $0 \leqslant |\alpha_n| < 1$, and the product $\prod |\alpha_n|^{p_n}$ is convergent.

Theorem ([15], pp. 98–99). *An inner function $\varphi(z)$ is uniquely representable as the product*

$$\varphi(z) = B(z)\exp\left\{-\int_{-\pi}^{\pi}\frac{e^{i\lambda}+z}{e^{i\lambda}-z}\mu(d\lambda)\right\},$$

where $B(z)$ is a Blaschke function and $\mu(d\lambda)$ is a singular measure.

It follows readily from this result ([15], p. 123) that any nonempty family of inner functions has a greatest common (inner) divisor.

We shall denote by D the class of functions $\varphi(z)$ (introduced by V. I. Smirnov—see [23]) analytic in a circle $|z| < 1$ that permit the representation

$$\varphi(z) = B(z)\exp\left\{-\int_{-\pi}^{\pi}\frac{e^{i\lambda}+z}{e^{i\lambda}-z}\mu(d\lambda)\right\} \times \exp\left\{\frac{1}{2\pi}\int_{-\pi}^{\pi}\frac{e^{i\lambda}+z}{e^{i\lambda}-z}\ln p(\lambda)\,d\lambda\right\}, \quad (1.4)$$

where $B(z)$ is the Blaschke product, $\mu(d\lambda)$ is a singular measure, and $p(\lambda) \geqslant 0$, $\ln p \in \mathscr{L}^1(-\pi, \pi)$. Therefore, class D consists of functions $\varphi(z)$ that can be expressed as the product of some inner function (*the inner part of* φ) and some exterior function (*the outer part of* φ).

For each function $\varphi(z)$ of class D there exist, for almost all λ, boundary values $\varphi(e^{i\lambda}) = \lim_{r\to 1}\varphi(re^{i\lambda})$ satisfying the condition $|\varphi(e^{i\lambda})| = p(\lambda)$, where $p(\lambda)$ is the function from the representation given in (1.4) for $\varphi(z)$.

Theorem ([15], p. 80). *Each function $\varphi \in \mathscr{H}^1$ is the product of two functions from $\mathscr{H}^2$.*

Theorem ([15], p. 81). *For $f(\lambda) > 0$ and $f \in \mathscr{L}^1$ we have $f = |\varphi|^2$ where $\varphi \in \mathscr{H}^2$ if and only if $\ln f \in \mathscr{L}^1$.*

With respect to this theorem, we note that one can take

$$\varphi(z) = \exp\left\{\frac{1}{4\pi}\int_{-\pi}^{\pi}\frac{e^{i\lambda}+z}{e^{i\lambda}-z}\ln f(\lambda)\,d\lambda\right\}.$$

Beurling's Theorem ([15], p. 145). *For $\varphi \in \mathscr{H}^2$ the functions $\{z^n\varphi\}$, $n = 0, 1, \ldots$, generate the total class $\mathscr{H}^2$ if and only if φ is an outer function.*

Let $\varphi(z)$ be a function analytic in a circle. In this case the function $\varphi(1/z)$ is analytic outside of a circle. Associating in particular each function analytic in a circle with a function analytic outside a circle, we have classes D and $\mathscr{H}^p$ of functions analytic outside of a circle. To distinguish between these two classes, we denote, if necessary, by D^+ and $\mathscr{H}^{p+}$ the classes inside a circle and by D^- and $\mathscr{H}^{p-}$ those outside a circle.

II.1.3 Functions Analytic in a Half-Plane

Let us denote by $\tilde{\mathscr{H}}^p$ and D the classes of functions analytic in the upper half-plane that are images of classes $\mathscr{H}^p$ and D in a circle under the conformal mapping of the circle onto the upper half-plane.

We shall denote by $\mathscr{H}^p$ the class of functions $\varphi(z)$ analytic in the upper half-plane for which

$$\int_{-\infty}^{\infty}|\varphi(x+iy)|^p\,dx \leqslant M < \infty, \qquad y \geqslant 0,$$

where the constant M is independent of y ([15], [17]).

A function $\varphi \in D$ is said to be *outer* if it can be expressed as

$$\varphi(z) = \exp\left\{\frac{1}{\pi i}\int_{-\infty}^{\infty}\frac{1+\lambda z}{\lambda - z}\frac{\ln p(\lambda)}{1+\lambda^2}\,d\lambda\right\},$$

where $p(\lambda)$ is a real function, $p(\lambda) \geqslant 0$, and $\ln p(\lambda)/(1+\lambda^2) \in \mathscr{L}^1(-\infty, \infty)$. We have equality for *outer* functions in the inequality for Poisson's integral:

$$\ln|\varphi(z)| \leqslant \frac{y}{\pi}\int_{-\infty}^{\infty}\frac{\ln|\varphi(\lambda)|}{(\lambda - x)^2 + y^2}\,d\lambda, \qquad z = x + iy,\ y > 0.$$

A function $\varphi \in D$ is said to be *inner* if $|\varphi(z)| \leqslant 1$ and $|\varphi(\lambda)| = 1$ ($z = \lambda + i\mu$, $\mu \geqslant 0$).

We have for functions of class D representations of the type (1.4):

$$\varphi(z) = e^{i\alpha z}B(z)\exp\left\{-\int_{-\infty}^{\infty}\frac{1+\lambda z}{\lambda - z}\mu(d\lambda)\right\} \times \exp\left\{\frac{1}{\pi i}\int_{-\infty}^{\infty}\frac{1-\lambda z}{\lambda - z}\frac{\ln p(\lambda)}{1+\lambda^2}\,d\lambda\right\},$$

where α is a real number, $B(z)$ is a Blaschke function, $\mu(d\lambda)$ is a singular finite measure, and $p(\lambda) \geqslant 0$, $\ln p(\lambda)/(1+\lambda^2) \in \mathscr{L}^1(-\infty, \infty)$.

The other assertions can be generalized to classes D and $\mathscr{H}^p$ in the upper half-plane in a similar way. In particular, we have the following:

Theorem of Lax ([15]). *For $\varphi \in \mathscr{H}^2$ the functions $\{e^{i\lambda t}\varphi(\lambda),\ t \geqslant 0\}$ generate all $\mathscr{H}^2$ if and only if φ is an outer function.*

From now on the following characteristic feature of the spaces $\mathscr{H}^2$ will be frequently used.

Paley–Wiener Theorem ([15], p. 187). *The function $\varphi \in \mathscr{H}^2$ (in the upper half-plane) if and only if*

$$\varphi(z) = \int_0^\infty e^{izt}c(t)\,dt, \qquad \operatorname{Im} z \geqslant 0,$$

where $c(t) \in \mathscr{L}^2(0, \infty)$.

If we consider classes D and $\mathscr{H}^p$ in the upper half-plane and classes D and $\mathscr{H}^p$ in the lower half-plane at the same time, we shall write D^+ and $\mathscr{H}^{p+}$ in the case of the upper half-plane and D^- and $\mathscr{H}^{p-}$ in the case of the lower half-plane. Note that

$$\mathscr{L}^2(-\infty, \infty) = \mathscr{H}^{2+} \oplus \mathscr{H}^{2-}.$$

II.2 The Spaces $L^+(F)$ and $L^-(F)$

Let $\xi(t)$ be a stationary wide-sense process with a spectral measure $F(d\lambda)$. Let $F = F_a + F_s$, where F_a is the absolutely continuous component and F_s is the singular component of the measure F. Set $f(\lambda) = dF_a/d\lambda$. In this case we shall call $f(\lambda)$ a spectral density even if $F \neq F_a$. For the sake of brevity, let us introduce some new notation:

$$\begin{aligned} L &= L(F) = L_{(-\infty,\infty)}(F), \\ L^- &= L^-(F) = L_{(-\infty,0]}(F), \\ L^+ &= L^+(F) = L_{[0,\infty)}(F). \end{aligned}$$

Theorem 1. *If $\xi(t)$ is a stationary random process with discrete time $t = 0, \pm 1, \ldots$, then:*

1. $L^+(F_s) = L^-(F_s) = L(F_s)$;
2. $L^+(F_a) = L^-(F_a) = L(F_a)$ *if and only if*

$$\int_{-\pi}^{\pi} |\ln f(\lambda)|\,d\lambda = \infty;$$

3. *if*

$$\int_{-\pi}^{\pi} |\ln f(\lambda)|\,d\lambda < \infty, \tag{2.1}$$

we can write $f(\lambda)$ *as* $f(\lambda) = |g(e^{i\lambda})|^2$ *where* g *is an outer function of class* $\mathscr{H}^2$ *in the circle* $|z| < 1$.

Further,

$$L^-(F_a) = D^- \cap L(F_a) = \frac{1}{\bar{g}} \mathscr{H}^{2-}, \qquad L^+(F_a) = D^+ \cap L(F_a) = \frac{1}{g} \mathscr{H}^{2+}.$$

Theorem 2. *If* $\xi(t)$ *is a stationary random process with continuous time* t, $-\infty < t < \infty$, *then*:

1. $L^+(F_s) = L^-(F_s) = L(F_s)$;
2. $L^+(F_a) = L^-(F_a) = L(F_a)$ *if and only if*

$$\int_{-\infty}^{\infty} \frac{|\ln f(\lambda)|}{1 + \lambda^2} \, d\lambda = \infty;$$

3. *if*

$$\int_{-\infty}^{\infty} \frac{|\ln f(\lambda)|}{1 + \lambda^2} \, d\lambda < \infty, \tag{2.2}$$

we can write $f(\lambda)$ *as* $f(\lambda) = |g(\lambda)|^2$, *where* g *is now an outer function of class* $\mathscr{H}^2$ *in the upper half-plane* $\operatorname{Im} z > 0$, $z = \lambda + i\mu$.

Further,

$$L^-(F_a) = D^- \cap L(F_a), \; L^+(F_a) = D^+ \cap L(F_a).$$

Assertions (1)–(3) of the two theorems are, in essence, equivalent to those due to A. N. Kolmogorov and M. G. Krein in the theory of prediction of stationary random processes.* Since these assertions play a fundamental role from now on we shall sketch the proofs briefly.

Proof. Let us first prove (1)–(3) of Theorem 1. Suppose that F is a singular measure on an interval $[-\pi, \pi]$, and that $L^- \neq L$. Then necessarily $e^{i\lambda} \notin L^-$. Denote by $\varphi(\lambda)$ the projection of the element $e^{i\lambda}$ on the subspace L^-. Then $e^{i\lambda} - \varphi(\lambda) \not\equiv 0$ and $e^{i\lambda} - \varphi(\lambda) \perp L^-$, and

$$\int_{-\pi}^{\pi} e^{in\lambda}(e^{i\lambda} - \varphi)F(d\lambda) = 0, \qquad n = 0, 1, \ldots .$$

Since the generalized measure $F_1(d\lambda) = (e^{i\lambda} - \varphi)F(d\lambda)$ is analytic, that is,

$$\int_{-\pi}^{\pi} e^{in\lambda} F_1(d\lambda) = 0, \qquad n = 0, 1, \ldots,$$

by the theorem of F. and M. Riesz ([15], p. 73) it must be absolutely continuous with respect to a measure $F(d\lambda)$. The contradiction obtained proves (1).

As to (2), suppose $\int_{-\pi}^{\pi} |\ln f(\lambda)| \, d\lambda = \infty$. Assume contrariwise that $L^- \neq L$. Then we have again that $e^{i\lambda} \notin L^-$, and if $\varphi(\lambda)$ is the projection of $e^{i\lambda}$ onto

* See, for example, [24].

L^-, we have $\psi(\lambda) = e^{i\lambda} - \varphi(\lambda) \not\equiv 0$ and

$$\int_{-\pi}^{\pi} e^{in\lambda}\psi(\lambda)f(\lambda)\,d\lambda = 0, \qquad n = 0, 1, \ldots . \tag{2.3}$$

Let us denote by $\|\varphi\|^{(p)}$ the norm in the space $\mathscr{L}^p(-\pi, \pi)$. We note that the function $\psi f \in \mathscr{L}^1(-\pi, \pi)$. In fact, by the Schwarz inequality,

$$\|\psi f\|^{(1)} \leqslant \|\psi\|_{F_a} \cdot (\|f\|^{(1)})^{1/2} < \infty,$$

from which it follows by (2.3) that $\psi f \in \mathscr{H}^1$. The logarithm of any function from $\mathscr{H}^1$ is summable; therefore,

$$\int_{-\pi}^{\pi} \ln|\psi(\lambda)f(\lambda)|\,d\lambda > -\infty.$$

Further, from the elementary inequality $\ln x < x$ we have

$$\int_{-\pi}^{\pi} \ln|\psi^2(\lambda)f(\lambda)|\,d\lambda \leqslant \|\psi\|_{F_a}^2 < \infty,$$

$$\int_{-\pi}^{\pi} \ln f(\lambda)\,d\lambda \leqslant \int_{-\pi}^{\pi} f(\lambda)\,d\lambda < \infty.$$

Together with the previous inequality the above two inequalities imply $\ln f \in \mathscr{L}^1$, contrary to the hypothesis.

The contradiction obtained proves the first part of (2); the second part of (2) will follow from (3), which we shall prove next. By virtue of (2.1) we can write $f(\lambda)$ as $f(\lambda) = |g(e^{i\lambda})|^2$, where

$$g(z) = \exp\left\{\frac{1}{4\pi}\int_{-\pi}^{\pi} \frac{e^{i\theta} + z}{e^{i\theta} - z} \ln|f(\theta)|\,d\theta\right\}, \qquad z = re^{i\lambda},$$

is the exterior function from $\mathscr{H}^2$. Let $\varphi(e^{i\lambda}) = \varphi \in L^+(F)$. This fact implies that there exists a sequence of polynomials $P_n(z)$ such that $\|\varphi - P_n\|_F \to 0$. But in this case it follows also that $\|\varphi g - P_n g\|^{(2)} = \|\phi - P_n\|_F \to 0$ for $n \to \infty$. Obviously, $P_n g \in \mathscr{H}^{2+}$. Hence the limiting function $\psi = \varphi g \in \mathscr{H}^{2+}$, that is, $\varphi = \psi/g$ where ψ, $g \in H^{2+}$. Taking the canonical representation given in (1.4) of the functions from $\mathscr{H}^{2+}$ and D^+, we can see that $\varphi \in D^+$.

Conversely, let $\varphi \in D^+ \cap L(F)$. Then $\psi = \varphi g \in \mathscr{H}^{2+}$. The function g is exterior, and by Beurling's theorem (see Section II.1) the aggregation of functions $\{gP\}$ in which P runs through all the polynomials is dense in $\mathscr{H}^{2+}$. This fact implies, in particular, that we can find a sequence of polynomials P_n for which, as $n \to \infty$, we have

$$\|\psi - gP_n\|_2 = \|(\varphi - P_n)g\|_2 = \|\varphi - P_n\|_F \to 0,$$

that is, that $\varphi \in L^+(F)$. The case $L^-(F)$ should be considered in a similar way.

The proofs of (1)–(3) of Theorem 2 coincide almost completely with the proofs of (1)–(3) of Theorem 1. In fact, we can prove with the aid of the conformal mapping of a circle onto a half-plane that the generalized theorem of F. and M. Riesz is true in this case as well (see [23], p. 209). Further, if $\varphi \in \mathscr{H}^2$ in the upper half-plane, then, necessarily, we have

$$\int_{-\infty}^{\infty} \frac{|\ln|\varphi(\lambda)||}{1+\lambda^2}\, d\lambda < \infty.$$

Finally, in proving (3) we should refer to the Lax theorem instead of the Beurling theorem. □

It is useful to note that, in proving (3) in Theorem 1 and Theorem 2, we obtain, in essence, the following: *if the conditions given by* (2.1) *or* (2.2) *are satisfied, then*

$$L^+ = \frac{1}{g}\mathscr{H}^{2+}, \qquad L^- = \frac{1}{\bar{g}}\mathscr{H}^{2-}. \tag{2.4}$$

II.3 The Construction of Spaces $L_T(F)$ When T Is a Finite Interval

We investigated in Section II.2 the spaces $L_T(F)$ (and therefore, the spaces $H(T)$ generated by values of the corresponding stationary process $\xi(t)$, $t \in T$, with the spectral measure $F(d\lambda)$), when T is an infinite interval. We shall investigate in this section the case where $T = [a, b]$ is a finite interval.

Since the case of discrete time is trivial—$L_T(F)$ consists of trigonometrical polynomials of the form $\sum_{a \leqslant t \leqslant b} a_t e^{i\lambda t}$—we shall deal only with processes with continuous time. Furthermore, we restrict ourselves to the investigation of a process $\xi(t)$ with an absolutely continuous spectral measure $F(d\lambda)$ and a spectral density $f(\lambda)$ satisfying the condition given by (2.2). As noted above, it suffices to consider intervals of the form $T = [-a, a]$. Let

$$L^\sigma(F) = \bigcap_{T=[-a,a],\, a>\sigma} L_T(F)$$

and

$$L^0(F) = \bigcap_T L_T(F).$$

The space $L^0(F)$ is defined by the behavior of the process $\xi(t)$ in an infinitely small neighborhood of zero; its isometric space $H^0 = \bigcap_T H_T$, in particular, contains all existing derivatives $\xi^{(k)}(0)$.

Let us agree to denote by D_σ the aggregate of entire analytic functions $\varphi(z)$, $z = \lambda + i\mu$, of finite degree $\leqslant \sigma$, i.e., entire functions such that

$$\overline{\lim_{R\to\infty}}\, R^{-1} \max_\theta \ln|\varphi(\mathrm{Re}^{i\theta})| \leqslant \sigma$$

(in particular, D_0 denotes the aggregate of entire functions of zero degree).

Theorem 3. *If the spectral density $f(\lambda)$ of the stationary process $\xi(t)$ satisfies the condition given by* (2.2), *then* it follows that*

$$L^\sigma(F) = D_\sigma \cap L(F), \qquad L^0(F) = D_0 \cap L(F). \tag{3.1}$$

Proof. We need to prove two inclusions:

$$L^\sigma(F) \subset D_\sigma \cap L(F) \quad \text{and} \quad L^\sigma(F) \supset D_\sigma \cap L(F).$$

We shall prove the first inclusion for all $\sigma \geqslant 0$, and we shall prove the second inclusion for $\sigma = 0$ only.†

1. *Proof of the inclusion $L^\sigma(F) \subset D_\sigma \cap L(F)$.* Given any function $\varphi \in L^\sigma(F)$, there exist functions

$$\varphi_n(\lambda) = \sum_j a_{jn} \exp\{it_{jn}\lambda\}, \qquad |t_{jn}| \leqslant \sigma + 1/n,$$

such that $\|\varphi - \varphi_n\|_F < 1/n$, $n = 1, 2, \ldots$. It is seen that all $\varphi_n \in D_{\sigma + 1/n}$.

Next let us prove that at each point of the complex half-plane we have

$$|\varphi_n(z)| \leqslant C_\varepsilon e^{(\sigma+\varepsilon)|z|} \tag{3.2}$$

for any $\varepsilon > 0$, the constants C_ε being dependent of ε only (and not of n).

The uniformly bounded family of analytic functions φ_n is compact. Furthermore, $\|\varphi_n - \varphi\|_F \to 0$, and hence $\varphi_n(z)$ is convergent to an entire function $\varphi(z)$ which, under the condition given by (3.2), is a function of finite degree less than σ. It can be seen that the restriction of $\varphi(z)$ on $\operatorname{Im} z = 0$ coincides with $\varphi(\lambda)$.

Therefore, we need only to prove the inequality in (3.2). To this end let us estimate $|\varphi_n(z)|$ along 45 degree rays through the origin, and make use of the Phragmen–Lindelof principle.‡

We shall estimate first $|\varphi_n(z)|$ for $|\mu| \geqslant 1$.

We note that, by virtue of (2.2), $f(\lambda) = |g(\lambda)|^2$, where $g(z)$ is an exterior function of class $\mathscr{H}^2$. Let us introduce the functions $\psi_n(z) = \varphi_n g \exp\{iz(\sigma + \delta)\}$,

* We do not distinguish between the functions from D_σ and the restriction of them on the line $\mu = 0$.

† The proof of the general case ($\sigma > 0$) is similar but rather cumbersome—the reader can find this proof in N. Levinson and H. McKean, "Weighted trigonometrical approximation on $\mathbb{R}^1$ with Application to the germ field of a stationary Gaussian noise, *Acta Math.* **112**, Nos. 1–2 (1964), 99–143. A stronger result was obtained earlier by M. G. Krein who gave the integral representation for entire functions from $L^\sigma(F)$—see "On the basic approximation problem of the theory of extrapolation of stationary random processes," *DAN SSSR* **94** (1954), 13–16.

The further results of this chapter are borrowed from the paper by Levinson and McKean cited above; we have altered their proofs slightly.

‡ See, for example, [20].

$0 < \delta \leqslant 1/n$. It is clear that $\psi_n \in \mathscr{H}^2$, and therefore for all $\mu > 0$ we have

$$\int_{-\infty}^{\infty} |\psi_n(\lambda + i\mu)|^2 \, d\mu \leqslant \int_{-\infty}^{\infty} |\psi_n(\lambda)|^2 \, d\lambda$$
$$= \|\varphi_n\|_F^2 \leqslant \left(\|\varphi\|_F + \frac{1}{n} \right)^2 \leqslant (\|\varphi\|_F + 1)^2 = C_1. \quad (3.3)$$

Further, since the function $g \in \mathscr{H}^2$ it can be represented by the Paley–Wiener theorem as

$$g(z) = \int_0^{\infty} e^{iuz} \hat{g}(u) \, du, \qquad z = \lambda + i\mu, \quad \mu > 0,$$

where $\hat{g}(u)$ is the Fourier transform of the function $g(\lambda)$. Hence for all $\mu > 0$

$$|g(\lambda + i\mu)| \leqslant \left(\int_0^{\infty} |\hat{g}(u)|^2 \, du \right)^{1/2} \left(\int_0^{\infty} e^{-2\mu u} \, du \right)^{1/2} = \frac{C_2}{\sqrt{\mu}}. \quad (3.4)$$

Denote by Γ_R the contour consisting of a segment of a line $|\lambda| \leqslant R$, $\mu = \frac{1}{2}$, and an arc resting on this segment of a circle with radius R centered at the point $z = i/2$, $\operatorname{Re} z \geqslant \frac{1}{2}$. By the Cauchy formula, if $\operatorname{Re} z_0 \geqslant \frac{1}{2}$, and the radius R is sufficiently large, we have

$$\psi_n(z_0) = \frac{1}{2\pi} \int_{\Gamma_R} \frac{\psi_n(z)}{z - z_0} \, dz.$$

Relying on the relation given in (3.4) we can easily show that the integral over the semicircle Γ_R vanishes starting from $\psi(z)/(z - z_0)$ as $R \to \infty$. Consequently, by virtue of the inequalities given in (3.3) for all $\mu \geqslant 1$ we obtain

$$|\psi_n(\lambda + i\mu)| \leqslant \frac{1}{2\pi} \int_{-\infty}^{\infty} \frac{\left| \psi_n\left(u + \dfrac{i}{2}\right) \right|}{\left| (u - \lambda) + i\left(\dfrac{1}{2} - \mu\right) \right|} \, du$$
$$\leqslant \frac{1}{2\pi} \left(\int_{-\infty}^{\infty} \left| \psi_n\left(u + \frac{i}{2}\right) \right|^2 du \int_{-\infty}^{\infty} \frac{du}{u^2 + \dfrac{1}{4}} \right)^{1/2} = C_3. \quad (3.5)$$

Next, writing $\ln|g(z)|$ as the Poisson integral (g is an exterior function, see Section II.1)

$$\ln|g(z)| = \frac{\mu}{\pi} \int_{-\infty}^{\infty} \frac{\ln|g(u)|}{(u - \lambda)^2 + \mu^2} \, du, \qquad z = \lambda + i\mu,$$

we find that if $z = \mathrm{Re}^{i\theta}$, $\theta = \pi/4, 3\pi/4$, then, as $R \to \infty$,

$$\frac{\ln|g(z)|}{|z|} \to 0. \quad (3.6)$$

In fact, for a fixed $T > 0$ and $z = \mathrm{Re}^{i\theta}$ we have

$$\frac{\mu}{\pi}\int_{-T}^{T}\frac{\ln|g(u)|}{(u-\lambda)^2+\mu^2}\,du = O\left(\frac{1}{\mu}\right) = O\left(\frac{1}{R}\right) \to 0.$$

At the same time we have

$$\begin{aligned}\frac{\mu}{\pi}\left|\int_{|u|>T}\frac{\ln|g(u)|}{(u-\lambda)^2+\mu^2}\,du\right| &\leqslant \frac{\mu^2+\lambda^2+1}{2\pi\mu}\int_{|u|>T}\frac{|\ln f(u)|}{1+u^2}\,du\\ &= \frac{R}{\pi\sqrt{2}}\frac{R^2+1}{R^2}\int_{|u|>T}\frac{|\ln f(u)|}{1+u^2}\,du\\ &= R\cdot o(1), \qquad T\to\infty,\end{aligned}$$

which proves (3.6).

It follows from (3.6) that for any $\varepsilon > 0$ and $z = \mathrm{Re}^{i\theta}$ $(\theta = \pi/4, 3\pi/4; R \geqslant 1)$ the inequality $|g(z)| \geqslant C_4 e^{-\varepsilon|z|}$ is satisfied, where the constant C_4 might depend on ε. Noting (3.5) we can obtain the estimate

$$|\varphi_n(z)| \leqslant C_5 e^{(\sigma+\varepsilon+\delta)|z|} \tag{3.7}$$

on the rays $z = \mathrm{Re}^{i\theta}$ $(\theta = \pi/4, 3\pi/4; R \geqslant 1)$.

Similarly, if we introduce the function $\psi_n^-(z) = \varphi_n\bar{g}\exp\{-iz(\sigma+\delta)\}$, where $\bar{g}(z) = \overline{g(\bar{z})} \in \mathscr{H}^{2-}$ for z from the lower half-plane, we shall obtain the estimate

$$|\varphi_n(z)| \leqslant C_5 e^{(\sigma+\varepsilon+\delta)|z|} \tag{3.8}$$

on the rays $z = \mathrm{Re}^{i\theta}$ $(\theta = 5\pi/4, 7\pi/4; R \geqslant 1)$.

We shall take next the estimate of $|\varphi_n(z)|$ on segments of the rays lying within a circle $|z| \leqslant 1$. We assume, as usual, that $\ln^+ a = \ln a$ if $a > 1$, and $\ln^+ a = 0$ if $a \leqslant 1$. Estimating the subharmonic function $\ln|\varphi_n(z)|$ with the aid of the Poisson integral, we have on the lines $|\mathrm{Im}\, z| = 1$ that

$$\begin{aligned}\ln|\varphi_n(z)| &\leqslant \frac{1}{\pi}\int_{-\infty}^{\infty}\frac{\ln|\varphi_n(u)|}{(u-\lambda)^2+1}\,du\\ &\leqslant \frac{1}{\pi}\sup_u\frac{u^2+1}{(u-\lambda)^2+1}\int_{-\infty}^{\infty}\frac{\ln^+|\varphi_n(u)|}{u^2+1}\,du\\ &\leqslant \frac{1}{\pi}\frac{\lambda^2+2}{2}\int_{-\infty}^{\infty}\frac{\ln^+|\varphi_n(u)|}{u^2+1}\,du\\ &= \frac{1}{\pi}\frac{\lambda^2+2}{2}\int_{-\infty}^{\infty}\frac{\ln^+\dfrac{|\varphi_n(u)||g(u)|}{|g(u)|}}{1+u^2}\,du\\ &\leqslant \frac{1}{\pi}\frac{\lambda^2+2}{2}\left(\int_{-\infty}^{\infty}\frac{|\ln|\varphi_n(u)g(u)||}{1+u^2}\,du + \int_{-\infty}^{\infty}\frac{|\ln|g(u)||}{1+u^2}\,du\right)\\ &\leqslant C_6(\lambda^2+2)\left(\|\varphi_n\|_F + \int_{-\infty}^{\infty}\frac{|\ln f(u)|}{1+u^2}\,du\right) \leqslant C_7(\lambda^2+2).\end{aligned}$$

Therefore, the function $e^{-C_7z^2}\varphi_n(z)$, analytic in a strip $|\text{Im } z| \leqslant 1$, is bounded in this strip and satisfies on its boundaries the inequality $|e^{-C_7z^2}\varphi_n(z)| \leqslant C_8$, where C_8 is independent of n. According to the Phragmen–Lindelof principle the last inequality holds true at each point of the strip indicated. In particular, all the functions $\varphi_n(z)$ are uniformly bounded in the circle $|z| \leqslant 1$ From this, and from (3.7) and (3.8), it follows that the inequality given in (3.2) holds true, which proves, as mentioned above, the first part of the theorem.

2. *Proof of the inclusion* $L^0(F) \supset D_0 \cap L(F)$. Let $\varphi(\lambda) \in D_0 \cap L(F)$. By the familiar Hadamard theorem* the function $\varphi(\lambda)$, the entire function of zero degree, can be written as the product

$$\varphi(z) = z^m e^b \prod_1^\infty \left(1 - \frac{z}{z_n}\right) e^{z/z_n}, \qquad \sum \frac{1}{|z_n|^2} < \infty, \tag{3.9}$$

where the $z_n \neq 0$ are zeros of $\varphi(z)$. The function $\varphi(z)$ can be rewritten as the sum $\varphi_1(z) + \varphi_2(z)$, where $\varphi_1(z) = \frac{1}{2}(\varphi(z) + \varphi(-z))$ and $\varphi_2(z) = \frac{1}{2}(\varphi(z) - \varphi(-z))$ are even and odd, respectively. In this case, we have as before that $\varphi_1, \varphi_2 \in D_0 \cap L(F)$. Therefore, we need only to prove the theorem for even functions and odd functions. In both cases the proofs are identical, and for the sake of definiteness we shall consider the even functions $\varphi(\lambda)$.

We need to prove that for any $\varepsilon > 0$ there is a function $\varphi_\varepsilon \in L^\varepsilon(F)$ for which $\|\varphi - \varphi_\varepsilon\|_F \leqslant \varepsilon$.

Let us note first that each square-summable entire function $\varphi(\lambda)$ of finite degree $\leqslant \varepsilon$ belongs to $L^\varepsilon(F)$. In fact, each function of this kind belongs to $L(F)$. By the Paley–Wiener theorem† the function φ has, relative to entire functions from $\mathscr{L}^2$, the Fourier transform $\hat{\varphi}$ equal to zero outside of $[-\varepsilon, \varepsilon]$. We infer that

$$\varphi(\lambda) = \int_{-\varepsilon}^{\varepsilon} e^{i\lambda u}\hat{\varphi}(u)\, du$$

and is finite, $\varphi \in L^\varepsilon(F)$.

Therefore, it suffices to construct a square-summable integral function of degree $\leqslant \varepsilon$ adequately approximating φ. We note that the Hadamard factorization given by (3.9) for the even function φ will be of the form

$$\varphi(\lambda) = \lambda^{2m} \prod_1^n \left(1 - \frac{\lambda^2}{z_n^2}\right), \qquad m \geqslant 0.$$

We shall define the function $\varphi_\varepsilon(\lambda)$ by the equality

$$\varphi_\varepsilon(\lambda) = \lambda^{2m} \prod_{|z_n| < d} \left(1 - \frac{\lambda^2}{z_n^2}\right) \prod_{n > d\delta} \left(1 - \frac{\lambda^2\delta^2}{n^2}\right),$$

where $\delta = \varepsilon/\pi$; the number $d = d(\varepsilon)$ will be defined later on.

* See, for example, [20], p. 525.

† See, for example, [15], p. 82.

We shall show next that φ_ε is a square-summable integral function of degree $\leqslant \varepsilon$ (and, therefore, $\varphi_\varepsilon \in L^2(F)$). The Euler formula

$$\sin \pi\lambda = \pi\lambda \prod_1^\infty \left(1 - \frac{\lambda^2}{n^2}\right)$$

enables us to write φ_ε as

$$\varphi_\varepsilon(\lambda) = \frac{\sin \pi\,\delta\lambda}{\pi\delta}\,\frac{\lambda^{2m-1} \prod\limits_{|z_n|<d} \left(1 - \dfrac{\lambda^2}{z_n^2}\right)}{\prod\limits_{n \leqslant d\delta} \left(1 - \dfrac{\lambda^2\delta^2}{n^2}\right)}. \tag{3.10}$$

Therefore $\varphi_\varepsilon(\lambda)$ is an entire function of degree $\pi\delta = \varepsilon$. We estimate next the ratio of polynomials in the right-hand side of (3.10) for large λ. Let us introduce the monotone nondecreasing function $N_\varphi(R)$ equal to the number of roots of the function $\varphi(z)$ in a circle $|z| \leqslant R$. The function $N_\varphi(R)$ is closely related to the growth order of the function $\varphi(z)$. In particular, $N\varphi(R) = o(R)$, $R \to \infty$,* for functions of zero degree. Therefore, the degree of the polynomial in the numerator of (3.10) is equal to $o(d)$, whereas the degree of the polynomial in the denominator is equal to $\sim 2d\delta$, and $\varphi_\varepsilon \in \mathscr{L}_2(-\infty, \infty)$ for sufficiently large d.

It remains only to estimate $\|\varphi_\varepsilon - \varphi\|_F$. To this end we prove the following lemma characterizing the proximity of φ_ε to φ. □

Lemma 1. *For given $\varepsilon > 0$, $A < \infty$, we can find a number d_0 such that for all $d \geqslant d_0$ we have the inequalities*

$$\max_{|\lambda| \leqslant A} |\varphi(\lambda) - \varphi_\varepsilon(\lambda)| \leqslant \frac{\varepsilon}{\sqrt{2}} \left(\int_{-A}^{A} f(\lambda)\, d\lambda\right)^{-1/2},$$

$$\max_{A \leqslant |\lambda| \leqslant d/2} \left|\frac{\varphi_\varepsilon(\lambda)}{\varphi(\lambda)}\right| \leqslant 1, \qquad \max_{|\lambda| \geqslant d/2} |\varphi(\lambda)| \leqslant 1. \tag{3.11}$$

The first inequality in (3.11) is obvious. To prove the second inequality in (3.11) we shall use the elementary inequality

$$e^{-2x} \leqslant 1 - x \leqslant e^{-x}, \qquad 0 \leqslant x < \frac{1}{4},$$

and write

$$\left|\frac{\varphi_\varepsilon}{\varphi}\right| = \frac{\prod\limits_{n > d\delta} \left(1 - \dfrac{\lambda^2\delta^2}{n^2}\right)}{\prod\limits_{|z_n| \geqslant d} \left|1 - \dfrac{\lambda^2}{z_n^2}\right|} \leqslant \frac{\prod\limits_{n > d\delta} \left|1 - \dfrac{\lambda^2\delta^2}{n^2}\right|}{\prod\limits_{|z_n| \geqslant d} \left(1 - \dfrac{\lambda^2}{|z_n|^2}\right)}$$

$$\leqslant \exp\left\{-\lambda^2\delta^2 \sum_{n > d\delta} \frac{1}{n^2}\right\} \exp\left\{2\lambda^2 \sum_{|z_n| \geqslant d} \frac{1}{|z_n|^2}\right\}. \tag{3.12}$$

* See [20], p. 521.

To estimate the first multiplier in (3.12) we note that

$$\sum_{n>d\delta} \frac{1}{n^2} \geqslant \int_{d\delta+1} \frac{dx}{x^2} = \frac{1}{d\delta+1}.$$

To estimate $\sum_{|z_n|\geqslant d} (1/|z_n|^2)$ we take again the distribution function of zeros, $N_\varphi(R)$. Noting that $N_\varphi(R) = o(R)$ we have

$$\begin{aligned}\sum_{|z_n|\geqslant d} \frac{1}{|z_n|^2} &= \int_d^\infty \frac{dN_\varphi(R)}{R^2} \\ &= -\frac{N_\varphi(d)}{d^2} + 2\int_d^\infty \frac{N_\varphi(R)}{R^3}\,dR \\ &\leqslant 2\sup_{R\geqslant d} \frac{N_\varphi(R)}{R} \int_d^\infty \frac{dR}{R^2} = \frac{2}{d}\sup_{R\geqslant d} \frac{N_\varphi(R)}{R} \leqslant \frac{\delta}{2d},\end{aligned}$$

for d sufficiently large. Putting the estimates obtained into (3.12), we have that

$$\max_{|\lambda|\leqslant d/2} \left|\frac{\varphi_\varepsilon(\lambda)}{\varphi(\lambda)}\right| \leqslant \max_{|\lambda|\leqslant d/2} \exp\left\{-\lambda^2\left(\frac{\delta^2}{d\delta+1} - \frac{\delta}{2d}\right)\right\} = 1,$$

letting $d \geqslant 1/\delta$, thus proving the second inequality in (3.11).

Let us prove the last inequality in the lemma. Let $|\lambda| \geqslant d/2$. Estimating the second multiplier in the expression for φ_ε, we find that

$$\begin{aligned}\ln\left(\prod_{|z_n|<d} \left|1 - \frac{\lambda^2}{z_n^2}\right|\right) &= \sum_{|z_n|<d} \ln\left|1 - \frac{\lambda^2}{z_n^2}\right| \\ &= \int_0^d \ln\left|1 - \frac{\lambda^2}{R^2}\right| dN_\varphi(R) \\ &\leqslant \int_0^d \ln\left(1 + \frac{\lambda^2}{R^2}\right) dN_\varphi(R) \\ &= N_\varphi(d)\ln\left(1 + \frac{\lambda^2}{d^2}\right) + \int_0^d \frac{2\lambda^2}{\lambda^2 + R^2}\frac{N_\varphi(R)}{R}\,dR \\ &\leqslant 2N(d)\ln\frac{5|\lambda|}{d} + 2\int_0^d \frac{N_\varphi(R)}{R}\,dR \\ &= o(d\ln d) + o\left(d\ln\frac{5|\lambda|}{d}\right). \end{aligned} \tag{3.13}$$

To estimate the last multiplier in the formula for φ_ε we shall consider separately the case $d/2 \leqslant |\lambda| \leqslant d$ and the case $|\lambda| > d$. If $\lambda \leqslant d$, then

$$\begin{aligned}\prod_{n>d\delta}\left(1 - \frac{\lambda^2\delta^2}{n^2}\right) &\leqslant \exp\left\{-\lambda^2\delta^2 \sum_{n>d\delta} \frac{1}{n^2}\right\} \\ &\leqslant \exp\left\{-\frac{\lambda^2\delta^2}{d\delta+1}\right\} \leqslant e^{-(\lambda\delta/2)},\end{aligned}$$

if only $d\delta \geqslant 1$. If $|\lambda| > d$, then, using again the Euler decomposition, we find that

$$\prod_{n > d\delta} \left|1 - \frac{\lambda^2\delta^2}{n^2}\right| = \frac{\sin \pi\delta\lambda}{\pi\delta\lambda \prod_{n < d\delta} \left|1 - \frac{\lambda^2\delta^2}{n^2}\right|}.$$

For $n < d\delta/\sqrt{2}$ we shall have

$$\left|1 - \frac{\lambda^2\delta^2}{n^2}\right| > \frac{\lambda^2\delta^2}{2n^2};$$

therefore, by the Stirling formula for $n!$, we have

$$\prod_{n > d\delta} \left|1 - \frac{\lambda^2\delta^2}{n^2}\right| \leqslant \frac{1}{\pi\delta} \exp\left\{-d\delta\sqrt{2}\ln\frac{|\lambda|\delta}{\sqrt{2}} + d\delta\sqrt{2}\ln\frac{d\delta}{\sqrt{2}}\right\}$$

$$= \frac{1}{\pi\delta} \exp\left\{-d\delta\sqrt{2}\ln\frac{|\lambda|}{d}\right\}.$$

Comparing the estimates thus obtained with (3.13) we can see that, if the d is sufficiently large and $|\lambda| \geqslant d/2$, then $|\varphi_\varepsilon(\lambda)| \leqslant 1$, thus proving the lemma.

Using this lemma we can immediately obtain the estimate for $\|\varphi - \varphi_\varepsilon\|_F$. By (3.11) we have

$$\|\varphi - \varphi_\varepsilon\|_F^2 \leqslant \int_{-A}^{A} |\varphi(\lambda) - \varphi_\varepsilon(\lambda)|^2 f(\lambda)\, d\lambda$$

$$+ 5\int_{|\lambda| \geqslant A} |\varphi(\lambda)|^2 f(\lambda)\, d\lambda + 2\int_{|\lambda| > d/2} f(\lambda)\, d\lambda < \varepsilon^2,$$

if the A and d are chosen to be sufficiently large. Therefore Theorem 3 is proved. □

II.4 The Projection of $L^+(F)$ on $L^-(F)$

It is a well-known fact* that the projections of the spaces $L_T(F)$, $T = [a, \infty)$, $a \geqslant 0$, on the space $L^-(F)$ play an essential role in prediction theory. They appear in investigating the conditions for random processes to be regular (see Chapter IV). We prove here a few theorems on the construction of the space $L^{+|-}(F)$ being the projection of $L^+(F)$ on $L^-(F)$.

By Theorems 1 and 2 it suffices to consider random processes with spectral density $f(\lambda)$. The latter satisfies the conditions given by (2.1) for processes with discrete time or those given by (2.2) for processes with continuous time. We recall that the spectral density $f(\lambda)$ permits the factorization

$$f(\lambda) = |g(e^{i\lambda})|^2 = g(e^{i\lambda}) \cdot \bar{g}(e^{i\lambda})$$

(time t is discrete), where $g(z)$ is a function of class $\mathscr{H}^{2+}$ with $|z| < 1$, and $g(z) = g(1/\bar{z})$ is a function of class $\mathscr{H}^{2-}$ with $|z| > 1$. Similarly, for processes

* See, for example, [24].

with continuous time we infer that

$$f(\lambda) = |g(\lambda)|^2 = g(\lambda)\cdot\bar{g}(\lambda),$$

$g \in \mathscr{H}^{2+}$, $\operatorname{Im} z > 0$; $\bar{g}(z) = \overline{g(\bar{z})}$, $\operatorname{Im} z < 0$, $\bar{g} \in \mathscr{H}^{2-}$.

The inclusions

$$L^-(F) \supseteq L^{+|-}(F) \supseteq L^-(F) \cap L^+(F) \supseteq L^0(F) \tag{4.1}$$

are obvious.

Our objective is to find out when each of these inclusions becomes an equality.

We describe first the analytic form of operators $\mathscr{P}^-$ (denoted also by $\mathscr{P}$) and $\mathscr{P}^+$ projecting $L_T(F)$ on $L^-(F)$ and $L^+(F)$, respectively. Denote by π^- $(=\pi)$ and π^+ the projection operators in $\mathscr{L}^2(-\infty, \infty)$ $(\mathscr{L}^2(-\pi, \pi))$ on $\mathscr{H}^{2-}$ and $\mathscr{H}^{2+}$, respectively. These operators can be defined as follows. Let $\varphi \in \mathscr{L}^2(-\infty, \infty)$ and let $\varphi(\lambda) = \int_{-\infty}^{\infty} e^{i\lambda u}\hat{\varphi}(u)\,du$ where $\hat{\varphi} \in \mathscr{L}^2(-\infty, \infty)$. Then we have

$$\pi^-\varphi(\lambda) = \int_{-\infty}^{0} e^{i\lambda u}\hat{\varphi}(u)\,du, \qquad \pi^+\varphi(\lambda) = \int_0^{\infty} e^{i\lambda u}\varphi(u)\,du.$$

Similarly, if $\varphi \in \mathscr{L}^2(-\pi, \pi)$, $\varphi(\lambda) = \sum_{-\infty}^{\infty} e^{i\lambda j}a_j$ where $\sum |a_j|^2 < \infty$, then we have

$$\pi^-\varphi = \sum_{-\infty}^{0} e^{i\lambda j}a_j, \qquad \pi^+\varphi = \sum_{0}^{\infty} e^{i\lambda j}a_j.$$

Lemma 2. *The following relations hold*:

$$\mathscr{P}^-\varphi(\lambda) = \bar{g}^{-1}(\lambda)\pi^-\bar{g}(\lambda)\varphi(\lambda), \qquad \mathscr{P}^+ = g^{-1}\pi^+g.$$

Proof. We consider $\mathscr{P}^- = \mathscr{P}$ only.

Let us consider here the case of continuous time t (the case of discrete time is considered in a similar way). We need to prove that

$$\mathscr{P}^* = \mathscr{P}, \qquad \mathscr{P}^2 = \mathscr{P}, \qquad \mathscr{P}L(F) \subseteq L^-(F), \qquad \mathscr{P}L^-(F) = L^-(F).$$

The first two equalities imply that $\mathscr{P}$ is a projection.* The two equalities on the left imply that $\mathscr{P}$ is the projection on $L^-(F)$. Since π is a projection it follows that $\pi^* = \pi$, $\pi^2 = \pi$. Hence

$$\begin{aligned}\langle \mathscr{P}\varphi, \psi\rangle_F &= (\pi(\bar{g}\varphi), \bar{g}\psi)_2\\ &= (\bar{g}\varphi, \pi(\bar{g}\psi))_2\\ &= (g\bar{g}\varphi, \bar{g}^{-1}\pi\bar{g}\psi)_2 = \langle \varphi, \mathscr{P}\psi\rangle_F.\end{aligned}$$

The equality $\mathscr{P}^2 = \mathscr{P}$ follows obviously from $\pi^2 = \pi$.

* See, for example, [2], p. 111.

Further, we showed in Section II.2 that $\bar{g}\varphi \in \mathscr{H}^{2-}$ if $\varphi \in L^-(F)$ and that $\psi\bar{g}^{-1} \in L^-(F)$ if $\psi \in \mathscr{H}^{2-}$. $\mathscr{P}\varphi = \bar{g}^{-1}\pi\bar{g}\varphi \in L^-(F)$ since $\pi(\bar{g}\varphi) \in \mathscr{H}^{2-}$. Finally,

$$\mathscr{P}\varphi = \bar{g}^{-1}\pi(\bar{g}\varphi) = \bar{g}^{-1}\bar{g}\varphi = \varphi \quad \text{if } \varphi \in L^-(F).$$

The lemma is proved.

Theorem 4. *$L^-(F) \neq L^{+|-}(F)$ if and only if $\theta = g/\bar{g}$ coincides with the ratio of two inner functions; $L^{+|-}(F) = L^-(F) \cap L^+(F)$ if and only if $\theta = g/\bar{g}$ is an inner function.*

Proof. This theorem is to be proved for processes with discrete time in a fashion similar to that for processes with continuous time. We give here the proof of the first part of the theorem for the case of discrete time and the proof of the second part of the theorem for the case of continuous time.

Let $\theta = s_1(e^{i\lambda})/s_2(e^{i\lambda})$ where $s_1(z)$ and $s_2(z)$ are inner functions; therefore $|s_1(e^{i\lambda})| = |s_2(e^{i\lambda})| \equiv 1$. Assume $s_1(0) = s_2(0) = 0$; otherwise, we could replace s_1 and s_2 by zs_1 and zs_2 without changing θ. Consider the function $\varphi = \bar{g}^{-1}(s_1\bar{g})$ and show that $\varphi \in L^-(F)$. But $\varphi \perp L^{+|-}(F)$ ($\varphi \neq 0$, naturally); therefore $L^- \neq L^{+|-}$. Obviously, $\bar{s}_1\bar{g} \in \mathscr{H}^{2-}$ and hence $\varphi \in L^-$. Further, we can write φ as

$$\varphi = \frac{1}{g}\frac{g}{\bar{g}}(\bar{s}_1\bar{g}) = \frac{1}{g}\frac{|s_1|^2}{s_2}\bar{g} = \frac{1}{g}(\bar{s}_2\bar{g}),$$

and since $s_2(0) = 0$ we infer that $\varphi g \in e^{-i\lambda}\mathscr{H}^{2-}$. Next let ψ be an arbitrary element of $L^+(F)$. We have

$$\begin{aligned}\langle\varphi, \mathscr{P}\psi\rangle_F &= (\varphi\bar{g}, \pi(\bar{g}\psi))_2 \\ &= (\pi(\varphi\bar{g}), \bar{g}\psi)_2 = (\varphi\bar{g}, \bar{g}\psi)_2 = (\varphi g, \psi g)_2 = 0.\end{aligned}$$

Since $\varphi g \in e^{-i\lambda}\mathscr{H}^{2-}$ it follows that $\psi g \in \mathscr{H}^{2+}$.

Conversely, let $L^- \neq L^{+|-}$ and let $\varphi \in L^-$, but let $\varphi \perp L^{+|-}$. Write the function φ as $\bar{\varphi}_1/\bar{g}$, $\varphi_1 \in \mathscr{H}^{2+}$. The function $\varphi_1 \in \mathscr{H}^2$ can be expressed as the product $\varphi_1 = \theta_1\gamma_1$ where θ_1 is an inner function and $\gamma_1 \in \mathscr{H}^2$ is an outer function (see Section II.1). We note that the function φg is orthogonal to $\mathscr{H}^{2+}$ in $\mathscr{L}^2(-\pi, \pi)$. In fact, if $\psi \in \mathscr{H}^{2+}$, we have

$$\begin{aligned}(\varphi g, \psi)_2 &= \left(\varphi\bar{g}, \frac{\bar{g}}{g}\psi\right)_2 \\ &= \left(\pi(\varphi\bar{g}), \frac{\bar{g}}{g}\psi\right)_2 \\ &= \left(\frac{1}{\bar{g}}\pi\bar{g}\varphi, \frac{\psi}{g}|g|^2\right)_2 = \left\langle\mathscr{P}\varphi, \frac{\psi}{g}\right\rangle_F = \left\langle\varphi, \mathscr{P}\frac{\psi}{g}\right\rangle_F = 0\end{aligned}$$

since $\psi/g \in L^+(F)$. Hence $\varphi g \in \mathscr{H}^{2-}$ and, therefore, $\varphi g = \bar{\theta}_2 \cdot \bar{\gamma}_2$, where θ_2 and γ_2 are an inner function and an outer function, respectively. Thus we have

$$\varphi g = \frac{g}{\bar{g}} \bar{\varphi}_1 = \frac{g}{\bar{g}} \bar{\theta}_1 \bar{\gamma}_1 = \bar{\theta}_2 \bar{\gamma}_2.$$

Since $|\theta_1| = |\theta_2| \equiv 1$ (θ_1 and θ_2 are inner), it follows that $|\gamma_1| = |\gamma_2|$. The outer functions $\gamma_1(z)$ and $\gamma_2(z)$ coinciding on $|z| = 1$ are equal; therefore

$$\frac{g}{\bar{g}} = \frac{\bar{\theta}_2}{\bar{\theta}_1} = \frac{\theta_1}{\theta_2}.$$

The first part of the theorem is proved.

Next we prove the second part of the theorem. Note first that $L^-(F) \neq L(F)$ by virtue of the assumptions made before and, therefore, $L^-(F) \neq L^-(F) \cap L^+(F)$. Hence, according to what was proved above, in investigating the equality $L^- \cap L^+ = L^{+|-}$ we may assume $\theta = \theta_1/\theta_2$, where θ_1 and θ_2 are interior functions without common divisor.

Our proof is based on the following:

$$L^{+|-}(F) = L^-(F) \cap \frac{1}{\theta_2} L^+(F), \tag{4.2}$$

from which it immediately follows that $L^{+|-} = L^- \cap L^+$ if and only if $\theta_2 \equiv 1$, that is, if $\theta = \theta_1/\theta_2 \equiv \theta$ is an inner function.

To prove (4.2) we consider the orthogonal complement M of the space $L^{+|-}$ in the space L^- and show that

$$M = L^- \cap \frac{\bar{g}}{g} L^- = \frac{1}{\bar{g}} \mathscr{H}^{2-} \cap \frac{1}{g} \mathscr{H}^{2-}. \tag{4.3}$$

In fact, let us write the elementary element $\phi \in M \subset L^-$ as $\varphi = \bar{\varphi}_1/\bar{g}$ where $\varphi_1 \in \mathscr{H}^2$. If φ is orthogonal to

$$L^{+|-} = \mathscr{P}L^+ = \frac{1}{\bar{g}} \pi \frac{\bar{g}}{g} \mathscr{H}^{2+},$$

then $\bar{\varphi}_1$ is orthogonal to $\pi(\bar{g}/g)\mathscr{H}^{2+}$ in a space $\mathscr{L}^2(-\infty, \infty)$. This implies, since $\pi\bar{\varphi}_1 = \bar{\varphi}_1$, that $\bar{\varphi}_1$ is orthogonal to $(\bar{g}/g)\mathscr{H}^{2+}$ and $(g/\bar{g})\bar{\varphi}_1 = \varphi \cdot g$ is orthogonal to the whole of $\mathscr{H}^{2+}$. Therefore, $\varphi g \in \mathscr{H}^{2-}$, which is equivalent to (4.3).

We now expand the function $\bar{\varphi}_1$ (belonging to $\mathscr{H}^{2-}$ as well as $\theta^{-1}\mathscr{H}^{2-}$ by virtue of (4.3)) as a product of its interior and exterior parts. We have

$$\bar{\varphi}_1 = \bar{\gamma}_3 \bar{\theta}_3 = \frac{\theta_2}{\theta_1} \bar{\gamma}_4 \bar{\theta}_4,$$

where the γ_i are the outer multipliers and the θ_i are inner multipliers. It is seen that $\gamma_3 = \gamma_4$; therefore

$$\theta_1 \theta_4 = \theta_2 \theta_3$$

and, since θ_2 and θ_1 have no common multipliers, the function θ_2 must divide θ_4. Therefore (4.3) can be rewritten as

$$M = \frac{1}{\bar{g}}\mathscr{H}^{2-} \cap \frac{1}{g\theta_2}\mathscr{H}^{-2} = L^- \cap \frac{\bar{g}}{g\theta_2}L^-. \tag{4.4}$$

Next, relying on the relation given by (4.4), we find the orthogonal complement of M in L^- (by definition, equal to $L^{+|-}$). Note that

$$\frac{\bar{g}}{g}\frac{1}{\theta_2}L^- = \bar{\theta}_1 L^- = \frac{1}{\bar{g}}\bar{\theta}_1\mathscr{H}^{-2} \subset \frac{1}{\bar{g}}\mathscr{H}^{2-} = L^-.$$

Hence it suffices to find the orthogonal complement of

$$\frac{\bar{g}}{g}\frac{1}{\theta_2}L^-$$

in $L(F)$ or, what is the same, to find the orthogonal complement of the space $(1/\theta_2)\mathscr{H}^{2-}$ in the whole space $\mathscr{L}^2(-\infty, \infty)$. The latter is, obviously, $(1/\theta_2)\mathscr{H}^{2+}$, to which the space $(1/\theta_2)L^+(F)$ corresponds in $L(F)$. Hence (4.2) and the theorem are proved. □

Let us find out when $L^{+|-}(F) = L^0(F)$ (this fact implies that we are dealing with the case of continuous time). According to Theorem 3 there is a complete description of the space $L^{+|-}$ for the equality indicated.

Theorem 5. *Suppose the condition given by (2.2) is satisfied. We have the equality $L^{+|-}(F) = L^0(F)$ if and only if $1/f(\lambda)$ is an entire analytic function of zero degree.*

Proof. Let $L^{+|-}(F) = L^0(F)$. Consider two cases which follow.

1. $f/(1+\lambda^2) = f_1 \in \mathscr{L}^1(-\infty, \infty)$. Consider a new spectral measure with spectral density f_1. Obviously, $f_1 = |g_1|^2$ where $g_1 = g/(\lambda + i) \in \mathscr{H}^{2+}$. Let us prove that $1/\bar{g} \in L^0(F_1)$. From this and Theorem 3 it follows that $1/g$ and, therefore, $1/f = 1/(g\bar{g})$ are entire functions of zero degree.

By virtue of (4.1) we have the equality $L^{+|-}(F) = L^-(F) \cap L^+(F)$, and, by Theorem 4, $g/\bar{g} = \theta$ is an interior function. It is clear that in this case $\theta_1 = g_1/\bar{g}_1 = \theta((\lambda - i)/(\lambda + i))$ is also an interior function. Consider the function $\varphi = \theta/g_1(\lambda + i) \in L^+(F_1)$ and find its projection on $L^-(F_1)$. Assuming $\mathscr{P}$ to be the projector in $L(F_1)$ on $L^-(F_1)$, we have

$$\begin{aligned}\mathscr{P}\varphi &= \frac{1}{\bar{g}_1}\pi\bar{g}_1\varphi \\ &= \frac{1}{\bar{g}_1}\pi\frac{\bar{g}_1}{g_1}\frac{\theta}{\lambda + i} \\ &= \frac{1}{\bar{g}_1}\pi\frac{\lambda + i}{\lambda - i}\frac{1}{\theta_1}\frac{\theta_1}{\lambda + i} = \frac{1}{\bar{g}_1(\lambda - i)} = \frac{1}{\bar{g}},\end{aligned}$$

since $1/(\lambda - i) = i\int_{-\infty}^{0} e^{iu(\lambda - i)}\,du$ belongs to $\mathscr{H}^{2-}$ and $\pi(1/(\lambda - i)) = 1/(\lambda - i)$. We have thus proved that $1/\overline{g} = \mathscr{P}\varphi \in L^{+|-}(F_1) = L^0(F_1)$.

2. $f_1 \in \mathscr{L}'(-\infty, \infty)$. Here (see Section II.3) the space $L^0(F)$ as well as the space $L^{+|-}(F)$ (equal to $L^0(F)$) contain only constants, $\dim L^{+|-} = 1$. It will follow from Theorem 6 (given below) that $f = \text{constant}/(1 + \lambda^2)$ necessarily.

The first part of the theorem is proved. To prove the second part of the theorem suppose that $1/f(\lambda) = b(\lambda)$ is an entire function of zero degree. Let us prove first that $L^{+|-} = L^+ \cap L^-$. According to Theorem 4 we need only to show that the function $\theta = g/\overline{g}$ is interior.

By a theorem due to N. I. Akhiezer (see [19], p. 567) the function $b(\lambda)$, nonnegative on the real axis, can be expressed as

$$b(\lambda) = \omega(\lambda)\overline{\omega}(\lambda) = |\omega(\lambda)|^2,$$

where $\omega(z)$ and $\overline{\omega}(z)$ are entire functions of zero degree, the function ω is an exterior function without zeros in the upper half-plane, and $\overline{\omega}(z) = \overline{\omega(\overline{z})}$ is an exterior function in the lower half-plane. Then $g = 1/\omega$ and $\overline{g} = 1/\overline{\omega}$. Next, let z_j be zeros of the function $b(z)$ lying in $\operatorname{Im} z < 0$. For $\operatorname{Im} z \geqslant 0$, $g/\overline{g} = \overline{\omega}/\omega$ is an analytic function with

$$|\theta(z)| = \left|\frac{g(z)}{\overline{g}(z)}\right| = \left|\frac{\overline{\omega}(z)}{\omega(z)}\right| = \prod_n \left|\frac{1 - \dfrac{z}{z_n}}{1 - \dfrac{z}{\overline{z}_n}}\right| \leqslant 1, \qquad |\theta(\lambda)| = 1,$$

the latter proving that θ is an interior function.

It remains to show that each element $\varphi \in L^+ \cap L^-$ is an entire function of zero degree. Let us prove first that φ is an entire function. By virtue of (2.4) we have

$$\varphi(\lambda) = \omega(\lambda)h^+(\lambda) = \overline{\omega}(\lambda)h^-(\lambda),$$

where $h^+ \in \mathscr{H}^{2+}$, $h^- \in \mathscr{H}^{2-}$. Therefore, $\varphi(\lambda)$ is the common boundary value of the functions $\omega(z)h^+(z)$ and $\overline{\omega}(z)h^-(z)$ analytic in $\operatorname{Im} z > 0$ and $\operatorname{Im} z < 0$, respectively. Let us prove that it is possible to extend them across the real line.

Let $\Phi(z) = \int_0^z \varphi(\xi)\,d\xi$ where the integral is taken over a segment of the line connecting the points 0 and z, and $\varphi(\xi)$ is either $\omega(\xi)h^+(\xi)$ or $\overline{\omega}(\xi)h^-(\xi)$. The function Φ is analytic in the upper and lower half-planes.

For the functions $h^+ \in \mathscr{H}^{2+}$ we infer that

$$\int_{-\infty}^{\infty} |h^+(\lambda + i\mu)|^2\,d\lambda \leqslant \int_{-\infty}^{\infty} |h^+(\lambda)|^2\,d\lambda = C_1 < \infty, \qquad \mu > 0. \tag{4.5}$$

Hence for all a, $|a| < \infty$, we have

$$\begin{aligned}\int_0^a |\varphi(\lambda + i\mu)|\,d\lambda &= \int_0^a |\omega(\lambda + i\mu)h^+(\lambda + i\mu)|\,d\mu \\ &\leqslant \max_{|z| \leqslant a} |\omega(z)|\sqrt{a}C_1^{1/2} \leqslant C_2 e^a.\end{aligned} \tag{4.6}$$

By the Paley–Wiener theorem we have

$$h^+(z) = \int_0^\infty e^{izu}\hat{h}(u)\,du,$$

$$\hat{h} \in \mathscr{L}^2(-\infty, \infty), \qquad \operatorname{Im} z > 0.$$

Therefore

$$|h^+(\lambda + i\mu)| \leqslant \left(\int_0^\infty e^{-2\mu u}\,du\right)^{1/2} C_1^{1/2} \leqslant \sqrt{\frac{C_1}{2\mu}}. \tag{4.7}$$

Hence

$$\int_0^\varepsilon |\varphi(a + i\mu)|\,d\mu = \int_0^\varepsilon |\omega(a + i\mu)h^+(a + i\mu)|\,d\mu \leqslant C_3 e^a\sqrt{\varepsilon}. \tag{4.8}$$

For almost all λ we obtain

$$\lim_{\mu\to 0} h^+(\lambda + i\mu) = h^+(\lambda).$$

From this and (4.6) it follows that

$$\lim_{\mu\to 0}\int_{i\mu}^{\lambda + i\mu} \varphi(\xi)\,d\xi = \int_0^\lambda \varphi(\xi)\,d\xi. \tag{4.9}$$

The integral of the analytic function $\varphi(\xi)$ over the side of a triangle with vertices at points 0 and $z = \lambda + i\mu$ ($\lambda = 0$). By virtue of (4.8) and (4.9),

$$\lim_{\mu\to 0} \Phi(z) = \lim_{\mu\to 0}\int_0^z \varphi(\xi)\,d\xi = \int_0^\lambda \varphi(\xi)\,d\xi = \Phi(\lambda), \qquad \operatorname{Im} z > 0.$$

A similar result is correct also for z in the lower half-plane. Therefore, the function $\Phi(z)$ is analytic in the upper and lower half-planes and is continuous in the entire complex plane. The inverse of Cauchy's theorem, known as Morera's theorem (see [20], p. 186) allows us to assert that $\Phi(z)$ is analytic in the entire plane. In this case the derivative $\varphi(z) = \Phi(z)$ is an entire function.

We need only to estimate the growth rate $|\varphi(\mathrm{Re}^{i\theta})|$, $R \to \infty$. We have $|\omega(z)| \leqslant C_\varepsilon e^{\varepsilon R}$, $|\overline{\omega}(z)| \leqslant C_\varepsilon e^{\varepsilon R}$ on a circle $|z| = R$. As regards $h^+(z)$ and $h^-(z)$, we have by the Paley–Wiener theorem for $z = \mathrm{Re}^{i\theta}$, $0 < \theta < \pi$,

$$\begin{aligned}|h^+(z)| &= \left|\int_0^\infty e^{izu}\hat{h}(u)\,du\right| \\ &\leqslant \int_0^\infty e^{-R\sin\theta u}|\hat{h}(u)|\,du \\ &\leqslant \frac{1}{\sqrt{2R\sin\theta}}\left(\int_0^\infty |\hat{h}(u)|^2\,du\right)^{1/2} = \sqrt{\frac{C_1}{2R\sin\theta}}.\end{aligned}$$

Similarly, we have on $z = \mathrm{Re}^{i\theta}$, $\pi < \theta < 2\pi$,

$$|h^-(z)| \leqslant \sqrt{\frac{C_1}{2R|\sin\theta|}}.$$

Hence it follows that for large R

$$\int_0^{2\pi} \ln^+|\varphi(\mathrm{Re}^{i\theta})|\,d\theta \leqslant \varepsilon R. \tag{4.10}$$

By the Poisson–Jensen formula (see [20], p. 456)

$$\ln|\varphi(\mathrm{Re}^{i\theta})| = \frac{1}{2\pi}\int_0^{2\pi} \ln|\varphi(\rho e^{i\alpha})| \frac{\rho^2 - R^2}{\rho^2 + R^2 - 2R\rho\cos(\alpha - \theta)}\, d\alpha$$
$$+ \sum \ln \frac{\rho|z - a_u|}{|\rho^2 - \bar{a}_u z|} - \alpha_0 \ln \frac{\rho}{R},$$

where $z = \mathrm{Re}^{i\theta}$, $\rho > R$, the sum to the right is extended to all zeros $a_u \neq 0$ of the function $\varphi(z)$ lying in the circle $|z| \leqslant \rho$, and α_0 is the zero multiplicity $\varphi(z)$ at the point $z = 0$. If here we set $\rho = 2R$ and note that

$$\ln \frac{\rho|z - a_u|}{|\rho^2 - \bar{a}_u z|} \leqslant 0,$$

we shall have by virtue of (4.10)

$$\ln|\varphi(\mathrm{Re}^{i\theta})| \leqslant \frac{3}{2\pi}\int_0^{2\pi} \ln^+|\varphi(2\mathrm{Re}^{i\alpha})|\, d\alpha \leqslant \varepsilon R.$$

The theorem is proved. □

Theorem 6. *Let a spectral measure F be absolutely continuous. The space $L^{+|-}(F)$ is of a finite dimension n if and only if the spectral density is either a rational function λ of degree $2n$ for processes with continuous time or a rational function $e^{i\lambda}$ of degree $2n$ for processes with discrete time.*

Proof. 1. $\xi(t)$ is a process with continuous time. Let $\dim L^{+|-}(F) = n < \infty$. Denote by $\varphi_1, \ldots, \varphi_n$ some basis in $L^{+|-}(F)$. Let $\eta_1, \ldots, \eta_n$ be an isometric basis in an isometric space of random variables $H^{+|-}$. If we denote by $\mathscr{P}$ the projection on H^-, then the correlation function

$$\begin{aligned} B(t+s) &= \mathsf{M}\xi(t)\overline{\xi(-s)} \\ &= (\xi(t), \xi(-s)) \\ &= (\mathscr{P}\xi(t), \xi(-s)) = \textstyle\sum c_j(t)(\eta_j, \xi(-s)), \qquad t, s \geqslant 0. \end{aligned}$$

Denoting $(\eta_j, \xi(-s))$ by $\mu_j(s)$ we have

$$B(t+s) = \sum_{j=1}^{n} c_j(t)\mu_j(s). \tag{4.11}$$

Further, for any numbers $0 < t_0 < \ldots < t_n < \infty$, the random variables $\mathscr{P}\xi(t_0), \ldots, \mathscr{P}\xi(t_n)$ are linearly dependent, so that there are numbers $a_0, a_1, \ldots, a_n$, for which we have

$$\textstyle\sum a_j B(t_j + s) = (\sum a_j \mathscr{P}\xi(t_j), \xi(-s)) \equiv 0. \tag{4.12}$$

It is sufficient to have the equalities given by (4.11) and (4.12) in order to find the correlation function $B(t)$. Relying on (4.11) we show first that $B(t)$ is infinitely differentiable for $t > 0$. Choose n infinitely differentiable functions $g_1(s), \ldots, g_n(s)$ with support inside of $(0, \infty)$ so that $\det\|\int_0^\infty g_i(s)\mu_j(s)\, ds\| \neq 0$.

By virtue of (4.11) we infer that

$$\int_0^\infty B(t+s)g_i(s)\,ds = \sum_1^n c_j(t)\int_0^\infty g_i(s)\mu_j(s)\,ds.$$

The left sides of these equalities, equal to $\int_t^\infty B(u)g_i(u-t)\,du$, are infinitely differentiable in t. Hence all $c_j(t)$ are infinitely differentiable as well as $B(t)$, $t > 0$, over (4.11). Differentiating the identity given by (4.12), we arrive at

$$\sum_{j=0}^n a_j \frac{d^u}{ds^u} B(t_j + s) = 0, \qquad u = 0, 1, 2, \ldots, n.$$

Consequently, one can find a polynomial $P(z)$ of degree less than n such that

$$P\left(\frac{d}{ds}\right) B(s) = 0, \qquad s > 0. \tag{4.13}$$

It is known (see [22], p. 58) that the fundamental system of solutions of (4.13) is made up of functions of the form

$$e^{\lambda_1 s}R_1(s), \ldots, e^{\lambda_k s}R_k(s), \tag{4.14}$$

where all numbers λ_j are different and the $R_j(s)$ are polynomials of degree $n_j - 1$, $n_1 + \ldots + n_k$ being the degree of the polynomial $P(\leqslant n)$. The correlation function $B(s)$ is a solution of (4.13) and hence it is the linear combination of the functions given by (4.14). By the Riemann–Lebesgue theorem we have as $s \to \infty$ that

$$B(s) = \int_{-\infty}^\infty e^{i\lambda s} f(\lambda)\,d\lambda \to 0.$$

Therefore, each $\operatorname{Re}\lambda_j < 0$. Finally, $B(s) = \overline{B(-s)}$ for $s < 0$.

Direct calculation yields

$$\int_0^\infty e^{-i\lambda s}e^{\lambda_j s}R_j(s)\,ds = \tilde{R}_j\left(\frac{1}{\lambda_j - i\lambda}\right), \tag{4.15}$$

where $\tilde{R}_j$ is rational of degree less than $2n$. Hence

$$f(\lambda) = \frac{1}{2\pi}\int_{-\infty}^\infty e^{-i\lambda s}B(s)\,ds$$

is a rational function of degree less than $2n$.

To make sure that the degree of f is equal to $2n$ exactly, we need only to prove the second part of the theorem: if f is a rational function of degree $2n$, then $\dim L^{+|-}(F) \leqslant n$.

The last assertion is well known in prediction theory (see [24], p. 174). To prove this assertion, we expand the rational function $f(\lambda)$ as the sum of simple fractions $1/(\lambda_j + i\lambda)^\alpha$, $\operatorname{Re}\lambda_j > 0$. The integer α does not exceed the multiplicity n_j of the conjugate poles $\pm i\lambda_j$. Applying the Fourier transform to (4.15) we find that the correlation function $B(s)$, $s > 0$, is again the sum of functions of the form (4.14). Therefore, $B(s)$ is a solution of a linear differential equation of the form (4.13) of degree less than n. Any $n+1$ solutions $B(t_0+s), \ldots, B(t_n+s)$, $t_j>0$, of (4.13) are linearly dependent, so that there are numbers a_j for which

$$\sum_0^n a_j B(t_j+s)=0, \qquad s>0.$$

In this case we infer that

$$\left(\sum a_j \mathscr{P}\xi(t_j), \xi(-s)\right)=\sum_0^n a_j(\xi(t_j), \xi(-s))=\sum_0^n a_j B(t_j+s)=0$$

for all $s>0$; therefore, $\sum_0^n a_j \mathscr{P}\xi(t_j)=0$, i.e., any $n+1$ vectors from $H^{+|-}$ (respectively, from $L^{+|-}$) are linearly dependent. This concludes the case of continuous time.

2. *$\xi(t)$ is a process with discrete time.* The proof in this case is similar to that given above and even simpler. The equality given by (4.12) holds true as before with t_j and s being integers. Define the operator Δ of taking the difference: $\Delta B(s)=B(s+1)-B(s)$. It follows from (4.12) that

$$\sum_0^n a_j \Delta^k B(t_j+s)=0,$$

from which follows the equation in finite differences (an analog of (4.13))

$$P(\Delta)B(s)=0. \tag{4.16}$$

We know from the theory of finite difference equations* that each solution of (4.16) is the linear combination of n linearly independent fundamental solutions of the form (4.14), with s being an integer. The further considerations are the same as those appearing in the discussion of the case of continuous time and, therefore, will be omitted. The theorem is proved. □

This theorem is adequate to explain the role that the rational spectral densities play in prediction theory. Let $f(\lambda)$ be a rational function of λ (or $e^{i\lambda}$) of degree $2n$. We can write $f(\lambda)$ as $f=|P/Q|^2$, where $P/Q=g \in \mathscr{H}^{2+}$, and Q and P are polynomials of degrees n and $n_1 \leqslant n-1$, respectively. It is easy to show that the basis of the space $L^{+|-}$ forms functions $\lambda^s/P(\lambda)$, $s=0,1,\ldots,n-1$, for processes with continuous time, and it forms functions $e^{ij\lambda}/P(e^{i\lambda})$, $j=1,\ldots,n$, for processes with discrete time. In particular, we have (it follows from Theorem 5 as well) that $L^{+|-}=L^0$ only if P is constant.

Note that the correlation function $B(t)$ is the sum of the functions given by (4.14) even if we do not assume the spectral measure F to be absolutely continuous. We might have, however, that $\operatorname{Re}\lambda_j=0$ for some λ_j (surely, $R_j(s) \equiv$ constant for these λ_j). The Fourier transform of the function $e^{\lambda_j s}$, $\operatorname{Re}\lambda_j=0$, is a δ-measure with jump at $\operatorname{Im}\lambda_j$. Recalling item 1 in Theorems 1

* See, for example, [9].

and 2 we arrive at the following general assertion:

$$\dim L^{+|-}(F) = n < \infty$$

if and only if $F = F_a + F_s$, where the derivative of the absolutely continuous part F'_a is a rational function $\lambda\,(e^{i\lambda})$ of degree $2n_1$. The singular measure F_s is concentrated at n_2 points, $n_1 + n_2 = n$.

II.5 The Structure of the σ-algebra of Events $\mathfrak{A}(T)$

In this section we show how some results on subspaces $H(T)$ (or their isometric subspaces $L_T(F)$) lead (for Gaussian processes) to theorems on σ-algebras of $\mathfrak{A}(T)$. More results can be found in Chapter IV.

Let us consider the relations given in (4.1). We try to find analogs of these relations and other results of Section II.4 expressed in terms of σ-algebras of $\mathfrak{A}(T)$. It is clear that the σ-algebra of events $\mathfrak{A}^- = \mathfrak{A}(-\infty, 0)$ corresponds to a space $H^- = H(-\infty, 0)$ (or its isometric space $L^-(F)$); the σ-algebra of $\mathfrak{A}^+ = \mathfrak{A}(0, \infty)$ corresponds to the space $H^+ = H(0, \infty)$; the σ-algebra of $\mathfrak{A}^0 = \bigcap_{t>0} \mathfrak{A}(-t, t)$ corresponds to the space $H^0 = \bigcap_{t>0} H(-t, t) = \bigcap_{t>0} H(0, t)$. The analog of the space $H^{+|-}$ (or $L^{+|-}(F)$)—the projection in H of a subspace H^+ on H^-—presents some difficulties. Let us introduce first the following concept. Definition:* by a *splitting σ-algebra at the point t for a process $\xi(t)$* we mean any σ-algebra of $\mathfrak{A}_1^{(t)} \subseteq \mathfrak{A}(-\infty, t)$ relative to which the "past" $\mathfrak{A}(-\infty, t)$ and the "future" $\mathfrak{A}(t, \infty)$ of the process are conditionally independent, i.e., for any $A \in \mathfrak{A}(-\infty, t)$ and $B \in \mathfrak{A}(t, \quad)$

$$\mathsf{P}\{AB|\mathfrak{A}_1^{(t)}\} = \mathsf{P}\{A|\mathfrak{A}_1^{(t)}\}\mathsf{P}\{B|\mathfrak{A}_1^{(t)}\}.$$

It can easily be seen that the splitting σ-algebra at a point t always exists; such is, for example, the σ-algebra of $\mathfrak{A}(-\infty, t)$. Naturally, the minimal splitting σ-algebra at a point t is of some interest (we shall prove further that it always exists). For example, for Markov processes the minimal splitting σ-algebra at a point t is the σ-algebra generated by a random variable $\xi(t)$.

We need to consider only splitting σ-algebras at the point 0 for narrow-sense stationary processes—in particular, for stationary Gaussian processes. Denote by $\mathfrak{A}^{+|-}$ the minimal splitting σ-algebra at the point 0. It will follow from the below that this σ-algebra is a natural analog of the space $H^{+|-}$.

Theorem 7. *Let $\xi(t)$ be a stationary Gaussian process with spectral measure $F(d\lambda)$. We always have the inclusions*

$$\mathfrak{A}^- \supseteq \mathfrak{A}^{+|-} \supseteq \mathfrak{A}\ \ \cap \mathfrak{A}^- \supseteq \mathfrak{A}^0. \tag{5.1}$$

* The term *splitting σ-algebra* is due to H. P. Mckean, Jr., "Brownian motion with a several-dimensional time," *Teorija Verojatn. i Primen.* **8** (1963), 335–354 (English translation); see also Levinson and Mckean's paper, cited in Section II.3.

Either of the inclusion symbols $\supseteq$ *in* (5.1) *can be replaced by the equality sign if and only if we can replace* $\supseteq$ *in the relations*

$$H^- \supseteq H^{+|-} \supseteq H^- \cap H^+ \supseteq H^0 \tag{5.2}$$

by the equality sign, or, what is equivalent, in the relations given by (4.1):

$$L^-(F) \supseteq L^{+|-}(F) \supseteq L^-(F) \cap L^+(F) \supseteq L^0(F).$$

To prove the theorem we investigate first some general properties of splitting σ-algebras.

Lemma 3. *Let* $\xi(t)$ *be an arbitrary random process and let* $\mathfrak{A}^- = \mathfrak{A}(-\infty, 0)$, $\mathfrak{A}^+ = (0, \infty)$.

Then:

1. $\mathfrak{A}^-$ *is a splitting* σ*-algebra* (*at the point* 0)
2. *if* $\mathfrak{A}_1$ *is a splitting* σ*-algebra and if the event* $B \in \mathfrak{A}^+$, *we infer that* $\mathsf{P}\{B|\mathfrak{A}^-\} = \mathsf{P}\{B|\mathfrak{A}_1\}$;
3. *if* $\mathfrak{A}_1$ *is a splitting* σ*-algebra and if* $\mathfrak{A}^- \supseteq \mathfrak{A}_2 \supseteq \mathfrak{A}_1$, *then* $\mathfrak{A}_2$ *is also a splitting* σ*-algebra*;
4. *if the* σ*-algebras of* $\mathfrak{A}_1$, *and* $\mathfrak{A}_2$ *are splitting algebras, the* σ*-algebra of* $\mathfrak{A}_1 \cap \mathfrak{A}_2$ *is also a splitting algebra*;
5. *there exists a minimal splitting* σ*-algebra* $\mathfrak{A}^{+|-}$;
6. $\mathfrak{A}^{+|-} \supseteq \mathfrak{A}^+ \cap \mathfrak{A}^-$ *always.*

Proof. Denote by χ_A the indicator of the event A, i.e.,

$$\chi_A = \chi_A(\omega) = \begin{cases} 0, & \omega \notin A \\ 1, & \omega \in A. \end{cases}$$

1. By virtue of properties of conditional mathematical expectations for $A^- \in \mathfrak{A}^-$, $B \in \mathfrak{A}^+$, we have

$$\begin{aligned} \mathsf{P}\{AB|\mathfrak{A}^-\} &= \mathsf{M}\{\chi_A \cdot \chi_B|\mathfrak{A}^-\} \\ &= \chi_A \cdot \mathsf{M}\{\chi_B|\mathfrak{A}^-\} \\ &= \mathsf{M}\{\chi_A|\mathfrak{A}^-\}\mathsf{M}\{\chi_B|\mathfrak{A}^-\} \\ &= \mathsf{P}\{A|\mathfrak{A}^-\}\mathsf{P}\{B|\mathfrak{A}^-\}. \end{aligned}$$

2. By definition, the splitting σ-algebra of $\mathfrak{A}_1 \subset \mathfrak{A}^-$; therefore the random variable $\mathsf{P}\{B|\mathfrak{A}_1\}$ is measurable with respect to $\mathfrak{A}^-$. Hence it suffices to prove that the integrals of random variables $\mathsf{M}\{\chi_B|\mathfrak{A}_1\}$, $\mathsf{M}\{\chi_B|\mathfrak{A}^-\}$ over each set $A \in \mathfrak{A}^-$ coincide. We invoke the Radon–Nikodym theorem and infer that

$$\begin{aligned} \int_A \mathsf{M}\{\chi_B|\mathfrak{A}^-\}\, d\mathsf{P} &= \mathsf{M}\{\chi_A \cdot \chi_B\} \\ &= \mathsf{M}\{\mathsf{M}\{\chi_A \cdot \chi_B|\mathfrak{A}_1\}\} \\ &= \mathsf{M}\{\mathsf{M}\{\chi_A|\mathfrak{A}_1\} \cdot \mathsf{M}\{\chi_B|\mathfrak{A}_1\}\} \\ &= \mathsf{M}\{\chi_A \cdot \mathsf{M}\{\chi_B|\mathfrak{A}_1\}\} \\ &= \int_A \mathsf{M}\{\chi_B|\mathfrak{A}_1\}\, d\mathsf{P}. \end{aligned}$$

3. Let $A \in \mathfrak{A}^-$, $B \in \mathfrak{A}^+$. Since the both sides of the equality $\mathsf{M}\{\chi_A \cdot \chi_B|\mathfrak{A}_2\} = \mathsf{M}\{\chi_A|\mathfrak{A}_2\} \times \mathsf{M}\{\chi_B|\mathfrak{A}_2\}$ are measurable with respect to $\mathfrak{A}_2$, we need only to prove that the integrals of these random variables over each set $C \in \mathfrak{A}_2$ coincide. By (2) we have $\mathsf{M}\{\chi_B|\mathfrak{A}_1\} = \mathsf{M}\{\chi_B|\mathfrak{A}^-\}$, and therefore $\mathsf{M}\{\chi_B|\mathfrak{A}_1\} = \mathsf{M}\{\chi_B|\mathfrak{A}_2\}$. Hence

$$\begin{aligned}\int_C \mathsf{M}\{\chi_A\chi_B|\mathfrak{A}_2\}\, d\mathsf{P} &= \mathsf{M}\{\chi_A\chi_B\chi_C\}\\ &= \mathsf{M}\{\mathsf{M}\{\chi_A\chi_C|\mathfrak{A}_1\} \cdot \mathsf{M}\{\chi_B|\mathfrak{A}_1\}\}\\ &= \mathsf{M}\{\mathsf{M}\{\mathsf{M}\{\chi_A\chi_C|\mathfrak{A}_2\} \cdot \mathsf{M}\{\chi_B|\mathfrak{A}_2\}|\mathfrak{A}_1\}\}\\ &= \mathsf{M}\{\chi_C\mathsf{M}\{\chi_A|\mathfrak{A}_2\} \cdot \mathsf{M}\{\chi_B|\mathfrak{A}_2\}\}\\ &= \int_C \mathsf{M}\{\chi_A|\mathfrak{A}_2\} \cdot \mathsf{M}\{\chi_B|\mathfrak{A}_2\}\, d\mathsf{P}.\end{aligned}$$

4. Note first that for any random variable ζ ($\mathsf{M}\zeta = 0$, $\mathsf{M}|\zeta|^2 < \infty$) we have the equality

$$\mathsf{M}\{\zeta|\mathfrak{A}_1 \cap \mathfrak{A}_2\} = \mathsf{M}\{\mathsf{M}\{\ldots\mathsf{M}\{\zeta|\mathfrak{A}_1\}|\mathfrak{A}_2\}|\mathfrak{A}_1\ldots\}. \tag{5.3}$$

In fact, consider a Hilbert space $\boldsymbol{H}$ of all random variables with zero mean, finite variance, and scalar product $(\eta_1, \eta_2) = \mathsf{M}\eta_1\eta_2$. Denote by $\boldsymbol{H}_i$, $i = 1, 2$, the subspaces of $\boldsymbol{H}$ consisting of variables measurable with respect to $\mathfrak{A}_i$. The operators $\mathscr{P}_i = \mathsf{M}\{\cdot|\mathfrak{A}_i\}$ are projectors in $\boldsymbol{H}$ on $\boldsymbol{H}_i$. The left side of (5.3) is the projection ζ on $\boldsymbol{H}_1 \cap \boldsymbol{H}_2$. The right side of (5.3) is a result of application of the operator $\lim_n(\mathscr{P}_2\mathscr{P}_1)^n$ (equal to the projector on $\boldsymbol{H}_1 \cap \boldsymbol{H}_2$) to ζ.

Next let $A \in \mathfrak{A}^-$, $B \in \mathfrak{A}^+$. Then

$$\mathsf{M}\{\chi_A\chi_B|\mathfrak{A}_1\} = \mathsf{M}\{\chi_A|\mathfrak{A}_1\} \cdot \mathsf{M}\{\chi_B|\mathfrak{A}_1\}. \tag{5.4}$$

By (2) and (5.3) we have

$$\begin{aligned}\mathsf{M}\{\chi_B|\mathfrak{A}_1 \cap \mathfrak{A}_2\} &= \lim_n \underbrace{\mathsf{M}\{\mathsf{M}\{\mathsf{M}\{\ldots\mathsf{M}}_{n \text{ times}}\{\chi_B|\mathfrak{A}_1\}|\mathfrak{A}_2\}\mathfrak{A}_1\ldots\}\\ &= \mathsf{M}\{\chi_B|\mathfrak{A}_1\} = \mathsf{M}\{\chi_B|\mathfrak{A}_2\} = \mathsf{M}\{\chi_B|\mathfrak{A}^-\}.\end{aligned}$$

Hence, if we apply the operator $\mathsf{M}\{\cdot|\mathfrak{A}_1 \cap \mathfrak{A}_2\}$ to each side of (5.4), we find that

$$\begin{aligned}\mathsf{M}\{\chi_A\chi_B|\mathfrak{A}_1 \cap \mathfrak{A}_2\} &= \mathsf{M}\{\chi_A\mathsf{M}\{\chi_B|\mathfrak{A}_1\}|\mathfrak{A}_1 \cap \mathfrak{A}_2\}\\ &= \mathsf{M}\{\chi_A|\mathfrak{A}_1 \cap \mathfrak{A}_2\} \cdot \mathsf{M}\{\chi_B|\mathfrak{A}_1 \cap \mathfrak{A}_2\}.\end{aligned} \tag{5.5}$$

5. By (1) the set of all splitting σ-algebras is nonempty. Let $\mathfrak{A}^{+|-}$ be the intersection of all splitting σ-algebras. Let us prove that $\mathfrak{A}^{+|-}$ is a splitting algebra, and therefore is a minimal splitting σ-algebra. By (2) all probabilities $\mathsf{P}\{B|\mathfrak{A}_1\}$, where $B \in \mathfrak{A}^+$ and $\mathfrak{A}_1$ is an arbitrary splitting σ-algebra, coincide, and therefore they are equal to $\mathsf{P}\{B|\mathfrak{A}^{+|-}\}$. Taking the conditional mathematical expectation on both sides in (5.4) under the condition $\mathfrak{A}^{+|-}$,

we find (as in (5.5)) that

$$\mathsf{M}\{\chi_A\chi_B|\mathfrak{A}^{+|-}\} = \mathsf{M}\{\chi_A|\mathfrak{A}^{+|-}\}\mathsf{M}\{\chi_B|\mathfrak{A}^{+|-}\},$$

i.e., $\mathfrak{A}^{+|-}$ is a splitting σ-algebra.

6. The relation $\mathfrak{A}^{+|-} \supseteq \mathfrak{A}^- \cap \mathfrak{A}^+$ is obvious; the lemma is proved. □

Lemma 4. *Let $\xi(t)$ be a stationary narrow-sense stochastically continuous process; then we infer that*

$$\mathfrak{A}^{+|-} \supseteq \mathfrak{A}^+ \cap \mathfrak{A}^- \supseteq \mathfrak{A}^0. \tag{5.6}$$

Proof. Let us prove first the left inclusion. Let $B \in \mathfrak{A}^+ \cap \mathfrak{A}^-$. Since $\mathfrak{A}^{+|-}$ is a splitting σ-algebra, we have

$$\mathsf{P}\{B|\mathfrak{A}^{+|-}\} = \mathsf{P}\{BB|\mathfrak{A}^{+|-}\} = (\mathsf{P}\{B|\mathfrak{A}^{+|-}\})^2.$$

This equality implies that the random variable $\mathsf{P}\{B|\mathfrak{A}^{+|-}\}$ takes on two values only: 0 and 1. Let $A = \{\omega\colon \mathsf{P}\{B|\mathfrak{A}^{+|-}\} = 1\}$, and let $\bar{A}$ be the complement of A. It is seen that $A \in \mathfrak{A}^{+|-}$, $\bar{A} \in \mathfrak{A}^{+|-}$. Hence

$$\int_A \mathsf{P}\{B|\mathfrak{A}^{+|-}\}\, d\mathsf{P} = \mathsf{P}\{AB\} = \mathsf{P}\{A\},$$

$$\int_{\bar{A}} \mathsf{P}\{B|\mathfrak{A}^{+|-}\}\, d\mathsf{P} = \mathsf{P}\{\bar{A}B\} = 0,$$

from which it follows that B differs from A only by events of probability zero. Therefore, we may assume that $B = A \in \mathfrak{A}^{+|-}$.

Next let us prove the right inclusion given by (5.6). Set $\mathfrak{A}^{0+} = \bigcap_{t>0} \mathfrak{A}(0, t)$, $\mathfrak{A}^{0-} = \bigcap_{t>0} \mathfrak{A}(-t, 0)$. It suffices to prove that $\mathfrak{A}^{0+} = \mathfrak{A}^{0-} = \mathfrak{A}^0$. The stochastically continuous process $\xi(t)$ generates a group of transformations $\boldsymbol{T}^t = TA$, $A \in \mathfrak{A}(-\infty, \infty)$, that support the probability measure. These transformations determine, in turn, a group of continuous unitary operators $\boldsymbol{U}^t$ on $\boldsymbol{H}$.*

Let A be some event from $\mathfrak{A}^0$. It is obvious that $\boldsymbol{T}^t A \in \mathfrak{A}(0, t)$. Next let $\eta(t)$ be an indicator of the event $\boldsymbol{T}^t A$. It is seen that $\eta(t) = \boldsymbol{U}^t \chi_A$ is a stationary process continuous in the mean square. Hence we infer that

$$\lim_{t\to 0} \mathsf{M}|\eta(t) - \eta(0)|^2 = 0.$$

The random variable $\eta(t)$ is measurable with respect to $\mathfrak{A}(0, t)$ for all $t > 0$. Hence $\eta(0)$ is measurable with respect to $\mathfrak{A}^{0+}$. We prove in a similar way that $\eta(0)$ is measurable with respect to $\mathfrak{A}^{0-}$. Therefore, $A \in \mathfrak{A}^{0+}$, $A \in \mathfrak{A}^{0-}$ and $\mathfrak{A}^0 = \mathfrak{A}^{0+} = \mathfrak{A}^{0-}$.

Noting that the left inclusion given by (5.1) is trivial, we see that the first part of the theorem together with Lemma 4 have been proved. □

* See [24], pp. 206–211.

We need only to find out when we have equalities in (5.1). To this end we prove a few lemmas on the relationship between σ-algebras of $\mathfrak{A}(T)$ and spaces $H(T)$. Let us agree to denote by $A(\Xi)$ the minimal σ-algebra of events, this algebra being generated by a set of the random variables Ξ; for example, $\mathfrak{A}(T) = A(\xi(t), t \in T)$. It is seen further that $\mathfrak{A}^- = A(H^-)$.

Lemma 5. *For a stationary Gaussian process the minimal splitting σ-algebra is*

$$\mathfrak{A}^{+|-} = A(H^{+|-}).$$

Proof. We prove first that $\mathfrak{A}^{+|-} \supseteq A(H^{+|-})$. Denote by $\mathscr{P}^- = \mathscr{P}$ the projection of $H(-\infty, \infty)$ on H. It is seen that $\mathscr{P} = \mathsf{M}\{\cdot|\mathfrak{A}^-\}$. The random variable $\xi(t)$, $t \geqslant 0$, can be approximated arbitrarily closely in the mean square by linear combinations of indicators χ_B, $B \in \mathfrak{A}^+$. Hence we have by (2) of Lemma 3 that

$$\mathscr{P}\xi(t) = \mathsf{M}\{\xi(t)|\mathfrak{A}^-\} = \mathsf{M}\{\xi(t)|\mathfrak{A}^{+|-}\}.$$

Consequently, all elements of the space $H^{+|-}$ are measurable with respect to $\mathfrak{A}^{+|-}$ and $A(H^{+|-}) \subseteq \mathfrak{A}^{+|-}$.

Let us show now that $A(H^{+|-})$ splits σ-algebras of $\mathfrak{A}^+$ and $\mathfrak{A}^-$. Since $\mathfrak{A}^{+|-}$ is a minimal splitting σ-algebra, it follows that $\mathfrak{A}^{+|-} = A(H^{+|-})$. Let χ_A be an indicator of the event $A \in \mathfrak{A}^-$. Let the random variable η measurable with respect to $\mathfrak{A}^+$ be representable as the product $\eta = \eta_1\eta_2$, where η_1 is measurable with respect to $A(H^{+|-})$ and where η_2 is independent of a σ-algebra of $\mathfrak{A}^-$. Let us prove that

$$\mathsf{M}\{\chi_A\eta|A(H^{+|-})\} = \mathsf{M}\{\chi_A|A(H^{+|-})\} \cdot \mathsf{M}\{\eta|A(H^{+|-})\}. \tag{5.7}$$

In fact, we have

$$\begin{aligned}\mathsf{M}\{\chi_A\eta|A(H^{+|-})\} &= \eta_1\mathsf{M}\{\chi_A\eta_2|A(H^{+|-})\}\\ &= \mathsf{M}\{\eta_1|A(H^{+|-})\} \cdot \mathsf{M}\{\chi_A\mathsf{M}\{\eta_2|\mathfrak{A}^-\}|A(H^{+|-})\}\\ &= \mathsf{M}\{\eta_1|A(H^{+|-})\} \cdot \mathsf{M}\eta_2 \cdot \mathsf{M}\{\chi_A|A(H^{+|-})\}\\ &= \mathsf{M}\{\eta_1\eta_2|A(H^{+|-})\} \cdot \mathsf{M}\{\chi_A|A(H^{+|-})\}.\end{aligned}$$

Any random variable $\xi(t)$, $t \geqslant 0$, can be written as the sum $\xi(t) = \eta_1(t) + \eta_2(t)$, where $\eta_1(t) = \mathscr{P}\xi(t) = \mathsf{M}\{\xi(t)|\mathfrak{A}^-\} \in H^{+|-}$ (and therefore is measurable with respect to $A(H^{+|-})$) and where $\eta_2(t)$ is orthogonal to H^-. Orthogonal Gaussian variables are independent; therefore $\eta_2(t)$ is independent of all variables from H^-, and consequently independent of a σ-algebra of $\mathfrak{A}^-$.

Next let $Q = Q(\xi(t_1), \ldots, \xi(t_n))$ be an arbitrary polynomial of the variables $\xi(t_1), \ldots, \xi(t_n)$, $t_i \geqslant 0$. Writing each $\xi(t_i)$ as the sum $\eta_1(t_i) + \eta_2(t_i)$ we can rewrite the polynomial Q as the sum of the form $\sum_j \zeta_j$, where each $\zeta_j = \eta_{1j} \cdot \eta_{2j}$, η_{1j} is measurable with respect to $A(H^{+|-})$, and η_{2j} is independent of $\mathfrak{A}^-$. By virtue of (5.7) we infer for any polynomial Q that

$$\mathsf{M}\{\chi_A \cdot Q|A(H^{+|-})\} = \mathsf{M}\{\chi_A|A(H^{+|-})\} \cdot \mathsf{M}\{Q|A(H^{+|-})\}. \tag{5.8}$$

Let B be an event from $\mathfrak{A}^+$. As we proved in Section I.5, any random variable measurable with respect to $\mathfrak{A}^+$ and having a finite variance, and in particular a random variable χ_B, is the limit in the mean square of the polynomials Q of the type indicated above. By virtue of (5.8) we have

$$\mathsf{M}\{\chi_A \cdot \chi_B | A(H^{+|-})\} = \mathsf{M}\{\chi_A | A(H^{+|-})\} \cdot \mathsf{M}\{\chi_B | A(H^{+|-})\},$$

that is, $A(H^{+|-})$ is a splitting σ-algebra, thus proving the lemma.

Lemma 6. *For a stationary Gaussian process we have*

$$\mathfrak{A}^+ \cap \mathfrak{A}^- = A(H^+ \cap H^-).$$

Proof. It suffices to show that, for any polynomial P of random variables $\xi(t_1), \ldots, \xi(t_n)$,

$$\mathsf{M}\{P | \mathfrak{A}^+ \cap \mathfrak{A}^-\} = \mathsf{M}\{P | A(H^+ \cap H^-)\}. \tag{5.9}$$

We prove this equality by induction. $P = \xi(t)$ is an element of the space $H(-\infty, \infty)$ for $n = 1$. Further, as in (5.3), for all random variables ζ, $M\zeta^2 < \infty$, we have

$$\mathsf{M}\{\zeta | \mathfrak{A}^+ \cap \mathfrak{A}^-\} = \lim_n \underbrace{\mathsf{M}\{\mathsf{M}\{\ldots \mathsf{M}\{\zeta | \mathfrak{A}^+\} | \mathfrak{A}^-\}}_{2n \text{ times}} \ldots | \mathfrak{A}^-\}. \tag{5.10}$$

We see that $\mathfrak{A}^+ = A(H^+)$, $\mathfrak{A}^- = A(H^-)$; therefore, if h is an element of $H(-\infty, \infty)$, then $\mathsf{M}\{h | \mathfrak{A}^+\} \in H^+ \subset H(-\infty, \infty)$.* From this and (5.10) it follows also that $\mathsf{M}\{h | \mathfrak{A}^- \cap \mathfrak{A}^+\} \in H(-\infty, \infty)$ for all $h \in H(-\infty, \infty)$. Since the random variable $\mathsf{M}\{h | \mathfrak{A}^- \cap \mathfrak{A}^+\}$ is measurable with respect to $\mathfrak{A}^-$, it belongs to H^-; similarly, $\mathsf{M}\{h | \mathfrak{A}^- \cap \mathfrak{A}^+\} \in H^+$. Thus, we have proved (5.9) for polynomials of degree one.

The rest of the proof is based on considerations similar to those in Section I.5. Suppose that (5.9) is proved for all polynomials P of degree less than $n - 1$, $n \geqslant 2$. We prove (5.9) for polynomials of degree n. We need only to consider polynomials P of the form $\xi(t_1) \cdots \xi(t_n)$. Denote by $\mathscr{P}^+$ and $\mathscr{P}^-$ the operators $\mathsf{M}\{\cdot | \mathfrak{A}^+\}$ and $\mathsf{M}\{\cdot | \mathfrak{A}^-\}$ (these operators being projections of $\boldsymbol{H}$ on $\boldsymbol{H}^+$, $\boldsymbol{H}^-$, respectively). If $\xi(t_i) = \mathscr{P}^+ \xi(t_i) + \eta(t_i) = \xi_1(t_i) + \eta(t_i)$, all random variables $\eta(t_i)$ are orthogonal to H^+ and, therefore, independent of $\mathfrak{A}^+$. We have (see Section I.5)

$$\mathsf{M}\{P | \mathfrak{A}^+\} = \xi_1(t_1) \cdots \xi_1(t_n) + Q_1,$$

where Q_1 is the limit in the mean square of polynomials of degree less than $n - 1$. Further, letting $\xi_1(t_i) = \mathscr{P}^{-1} \xi_1(t_i) + \eta_1(t_i) = \xi_2(t_i) + \eta_1(t_i)$, we obtain

$$\mathsf{M}\{\xi_1(t_1) \cdots \xi_1(t_n) + Q_1 | \mathfrak{A}^-\} = \xi_2(t_1) \cdots \xi_2(t_n) + Q_2,$$

* See Section I.5.

where Q_2 is the limit in the mean square of polynomials of degree less than $n-1$. Proceeding in a similar manner, we finally find that

$$\begin{aligned}\mathsf{M}\{P|\mathfrak{A}^+ \cap \mathfrak{A}^-\} &= \prod_{i=1}^{n} (\ldots \mathscr{P}^- \mathscr{P}^+ \xi(t_i)) + Q \\ &= \prod_{i=1}^{n} \mathsf{M}\{\xi(t_i)|\mathfrak{A}^+ \cap \mathfrak{A}^-\} + Q \\ &= \prod_{1}^{n} \mathsf{M}\{\xi(t_i)|A(H^+ \cap H^-)\} + Q.\end{aligned}$$

In this equation the first term to the right is obviously measurable with respect to $A(H^+ \cap H^-)$; the second term to the right is measurable with respect to the same σ-algebra by the induction assumption. The equality given by (5.9) and the lemma are proved. □

Lemma 7. *For a stationary Gaussian process we have*

$$\mathfrak{A}^0 = A(H^0). \tag{5.11}$$

Proof. It is seen that $\mathfrak{A}^0 \supseteq A(H^0)$. The inverse inclusion is proved as in Lemma 6. By induction over polynomials P of $\xi(t_1) \ldots \xi(t_n)$ we prove that for any polynomial

$$\mathsf{M}\{P|\mathfrak{A}^0\} = \mathsf{M}\{P|A(H^0)\}, \tag{5.12}$$

from which it follows that the equality given by (5.11) holds true. Inductive transition from $n-1$ to n is carried out in the same way as in Lemma 6. Hence we need only to prove (5.12) for polynomials of degree one. Let $P = h \in H(-\infty, \infty)$. Then all random variables $\mathsf{M}\{h|\mathfrak{A}(0, 1/n)\}$ and their limits in the mean square $\mathsf{M}\{h|\mathfrak{A}^0\}$ are elements of the space $H(-\infty, \infty)$. Further, since all variables $\mathsf{M}\{h|\mathfrak{A}(0, s)\}$, $s < t$, belong to the space $H(0, s) \subset H(0, t)$, it follows that $\mathsf{M}\{h|\mathfrak{A}^0\}$ belongs to $H(0, t)$, i.e., $\mathsf{M}\{h|\mathfrak{A}^0\} \in H^0$ and therefore $\mathsf{M}\{h|\mathfrak{A}^0\} = \mathsf{M}\{h|A(H^0)\}$. The lemma is proved. □

The second part of the theorem involving the equality conditions in (5.1) can be proved immediately with the help of Lemmas 5–7. We prove for definiteness that $\mathfrak{A}^{+|-} = \mathfrak{A}^0$ if and only if $H^{+|-} = H^0$. Other cases can be handled in a similar way.

Proof of Theorem 7. Let $H^{+|-} = H^0$; then by virtue of the lemma we have that $\mathfrak{A}^{+|-} = A(H^{+|-}) = A(H^0) = \mathfrak{A}^0$. Conversely, let $H^{+|-} \neq H^0$. Then $H^{+|-} \supset H^0$ and there exists a random variable $h \in H^{+|-}$ that is orthogonal to the space H^0. Since the process is Gaussian, the random variable h is independent of all variables from H, and therefore independent of a σ-algebra of $A(H^0) = \mathfrak{A}^0$. Hence the event $\{h < 0\} \in A(H^{+|-}) = \mathfrak{A}^{+|-}$ is independent of the σ-algebra of $\mathfrak{A}^0$ and cannot belong to this algebra. The theorem is proved. □

The results of the previous section in conjunction with the theorem proved above enable us to express in spectral terms the conditions under which we have equalities instead of inclusions in (5.1). For example, let $\xi(t)$ be a stationary Gaussian process with spectral density $f(\lambda)$. Then a σ-algebra of $\mathfrak{A}^0$ coincides with a minimal splitting σ-algebra of $\mathfrak{A}^{+|-}$ if and only if $1/f(\lambda)$ is an entire function of zero degree. It should be mentioned here that the random process $\xi(t)$ is a Markov process if for all t the minimal splitting σ-algebra at the point t coincides with the algebra generated by the random variable $\xi(t)$. Hence the process $\xi(t)$ is a Markov process if and only if $1/f(\lambda)$ is a polynomial of degree 2 ($\xi(t)$ is a Markov process with m chains if $1/f(\lambda)$ is a polynomial of degree $2m$).

CHAPTER III

Equivalent Gaussian Distributions and their Densities

III.1 Preliminaries

III.1.1 Introduction

Let $\xi = \xi(t)$ be a Gaussian random function of the parameter $t \in T$ with values $\xi(t) = \xi(\omega, t)$, $\omega \in \Omega$, on a probability space $(\Omega, \mathfrak{A}, \mathsf{P})$. We assume that the σ-algebra $\mathfrak{A}$ is generated by $\xi(t) = \xi(\omega, t)$ on Ω as the parameter t runs through the set T; in particular, then, the probability measure P on the σ-algebra $\mathfrak{A} = \mathfrak{A}_\xi$ is Gaussian.

Let P_1 be another Gaussian measure on the σ-algebra $\mathfrak{A}$.* It is said to be *absolutely continuous with respect to* P if $\mathsf{P}_1(A) = 0$ for $\mathsf{P}(A) = 0$, $A \in \mathfrak{A}$. It is known that the absolutely continuous measure P_1 is representable as

$$\mathsf{P}_1(A) = \int_A p(\omega)\mathsf{P}(d\omega), \qquad A \in \mathfrak{A}, \tag{1.1}$$

where $p(\omega)$ is nonnegative definite function on Ω called a *density* and designated $p(\omega) = \mathsf{P}_1(d\omega)/\mathsf{P}(d\omega)$. Measures P_1 and P are said to be *equivalent* if they are mutually absolutely continuous. The measures P_1 and P are said to be *orthogonal* if there exist nonoverlapping sets A and $A_1 \in \mathfrak{A}$ (known as the *carriers* of the corresponding measures P and P_1) for which

$$\mathsf{P}(A) = 1, \qquad \mathsf{P}(A_1) = 0,$$

and

$$\mathsf{P}_1(A) - 0, \qquad \mathsf{P}_1(A_1) = 1. \tag{1.2}$$

The absolute continuity implies in this case that for any $\varepsilon > 0$ there is $\delta > 0$ such that

$$\mathsf{P}_1(A) \leqslant \varepsilon \qquad \text{for } \mathsf{P}(A) \leqslant \delta \tag{1.3}$$

* More precisely, the measure P_1 is such that the random function $\xi(t)$ on $(\Omega, \mathfrak{A}, \mathsf{P}_1)$ is Gaussian.

for all $A \in \mathfrak{A}$. This fact, for instance, follows immediately from the inequality

$$\mathsf{P}_1(A) = \int_A p(\omega)\mathsf{P}(d\omega) \leqslant N\mathsf{P}(A) + \int_{|p(\omega)| > N} p(\omega)\mathsf{P}(d\omega),$$

where

$$\int_{|p(\omega)| > N} p(\omega)\mathsf{P}(d\omega) \leqslant \frac{\varepsilon}{2}$$

for sufficiently large N; hence, for example, $\mathsf{P}_1(A) \leqslant \varepsilon$ for $\mathsf{P}(A) \leqslant \varepsilon/(2N)$.

Each measure P_1 is representable* as the sum of orthogonal measures P_1' and P_1'', P_1' being orthogonal to P and P_1'' being absolutely continuous with respect to P.

Therefore

$$\mathsf{P}_1(A) = \mathsf{P}_1'(A) + \int_A p''(\omega)\mathsf{P}(d\omega), \tag{1.4}$$

where

$$p''(\omega) = \mathsf{P}_1''(d\omega)/\mathsf{P}(d\omega).$$

Let us note that the measures P and P_1 are orthogonal if for a sequence of sets $A_n \in \mathfrak{A}$, $n = 1, 2, \ldots$, the relations

$$\lim_{n\to\infty} \mathsf{P}(A_n) = 0 \quad \text{and} \quad \lim_{n\to\infty} \mathsf{P}_1(A_n) = 1 \tag{1.5}$$

are satisfied. This, for instance, follows from the fact that $\mathsf{P}_1''(\Omega) > 0$ and $\mathsf{P}_1'(\Omega) < 1$ for all measures; hence, under the condition $\lim_{n\to\infty} \mathsf{P}(A_n) = 0$ we have

$$\lim_{n\to\infty} \mathsf{P}_1(A_n) = \lim_{n\to\infty} \mathsf{P}_1'(A_n) + \lim_{n\to\infty} \mathsf{P}_1''(A_n)$$
$$= \lim_{n\to\infty} \mathsf{P}_1'(A_n) \leqslant \mathsf{P}_1'(\Omega) < 1.$$

In various branches of random process theory, mathematical statistics, information theory, etc., we encounter the following kind of problems. Under which conditions are the specified measures P_1 and P equivalent (or orthogonal)? How can the density $p(\omega) = \mathsf{P}_1(d\omega)/\mathsf{P}(d\omega)$ of the equivalent measures be determined? How can the nonoverlapping "carriers" A and A_1 of orthogonal measures be described explicitly?

It is clear (see Section I.2) that in solving the problems listed above one can go from the Gaussian measures $\mathsf{P}(d\omega)$ and $\mathsf{P}_1(d\omega)$ in an initial space Ω to the corresponding Gaussian distributions $\mathsf{P}(dx)$ and $\mathsf{P}_1(dx)$ in a function space $(X, \mathfrak{B})$ of real functions $x = x(t)$, of a parameter $t \in T$, that contains all trajectories $\xi(\omega, \cdot) = \xi(\omega, t)$ and in which the σ-algebra $\mathfrak{B}$ is generated by cylindric sets of this space (see (2.1)–(2.3), Chapter I). In particular, one can always go over to probability distributions P and P_1 in a space $X = \mathbb{R}^T$ of real functions $x = x(t)$ of $t \in T$.

The variable $\eta = \varphi[\xi(\omega, \cdot)]$ on Ω is measurable with respect to the σ-algebra $\mathfrak{A}$ for any function $\varphi = \varphi(x)$ on X measurable with respect to the

* See, for instance, [10], p. 111.

σ-algebra $\mathfrak{B}$. If the function $\varphi = \varphi(x)$ is integrable then the variable η will also be integrable, with

$$\int_X \varphi(x)\mathsf{P}(dx) = \int_\Omega \eta[\xi(\omega, \cdot)]\mathsf{P}(d\omega) = \mathsf{M}\eta. \tag{1.6}$$

Furthermore, on the measurable set $\xi(\Omega)$ each measurable variable η can be represented as $\eta = \varphi[\xi(\omega, \cdot)]$ where $\varphi = \varphi(x)$ is some measurable variable on $(X, \mathfrak{B})$.

In fact, as was mentioned in Section I.2, any set A is the preimage under the mapping $\xi = \xi(\omega, \cdot)$ of Ω onto X: $A = \{\xi \in B\}$ where B is some set in the σ-algebra $\mathfrak{B}$. Hence, for any measurable sets $A_k = \{\xi \in B_k\}$ the variables $\eta = \sum_k c_k \chi_{A_k}(\omega)$ (where χ_A is the indicator of the set A, that is, $\chi_A(\omega) = 1$ for $\omega \in A$ and $\chi_A(\omega) = 0$ for $\omega \notin A$) are representable as $\eta = \sum_k c_k \chi_{B_k}[\xi(\omega, \cdot)] = \varphi[\xi(\omega, \cdot)]$. Each measurable variable $\eta = \eta(\omega)$ is the (uniform over $\omega \in \Omega$) limit of variables $\eta_n = \varphi_n[\xi(\omega, \cdot)]$ of the type indicated above. Obviously, $\eta = \varphi[\xi(\omega, \cdot)]$ where the function $\varphi = \varphi(x)$ on the measurable space $\xi(\Omega)$ is defined as the (uniform over $x \in \xi(\Omega)$) limit of the corresponding functions $\varphi_n = \varphi_n(x)$.

From (1.6) it follows, in particular, that if $p(x) = \mathsf{P}_1(dx)/\mathsf{P}(dx)$ is the density of the distributions P_1 and P, $p[\xi(\omega, \cdot)] = \mathsf{P}_1(d\omega)/\mathsf{P}(d\omega)$ will be the density of the primary probability measures on the space Ω since each set $A \in \mathfrak{A}$ is representable as $A = \{\xi \in B\}$ where $B \in \mathfrak{B}$ and, by virtue of (1.6),

$$\mathsf{P}_1(B) = \int_B p(x)\mathsf{P}(dx) = \int_A p[\xi(\omega, \cdot)]\mathsf{P}(d\omega) = \mathsf{P}_1(A).$$

Let us note that equivalent random functions $\xi(t)$ and $\tilde{\xi}(t)$ have the same probability distributions in the corresponding function space $X = \mathbb{R}^T$. Therefore, if, for example,

$$\tilde{\xi}(\omega, t) = \xi(\omega, t) \tag{1.7}$$

for almost all $\omega \in \Omega$ (with respect to $\mathsf{P}(d\omega)$ and $\mathsf{P}_1(d\omega)$), the probability measures $\mathsf{P}(d\omega)$ and $\mathsf{P}_1(d\omega)$ on the σ-algebra $\mathfrak{A}_\xi$ are equivalent or orthogonal if and only if they possess the same property on the σ-algebra $\mathfrak{A}_{\tilde{\xi}}$ (generated by $\tilde{\xi}(t)$, $t \in T$).

As to the condition given by (1.7), it should be mentioned that the completion* of the σ-algebras $\mathfrak{A}_\xi$ and $\mathfrak{A}_{\tilde{\xi}}$ by sets of measure zero (with respect to the equivalent P and P_1) is the σ-algebra $\mathfrak{A}^*$, the density $p(\omega) = \mathsf{P}_1(d\omega)/\mathsf{P}(d\omega)$ on $\mathfrak{A}_\xi$ (or $\mathfrak{A}_{\tilde{\xi}}$) being at the same time the density on the complete σ-algebra $\mathfrak{A}^*$, i.e., for any set $A^* \in \mathfrak{A}^*$

$$\int_{A^*} p(\omega)\mathsf{P}(d\omega) = \int_A p(\omega)\mathsf{P}(d\omega) = \mathsf{P}_1(A) = \mathsf{P}_1(A^*), \tag{1.8}$$

where the set $A \in \mathfrak{A}_\xi$ (or $A \in \mathfrak{A}_{\tilde{\xi}}$) is such that

$$\mathsf{P}(A^* \circ A) = \mathsf{P}_1(A^* \circ A) = 0$$

for the symmetric difference $A^* \circ A = (A^* \backslash A) \cup (A \backslash A^*)$.

* See, for instance, [10], p. 127.

As was noted in Section I.2, each Gaussian measure P is determined by its mean value $a(t)$, $t \in T$, and by the correlation function $B(s, t)$, $s, t \in T$. It is natural to solve the problems stated above (of the equivalence of Gaussian measures P and P_1) starting from the given mean values $a(t)$ and $a_1(t)$ and the correlation functions $B(s, t)$ and $B_1(s, t)$ of Gaussian distributions of P and P_1.

We can consider without loss of generality that $a(t) \equiv 0$ since we can always pass from the variables $\xi(t) = \xi(\omega, t)$ to the variables $\xi(t) - a(t)$, $t \in T$; this is so because in this case the σ-algebra $\mathfrak{A}$ and the primary Gaussian measures $\mathsf{P}(d\omega)$ and $\mathsf{P}_1(d\omega)$ do not change. In the corresponding function space X we go over to the Gaussian distributions $\mathsf{P}(dx)$ and $\mathsf{P}_1(dx)$ with the correlation functions $B(s, t)$ and $B_1(s, t)$ having mean values 0 and $a_1(t) - a(t)$, $t \in T$.

III.1.2 Examples Illustrating Orthogonal Distributions

Let $\xi(t)$, $0 \leqslant t \leqslant \tau$, be a Gaussian stationary process (with respect to probability measure P) with zero mean and correlation function $B(t)$. Let P_1 also be a probability measure with respect to which $\xi(t)$ is a Gaussian stationary process with zero mean but correlation function $B_1(t)$. The measures P and P_1 will be considered further on the σ-algebra $\mathfrak{A}$ generated by variables $\xi(t) = \xi(\omega, t)$ for $t \in T$, where $T = [0, \tau]$ is an interval on a real line.

Simple examples illustrating the orthogonal probability measures P and P_1 can be obtained if we take into account local properties of trajectories (see Section I.7).

For example, if

$$\overline{\lim_{h \to 0}} \frac{\Delta_{-h}\Delta_h B(0)}{\Delta_{-h}\Delta_h B_1(0)} = \infty$$

or

$$\underline{\lim_{h \to 0}} \frac{\Delta_{-h}\Delta_h B(0)}{\Delta_{-h}\Delta_h B_1(0)} = 0, \tag{1.9}$$

then the probability measures P and P_1 are orthogonal.

In fact, if the second condition given by (1.9) is satisfied, for example, then for a function $\delta(h)$ such that as $h \to 0$

$$\lim_{h \to 0} \frac{\Delta_{-h}\Delta_h B(0)}{\delta(h)} = 0 \quad \text{and} \quad \lim_{h \to 0} \frac{\Delta_{-h}\Delta_h B_1(0)}{\delta(h)} = \infty,$$

for fixed t with probability 1 we have the following relations:

$$\lim_{h \to 0} \frac{|\Delta_h \xi(t)|}{\delta^{1/2}(h)} = \begin{cases} 0 \text{ with respect to } \mathsf{P} \\ \infty \text{ with respect to } \mathsf{P}_1, \end{cases}$$

for a sufficiently rapidly decreasing sequence $h = h_n$, $n = 1, 2, \ldots$ (we recall here that $\Delta_{-h}\Delta_h B(0) = \mathsf{M}[\Delta_h \xi(t)]^2$). It is seen that the probability measures

P and P_1 have nonoverlapping carriers A and A_1 of the form

$$A = \left\{\omega\colon \lim_{h=h_n\to 0} \frac{|\Delta_h \xi(\omega, t)|}{\delta^{1/2}(h)} = 0\right\}$$

and

$$A_1 = \left\{\omega\colon \lim_{h=h_n\to 0} \frac{|\Delta_h \xi(\omega, t)|}{\delta^{1/2}(h)} = \infty\right\}.$$

Therefore, if the condition

$$\frac{\Delta_{-h}\Delta_h B(0)}{\Delta_{-h}\Delta_h B_1(0)} \asymp 1 \tag{1.10}$$

is violated,* the probability measures P and P_1 are orthogonal. Taking into account the properties of trajectories described in Theorem 5 of Chapter I, we arrive at general examples illustrating the distributions P and P_1.

Let

$$b(t) = B(t) - B_1(t).$$

For the sake of simplicity we assume (1.10) to be satisfied. Let us show that if the condition

$$\Delta_{-h}\Delta_h b(0) = o\{\Delta_{-h}\Delta_h B(0)\} \tag{1.11}$$

is violated, the Gaussian measures P and P_1 are orthogonal.

In fact, according to Theorem 5 of Chapter I the variables

$$\begin{aligned}\eta(h) &= \left[1 - \frac{1}{N}\sum_{k=0}^{N-1} \frac{[\Delta_h \xi(kh)]^2}{\Delta_{-h}\Delta_h B(0)}\right]\frac{\Delta_{-h}\Delta_h B(0)}{\Delta_{-h}\Delta_h b(0)} \\ &= \left[1 - \frac{1}{N}\sum_{k=0}^{N-1} \frac{[\Delta_h \xi(kh)]^2}{\Delta_{-h}\Delta_h B_1(0)}\right]\frac{\Delta_{-h}\Delta_h B_1(0)}{\Delta_{-h}\Delta_h b(0)} + 1\end{aligned}$$

where $N = [\tau/h]$, are such that for a sufficiently rapidly decreasing sequence $h = h_n$, $n = 1, 2, \ldots$, for which

$$\frac{\Delta_{-h}\Delta_h B(0)}{\Delta_{-h}\Delta_h b(0)} \asymp \frac{\Delta_{-h}\Delta_h B_1(0)}{\Delta_{-h}\Delta_h b(0)} \asymp 1,$$

with probability 1 we have

$$\lim_{h\to 0} \eta(h) = \begin{cases} 0 \text{ with respect to } \mathsf{P} \\ 1 \text{ with respect to } \mathsf{P}_1. \end{cases} \tag{1.12}$$

It is seen that the Gaussian measures P and P_1 will be orthogonal if the condition given by (1.11) is violated.

* The relationship $\alpha \asymp \beta$ for variables α and β implies that $0 < c_1 \leqslant \alpha/\beta \leqslant c_2 < \infty$ for some constant c_1 and c_2.

We shall show also that the Gaussian measures P and P_1 are orthogonal if the condition

$$\Delta_{-h}\Delta_h b(0) = O\left\{\frac{1}{N^2}\sum_{k,j=0}^{N-1}[\Delta_{-h}\Delta_h B((k-j)h)]^2\right\}^{1/2} \tag{1.13}$$

is violated.

Indeed, the variable $\eta(h)$ given above has the mean values 0 and 1 with respect to the distributions of P and P_1 respectively. The dispersions (see (7.14), Chapter I) are

$$\frac{\frac{2}{N^2}\sum_{k,j=0}^{N-1}[\Delta_{-h}\Delta_h B((k-j)h)]^2}{[\Delta_{-h}\Delta_h b(0)]^2} \quad \text{and} \quad \frac{\frac{2}{N^2}\sum_{k,j=0}^{N-1}[\Delta_{-h}\Delta_h B_1((k-j)h)]^2}{[\Delta_{-h}\Delta_h b(0)]^2}.$$

Therefore, if the condition (1.13) is violated there will be a sequence $h = h_n$, $n = 1, 2, \ldots$, for which the limit relations (1.12) are satisfied. Here the Gaussian measures P and P_1 have nonoverlapping carriers of the form

$$A = \left\{\omega\colon \lim_{h_n\to 0} \eta(\omega, h_n) = 0\right\}$$

and

$$A_1 = \left\{\omega\colon \lim_{h_n\to 0} \eta(\omega, h_n) = 1\right\}.$$

Let us recall that Lemma 4 in Chapter I provides an estimate of the expression $\sum_{k,j=0}^{N-1}[\Delta_{-h}\Delta_h B((k-j)h)]^2$. By virtue of this estimate and the results obtained above, the Gaussian measures P and P_1 will be orthogonal if the following conditions are violated:

$$\Delta_{-h}\Delta_h b(0) = O\{|h|^{1/2}|\Delta_{-h}\Delta_h B(0)|\} \tag{1.14}$$

if

$$\frac{h}{\Delta_{-h}\Delta_h B(0)} = O(1), \tag{1.15}$$

and

$$\Delta_{-h}\Delta_h b(0) = O\{h[\Delta_{-h}\Delta_h B(0)]^{1/2}\} \tag{1.16}$$

otherwise.

From (1.14) and (1.16) spectral conditions can be deduced for the Gaussian measures P and P_1 to be orthogonal. Let the stationary process $\xi(t)$ have the spectral density $f(\lambda)$ with respect to P and the spectral density $f_1(\lambda)$ with respect to P_1. Then (see Subsection I.7.1) under the condition

$$\lim_{\lambda\to\infty} f(\lambda)|\lambda|^\alpha = 0 \tag{1.17}$$

the variable $\Delta_{-h}\Delta_h B(0)$ is such that

$$\lim_{h\to 0}\frac{\Delta_{-h}\Delta_h B(0)}{h^{\alpha-1}} = 0;$$

under the condition

$$\lim_{\lambda\to\infty} [f(\lambda) - f_1(\lambda)]|\lambda|^\beta = \infty \tag{1.18}$$

(where $f(\lambda) \geqslant f_1(\lambda)$) the variable $\Delta_{-h}\Delta_h b(0)$ satisfies

$$\lim_{h\to 0} \frac{\Delta_{-h}\Delta_h b(0)}{|h|^{\beta-1}} = \infty.$$

It is seen that for $\beta \leqslant \alpha + \frac{1}{2}$ the condition

$$\Delta_{-h}\Delta_h b(0) = O\{|h|^{1/2}|h|^{\alpha-1}\}$$

is violated, and *a fortiori* the condition given by (1.14). Therefore, the Gaussian measures P and P_1 are orthogonal for the spectral densities $f(\lambda)$ and $f_1(\lambda)$ satisfying (1.17) and (1.18) where $1 < \alpha \leqslant 2$ and $\beta \leqslant \alpha + \frac{1}{2}$.

Let us consider one more example illustrating measures P and P_1 with correlation functions $B(t)$ and $B_1(\mathrm{t})$ satisfying the conditions of Theorem 6 of Chapter I: at a point t of an interval $T = (0, \tau)$ the derivatives $B'(t)$ and $B_1'(t)$ have "jumps" different in magnitudes so that

$$B'(t-0) - B'(t+0) \neq B_1'(t-0) - B'(t-0)$$

(for an example illustrating correlation functions of this type, see Section III.4). By Theorem 6 of Chapter I the nonoverlapping carriers of the Gaussian measures P and P_1 are the sets

$$A = \left\{\omega: \lim_{h_n\to 0} \eta(\omega, h_n) = B'(t-0) - B'(t+0)\right\}$$

and

$$A_1 = \left\{\omega: \lim_{h_n\to 0} \eta(\omega, h_n) = B_1'(t-0) - B_1'(t+0)\right\},$$

where

$$\eta(h) = h^{-1}\frac{1}{N}\sum_{k=0}^{N-1} \Delta_h\xi(kh)\Delta_h\xi(t+kh)$$

and where h_n, $n = 1, 2, \ldots$, is a sufficiently rapidly decreasing sequence.

III.1.3 Basic Information about Equivalent Gaussian Distributions

Let $\xi(t)$, $-\infty < t < \infty$, be a stationary Gaussian process, let

$$\xi(t) = \int e^{i\lambda t}\Phi(d\lambda) \tag{1.19}$$

be its spectral representation, and let P be a probability measure on the σ-algebra $\mathfrak{A}(T)$ generated by $\xi(t) = \xi(\omega, t)$ on Ω (where the parameter t runs through a set T on a real line). Let the mean of this stationary process be equal to zero, and let the spectral measure of this process be $F(d\lambda)$. Let P_1 be a probability measure with respect to which the stationary process with the zero mean and the spectral measure $F_1(d\lambda)$ forms

$$\xi_1(t) = \xi(t) - a(t),$$

where $a(t)$ is the mean value of the initial process $\xi(t)$:

$$a(t) = M_1\xi(t), \qquad t \in T.$$

In considering the Gaussian measures P and P_1 on the σ-algebra $\mathfrak{A}(T)$ one can go from the initial process given by (1.19) to their linear combinations of the form

$$\eta(\varphi) = \int \varphi(\lambda)\Phi(d\lambda), \tag{1.20}$$

where

$$\varphi(\lambda) = \sum_k c_k e^{i\lambda t_k}; \tag{1.21}$$

here $t_1, \ldots, t_n \in T$ and $c_1, \ldots, c_n$ are real coefficients.

Let us denote by L_T^0 the linear space of functions $\varphi(\lambda)$ of the type given by (1.21). The variables $\eta(\varphi)$, $\varphi \in L_T^0$, representable by (1.20), can be viewed as Gaussian functionals on the space L_T^0. Let

$$a(\varphi) = \mathsf{M}_1\eta(\varphi), \qquad \varphi \in L_T^0. \tag{1.22}$$

We note that $a(\varphi) = \sum_k c_k a(t_k)$ for $\varphi(\lambda) = \sum_k c_k e^{i\lambda t_k}$.

Let us regard L_T^0 as a subspace of a Hilbert space $L_T(F)$ with the scalar product

$$\langle\varphi, \psi\rangle_F = \int_{-\infty}^{\infty} \varphi(\lambda)\overline{\psi(\lambda)}F(d\lambda), \tag{1.23}$$

where $F(d\lambda)$ is the spectral measure of the stationary process $\xi(t)$ with respect to the distribution P. We recall that $L_T(F)$ is the closure of the space L_T^0 with respect to the scalar product given by (1.23). At the same time let us regard L_T^0 as a subspace in a Hilbert space $L_T(F_1)$ with a scalar product $\langle\varphi, \psi\rangle_{F_1}$ (here $F_1(d\lambda)$ is the spectral measure of the stationary process $\xi_1(t) = \xi(t) - a(t)$ with respect to the distribution P_1).

The correlation function of the random function $\eta(\varphi)$ given by (1.20) of the functional parameter $\varphi(\lambda) \in L_T^0$ is

$$B(\varphi, \psi) = \langle\varphi, \psi\rangle_F, \qquad \varphi, \psi \in L_T^0, \tag{1.24}$$

with respect to the distribution P, and

$$B_1(\varphi, \psi) = \langle\varphi, \psi\rangle_{F_1}, \qquad \varphi, \psi \in L_T^0, \tag{1.25}$$

with respect to the distribution P_1.

It is clear that if $\|\varphi\|_F = 0$, and $\|\varphi\|_{F_1} \neq 0$ for a function $\varphi(\lambda) \in L_T^0$, the measures P and P_1 are orthogonal, since

$$\mathsf{P}\{\eta(\varphi) = 0\} = 1 \quad \text{and} \quad \mathsf{P}_1\{\eta(\varphi) = 0\} = 0. \tag{1.26}$$

Furthermore, the probability measures P and P_1 are orthogonal if the following condition is violated:*

$$\|\varphi\|_F \asymp \|\varphi\|_{F_1}, \qquad \varphi \in L_T^0. \tag{1.27}$$

* We recall that the relationship $\|\varphi\|_F \asymp \|\varphi\|_{F_1}$ implies $0 < c_1 \leqslant \|\varphi\|_F \|\varphi\|_{F_1} \leqslant c_2 < \infty$, where c_1 and c_2 are constants.

In fact, if there exists, for instance, a sequence $\varphi_n(\lambda) \in L_T^0$, $n = 1, 2, \ldots$, such that

$$\|\varphi_n\|_F^2 = 1, \qquad \sigma_n^2 = \|\varphi_n\|_{F_1}^2 \to 0 \quad \text{as } n \to \infty,$$

then

$$\mathsf{P}\{|\eta(\varphi_n) - a(\varphi_n)| < \sqrt{\sigma_n} = \int_{|x - a(\varphi_n)| < \sqrt{\sigma_n}} \frac{1}{\sqrt{2\pi}} e^{-x^2/2}\, dx \to 0,$$

$$\mathsf{P}_1\{|\eta(\varphi_n) - a(\varphi_n)| < \sqrt{\sigma_n}\} = \int_{|x| < (1/\sqrt{\sigma_n})} \frac{1}{\sqrt{2\pi}} e^{-x^2/2}\, dx \to 1.$$

Similar relations hold true as well if

$$\|\varphi_n\|_{F_1}^2 = 1, \qquad \|\varphi_n\|_F^2 \to 0 \quad \text{as } n \to \infty;$$

as a consequence (see (1.5)) the probability measures P and P_1 are orthogonal.

(1.27) implies, in particular, that the Hilbert spaces $L_T(F)$ and $L_T(F_1)$ are equivalent:

$$L_T(F) = L_T(F_1),$$

(1.27) being satisfied for each $\varphi(\lambda) \in L_T(F)$ $(\varphi \in L_T(F_1))$.

Further, let us consider the mean value given by (1.22) of the random functional $\eta(\varphi)$, $\varphi \in L_T^0$, with respect to the probability measure P_1.

It is seen that the measures P and P_1 are orthogonal if $a(\varphi) \neq 0$ for $\|\varphi\|_F = 0$, since either $\|\varphi\|_{F_1} \neq 0$ and in this case (1.26) holds true, or $\|\varphi\|_{F_1} = 0$ and then

$$\begin{aligned} \mathsf{P}\{\eta(\varphi) = a(\varphi)\} &= 0, \\ \mathsf{P}_1\{\eta(\varphi) = a(\varphi)\} &= 1. \end{aligned} \tag{1.28}$$

The above implies that the mean $a(\varphi)$, $\varphi \in L_T^0$, is a linear functional on the Hilbert space $L_T(F)$ for nonorthogonal measures.

Let us show that for the nonorthogonal measures P and P_1 the mean $a(\varphi)$, $\varphi \in L_T^0$, is a linear bounded functional on the Hilbert space $L_T(F)$.

The condition given by (1.27) can be regarded as satisfied if $a(\varphi_n) \to \infty$ as $n \to \infty$ for a sequence $\varphi_n(\lambda) \in L_T^0$ such that $\sigma_n = \|\varphi_n\|_{F_1} \asymp \|\varphi_n\|_F = 1$ then

$$\mathsf{P}\{\eta(\varphi_n) > \sqrt{a(\varphi_n)}\} = \frac{1}{\sqrt{2\pi}} \int_{\sqrt{a(\varphi_n)}}^{\infty} e^{-x^2/2}\, dx \to 0,$$

$$\mathsf{P}_1\{\eta(\varphi_n) > \sqrt{a(\varphi_n)}\} = \frac{1}{\sqrt{2\pi}\sigma_n} \int_{-a(\varphi_n) + \sqrt{a(\varphi_n)}}^{\infty} e^{-x^2/2\sigma_n^2}\, dx \to 1;$$

this implies that the measures P and P_1 are orthogonal (see (1.5))

Let $\varphi_n \in L_T^0$ be a sequence of functions that is complete in the Hilbert space $L_T(F)$ as well as in the Hilbert space $L_T(F_1)$ (we recall that the subspace L_T^0 of functions of the type given by (1.21) is everywhere dense in $L_T(F)$ as well as in $L_T(F_1)$). Let also $\mathfrak{A}$ be the σ-algebra generated by variables $\eta_k = \eta(\varphi_k)$, $k = 1, 2, \ldots$, of the type given by (1.20).

Lemma 1. *The Gaussian measures* P *and* P_1 *are equivalent on the* σ*-algebra* $\mathfrak{A}(T)$ *if and only if they are equivalent on the* σ*-algebra* $\mathfrak{A} \subseteq \mathfrak{A}(T)$.

Proof. Let P and P_1 be equivalent on $\mathfrak{A}$. Each element $\varphi(\lambda) = e^{i\lambda t}$ in the space $L_T(F)$ is the limit of linear combinations of the form $\psi_n(\lambda) = \sum_k c_{kn}\varphi_k(\lambda)$, and therefore

$$\mathsf{M}|\xi(t) - \eta(\psi_n)|^2 = \|e^{i\lambda t} - \psi_n(\lambda)\|_F^2 \to 0$$

as $n \to \infty$. (1.27) is satisfied for the equivalent measures P and P_1. The mean value $a(\varphi)$ is a linear continuous functional; therefore

$$\mathsf{M}_1|\xi(t) - \eta(\psi_n)|^2 = \|e^{i\lambda t} - \psi_n(\lambda)\|_{F_1}^2 + |a(e^{i\lambda t}) - a(\psi_n)|^2 \to 0.$$

Hence, for a sufficiently rapidly increasing subsequence n_k, $k = 1, 2, \ldots,$

$$\xi(\omega, t) = \lim_{n \to \infty} \eta(\omega, \psi_{n_k})$$

almost everywhere with respect to the probability measure P as well as the probability measure P_1. It is obvious that the variables $\tilde{\xi}(\omega, t)$, defined by the limit relation

$$\tilde{\xi}(\omega, t) = \lim_{k \to \infty} \eta(\omega, \psi_{n_k})$$

for those $\omega \in \Omega$ for which this limit exists, and defined more precisely by the equality $\tilde{\xi}(\omega, t) = 0$ for the remaining $\omega \in \Omega$, will be equivalent to the initial variables $\xi(\omega, t)$ in the sense of (1.7). These variables are measurable with respect to the σ-algebra $\mathfrak{A}^*$, the completion of the initial σ-algebra $\mathfrak{A}$ by sets of probability zero (with respect to the distributions P and P_1 equivalent on $\mathfrak{A}$). It can be seen that the probability measures P and P_1 will be equivalent on $\mathfrak{A}^*$ and *a fortiori* on the σ-algebra $\mathfrak{A}_{\tilde{\xi}}$ generated by $\tilde{\xi}(t)$, $t \in T$. From (1.7) we can conclude that P and P_1 are equivalent on the σ-algebra $\mathfrak{A}(T)$, as was to be proved. □

Let $\mathfrak{A}$ be the σ-algebra generated by the variables $\eta_k(\omega) = \eta(\omega, \varphi_k)$ on Ω, where $\varphi_k(\lambda)$, $k = 1, 2, \ldots,$ is a system of functions from L_T^0 complete in the spaces $L_T(F)$ and $L_T(F_1)$. Let $\mathfrak{A}_n$ denote the σ-algebra generated by the variables η_k, $k = 1, \ldots, n$. Obviously, $\mathfrak{A}$ is a minimal σ-algebra containing the whole sequence $\mathfrak{A}_1 \subseteq \mathfrak{A}_2 \subseteq \cdots$, which can be written

$$\mathfrak{A} = \lim_{n \to \infty} \mathfrak{A}_n.$$

Let us consider the Gaussian measures P and P_1 equivalent on the σ-algebra $\mathfrak{A}$. Let

$$p(\omega) = \frac{\mathsf{P}_1(d\omega)}{\mathsf{P}(d\omega)}$$

be the density of these measures on $\mathfrak{A}$.

It is seen that P and P_1 are equivalent measures on any σ-algebra $\mathfrak{A}'$ contained in $\mathfrak{A}$; for the corresponding density $p'(\omega) = \mathsf{P}_1(d\omega)/\mathsf{P}(d\omega)$ measurable with respect to $\mathfrak{A}'$ we have

$$\mathsf{P}_1(A') = \int_{A'} p'(\omega)\mathsf{P}(d\omega) = \int_{A'} p(\omega)\mathsf{P}(d\omega)$$

for any set $A' \in \mathfrak{A}'$. It can be seen that the density $p'(\omega)$ coincides with the conditional mathematical expectation of the variable $p(\omega)$ (with respect to the σ-algebra $\mathfrak{A}'$):

$$p'(\omega) = \mathsf{M}\{p(\omega)/\mathfrak{A}'\}. \tag{1.29}$$

According to the well-known properties of conditional mathematical expectations,* for the densities $p_n(\omega) = \mathsf{P}_1(d\omega)/\mathsf{P}(d\omega)$ on the σ-algebras $\mathfrak{A}_n \subseteq \mathfrak{A}$ representable as

$$p_n(\omega) = \mathsf{M}\{p(\omega)/\mathfrak{A}_n\}, \qquad n = 1, 2, \ldots, \tag{1.30}$$

we have with probability one that

$$p(\omega) = \lim_{n\to\infty} p_n(\omega) \tag{1.31}$$

(the limit relation given by (1.31) being satisfied in the mean as well). Next, let us take the arbitrary density $p'(\omega)$ of the type given by (1.29). Applying the familiar inequality† for the convex function $\log x$, we have

$$\log p'(\omega) = \log \mathsf{M}\{p(\omega)/\mathfrak{A}'\} \geqslant \mathsf{M}\{\log p(\omega)/\mathfrak{A}'\},$$

and therefore

$$\mathsf{M} \log p' \geqslant \mathsf{M} \log p. \tag{1.32}$$

In accord with this inequality, we obtain for the sequence of densities of the type given by (1.30)

$$\mathsf{M} \log p_1 \geqslant \mathsf{M} \log p_2 \geqslant \cdots.$$

Since $\mathsf{M} \log p_n \geqslant \mathsf{M} \log p$ for $\mathsf{M} \log p > -\infty$, the finite limit

$$\lim_{n\to\infty} \mathsf{M} \log p_n \tag{1.33}$$

exists for the monotone sequence $\mathsf{M} \log p_n$, $n = 1, 2, \ldots$.

It will be shown below (see Theorem 1) that $\mathsf{M} \log p$ is indeed $> -\infty$. Therefore, the density $p(\omega) = \mathsf{P}_1(d\omega)/\mathsf{P}(d\omega)$ on the σ-algebra $\mathfrak{A}$ (and therefore on the α-algebra $\mathfrak{A}(T)$) can be determined (as we shall show later) from a limit relation of the form

$$\log p(\omega) = \lim_{n\to\infty} \log p_n(\omega)$$
$$= \lim_{n\to\infty} \mathsf{M} \log p_n(\omega) + \lim_{n\to\infty} [\log p_n(\omega) - \mathsf{M} \log p_n(\omega)]. \tag{1.34}$$

* See, for instance, [6], p. 287.

† See [6], p. 37.

III.2 Some Conditions for Gaussian Measures to be Equivalent

III.2.1 Conditions for Equivalence Involving the Entropy of Distributions

Let us consider the Gaussian measures P and P_1 on the σ-algebra $\mathfrak{A}(T)$ generated by variables $\eta(\varphi) = \eta(\omega, \varphi)$ of the type given by (1.20) on the probability space Ω for $\varphi(\lambda)$ in L_T^0. It will be assumed that the condition given by (1.27) is satisfied; if this condition is violated the measures P and P_1 will be orthogonal.

Let $\eta_k = \eta(\varphi_k)$, $k = 1, 2, \ldots$, be a sequence of variables of the type given by (1.20) where $\varphi_k \in L_T^0$ is a complete system of linearly independent functions in either Hilbert space $L_T(F)$ or $L_T(F_1)$. Let $\mathfrak{A}$ be the σ-algebra generated by these variables. On each of the σ-algebras $\mathfrak{A}_n$ generated by the variables $\eta_k = \eta_k(\omega)$ on Ω ($k = 1, \ldots, n$), the Gaussian measures P and P_1 are equivalent since by virtue of (1.27) assumed above the corresponding correlation matrices $\{B(k, j)\}$ and $\{B_1(k, j)\}$ are nonsingular.

We consider the Gaussian measure P_1 on the σ-algebra $\mathfrak{A}_n$ generated by the sets

$$A = \{[\eta_1, \ldots, \eta_n] \in \Gamma\},$$

where Γ are Borel sets of an n-dimensional vector space $\mathbb{R}^n$. Let $(a_1, \ldots, a_n)$ be the mean value with respect to the probability measure P_1, i.e., $a_k = \mathsf{M}_1\eta(\varphi_k)$, $k = 1, 2, \ldots$. Then

$$\mathsf{P}_1(A) = \int_{\mathbb{R}^n} f_1(x_1, \ldots, x_n)\, dx_1 \cdots dx_n,$$

where $f_1(x_1, \ldots, x_n)$, being the density of the corresponding Gaussian distribution in the space $\mathbb{R}^n$ of the vectors $x = (x_1, \ldots, x_n)$, has the form

$$f_1(x_1, \ldots, x_n) = \frac{1}{(2\pi)^{n/2} D^{1/2}} \exp\left\{ -\frac{1}{2} \sum (x_k - a_k)(x_j - a_j) C_1(k, j) \right\}$$

(D denotes the determinant of the matrix $\{B_1(k, j)\}$; $\{C_1(k, j)\}$ is the inverse of the matrix $\{B_1(k, j)\}$).

Let $f(x_1, \ldots, x_n)$ be the density of the corresponding distribution P in the vector space $\mathbb{R}^n$ (similar to the density $f_1(x_1, \ldots, x_n)$, but with the zero mean and the correlation function $\{B(k, j)\}$). Then the distribution density is

$$\frac{\mathsf{P}_1(dx)}{\mathsf{P}(dx)} = \frac{f_1(x_1, \ldots, x_n)}{f(x_1, \ldots, x_n)}, \qquad x \in \mathbb{R}^n.$$

Therefore, the corresponding density $p_n(\omega) = \mathsf{P}_1(d\omega)/\mathsf{P}(d\omega)$ on the initial space Ω is

$$p_n(\omega) = \frac{f_1[\eta_1(\omega), \ldots, \eta_n(\omega)]}{f[\eta_1(\omega), \ldots, \eta_n(\omega)]}$$

and can be expressed as

$$\log p_n(\omega) = \log \frac{D}{D_1}$$
$$-\frac{1}{2} \sum_{j,\,k=1}^{n} [C_1(j, k)(\eta_j(\omega) - a_j)(\eta_k(\omega) - a_k) - C(j, k)\eta_j(\omega)\eta_k(\omega)], \tag{2.1}$$

where D and D_1 are the determinants of the matrices $\{B(j, k)\}$ and $\{B_1(j, k)\}$, $k, j = 1, \ldots, n$; and $\{C(j, k)\}$ and $\{C_1(j, k)\}$ are the matrices inverse to $\{B(j, k)\}$ and $\{B_1(j, k)\}$, respectively.

In the subspace generated by $\varphi_1, \ldots, \varphi_n$ one can find a basis $\varphi_{1n}, \ldots, \varphi_{nn}$ in which the positive bilinear forms $B(\varphi, \psi)$ and $B_1(\varphi, \psi)$ can be transformed into diagonal forms:

$$B(\varphi_{kn}, \varphi_{jn}) = \begin{cases} 1 & \text{for } k = j \\ 0 & \text{for } k \neq j; \end{cases}$$

$$B_1(\varphi_{kn}, \varphi_{jn}) = \begin{cases} \sigma_{kn}^2 & \text{for } k = j \\ 0 & \text{for } k \neq j. \end{cases}$$

The density $p_n(\omega) = \mathsf{P}_1(d\omega)/\mathsf{P}(d\omega)$ on the σ-algebra $\mathfrak{A}_n$ can be described by the following formula (compare with (2.1)):

$$p_n(\omega) = \frac{1}{\prod_{k=1}^{n} \sigma_{kn}} \exp\left\{-\frac{1}{2} \sum_{k=1}^{n} \left[\frac{(\eta_{kn}(\omega) - a_{kn})^2}{\sigma_{kn}^2} - \eta_{kn}^2(\omega)\right]\right\}, \tag{2.2}$$

where

$$\eta_{kn} = \eta(\varphi_{kn}), \qquad a_{kn} = \mathsf{M}_1\eta(\varphi_{kn}), \qquad k = 1, \ldots, n.$$

It is easy to show by computation that

$$\begin{aligned} \mathsf{M} \log p_n &= \frac{1}{2} \sum_{k=1}^{n} \left(\log \frac{1}{\sigma_{kn}^2} - \frac{1}{\sigma_{kn}^2} + 1 - \frac{a_{kn}^2}{\sigma_{kn}^2}\right), \\ \mathsf{M}_1 \log p_n &= \frac{1}{2} \sum_{k=1}^{n} (-\log \sigma_{kn}^2 + \sigma_{kn}^2 - 1 + a_{kn}^2), \\ \mathsf{D} \log p_n &= \frac{1}{2} \sum_{k=1}^{n} \frac{(1 - \sigma_{kn}^2)^2 + 2a_{kn}^2}{\sigma_{kn}^4}, \\ \mathsf{D}_1 \log p_n &= \frac{1}{2} \sum_{k=1}^{n} [(1 - \sigma_{kn}^2)^2 + 2\sigma_{kn}^2 a_{kn}^2], \end{aligned} \tag{2.3}$$

where D_η and $\mathsf{D}_1\eta$ denote the dispersions with respect to P and P_1.

Let us define the so-called *entropy distance* $r(\mathfrak{A}')$ between the equivalent Gaussian measures P and P_1 on the σ-algebra $\mathfrak{A}' \subseteq \mathfrak{A}$ by

$$r(\mathfrak{A}') = -\left[\mathsf{M} \log \frac{\mathsf{P}_1(d\omega)}{\mathsf{P}(d\omega)} + \mathsf{M}_1 \log \frac{\mathsf{P}(d\omega)}{\mathsf{P}_1(d\omega)}\right]. \tag{2.4}$$

According to the general inequality given by (1.32)

$$r(\mathfrak{A}') \leqslant r(\mathfrak{A}'') \quad \text{for} \quad \mathfrak{A}' \subseteq \mathfrak{A}''. \tag{2.5}$$

Let $r_n = r(\mathfrak{A}_n)$. Since $\mathfrak{A}_1 \subseteq \mathfrak{A}_2 \subseteq \dots$, the sequence r_n, $n = 1, 2, \dots$, is a monotone increasing sequence.

It is seen from (2.3) that if

$$\inf_{k,n} \sigma_{kn}^2 = 0 \quad \text{or} \quad \sup_{k,n} \sigma_{kn}^2 = \infty, \tag{2.6}$$

then

$$\lim_{n\to\infty} r_n = \infty. \tag{2.7}$$

If

$$\sigma_{kn}^2 \asymp 1, \tag{2.8}$$

which is obviously equivalent to (1.27), then

$$-\log \frac{1}{\sigma_{kn}^2} + \frac{1}{\sigma_{kn}^2} - 1 \asymp -\log \sigma_{kn}^2 + \sigma_{kn}^2 - 1 \asymp (1 - \sigma_{kn}^2)^2,$$

and hence

$$\mathsf{M}[-\log p_n] \asymp \mathsf{M}_1 \log p_n \asymp \mathsf{D} \log p_n \asymp \mathsf{D}_1 \log p_n \asymp r_n \asymp \sum_{k=1}^{n} [(1 - \sigma_{kn}^2)^2 + a_{kn}^2]. \tag{2.9}$$

Lemma 2. *If the relation given by (2.7) is satisfied, the Gaussian measures* P *and* P_1 *on the* σ*-algebra* $\mathfrak{A}$ *will be orthogonal.*

Proof. The case where (2.6) is satisfied has already been considered above (see (1.27)).

Let (2.8) (as well as (2.9)) be satisfied. Then for any sequence of sets $A_n \in \mathfrak{A}_n$ of the form

$$A_n = \left\{\log p_n - \mathsf{M} \log p_n \geqslant \frac{1}{2} r_n\right\}$$
$$= \Omega \Big\backslash \left\{-\log p_n + \mathsf{M}_1 \log p_n \geqslant \frac{1}{2} r_n\right\}$$

we have, using Chebyshev's inequality, the following limit relations:

$$\mathsf{P}(A_n) \leqslant \frac{\mathsf{D} \log p_n}{\frac{1}{4} r_n^2} \asymp \frac{1}{r_n} \to 0$$

and

$$\mathsf{P}_1(A_n) = 1 - \mathsf{P}_1(\Omega - A_n) \geqslant 1 - \frac{\mathsf{D}_1 \log p_n}{\frac{1}{4} r_n^2} \asymp 1 - \frac{1}{r_n} \to 1.$$

This implies (see (1.5)) that the probability measures P and P_1 are orthogonal. The lemma is proved. □

Lemma 3. *If*

$$\lim_{n\to\infty} r_n < \infty, \tag{2.10}$$

the probability measures P *and* P_1 *are equivalent on the σ-algebra* $\mathfrak{A}$.

Proof. Let us assume, for example, that the measure P_1 is not absolutely continuous with respect to P. This means that for some set $A \in \mathfrak{A}$

$$\mathsf{P}(A) = 0, \qquad \mathsf{P}_1(A) \neq 0.$$

There obviously exists a sequence of sets $A_n \in \mathfrak{A}_n$ such that $\mathsf{P}_2(A \circ A_n) \to 0$ as $n \to \infty$, where the measure $\mathsf{P}_2 = \mathsf{P} + \mathsf{P}_1$, and therefore for the sets $A_n \in \mathfrak{A}$

$$\mathsf{P}(A \circ A_n) \to 0, \qquad \mathsf{P}_1(A \circ A_n) \to 0.$$

We shall consider the σ-algebra $\mathfrak{A}'_n$ containing only the set A_n and its complement as well as the empty set and the whole space. The density $p'_n(\omega) = \mathsf{P}_1(d\omega)/\mathsf{P}(d\omega)$ on $\mathfrak{A}'_n$ is equal to $\mathsf{P}_1(A_n)/\mathsf{P}(A_n)$ for $\omega \in A_n$ and equal to $(1 - \mathsf{P}_1(A_n))/(1 - \mathsf{P}(A_n))$ for $\omega \notin A_n$. We have also

$$\mathsf{M}_1 \log p'_n = \log \frac{\mathsf{P}_1(A_n)}{\mathsf{P}(A_n)} \mathsf{P}_1(A_n) + \log \frac{1 - \mathsf{P}_1(A_n)}{1 - \mathsf{P}(A_n)} [1 - \mathsf{P}_1(A_n)].$$

Since $\mathsf{P}_1(A_n) \to \mathsf{P}_1(A) > 0$, and $\mathsf{P}(A_n) \to \mathsf{P}(A) = 0$, $\mathsf{M}_1 \log p'_n \to \infty$ as $n \to \infty$. By the general inequality given by (1.32) we infer that

$$-\mathsf{M}_1 \log p'_n \leqslant -\mathsf{M}_1 \log p_n \leqslant r_n,$$

where, as above, $p_n(\omega) = \mathsf{P}_1(d\omega)/\mathsf{P}(d\omega)$ is the density on the σ-algebra $\mathfrak{A}_n \supseteq \mathfrak{A}'_n$. Therefore, if the measure P_1 is not absolutely continuous with respect to P, $r_n \to \infty$ as $n \to \infty$. Similar arguments lead to the same result if the measure P is not absolutely continuous with respect to P_1. Therefore, the measures P and P_1 are equivalent on the σ-algebra $\mathfrak{A}$ under the condition given by (2.10), as was to be proved. □

Lemmas 2 and 3 proved above directly imply the following.*

Theorem 1. *The Gaussian measures* P *and* P_1 *are either equivalent or orthogonal. These measures are equivalent on the σ-algebra* $\mathfrak{A}(T)$ *if and only if the condition given by* (2.10) *is satisfied.*

Let P_1^0 be a Gaussian measure with the zero mean and the same correlation function as that of P_1. From (2.9) and Theorem 1 we easily deduce the following.

* This result is due to T. Hajek, "On a property of normal distributions of an arbitrary stochastic process," *Czechoslov. Matemat. J.* **8** (1958), 610–618.

Theorem 2. *The Gaussian measures* P *and* P_1 *are equivalent if and only if the pairs* P *and* P_1^0, *and* P_1^0 *and* P_1 *are equivalent; for the equivalent measures* P *and* P_1 *the density* $\mathsf{P}_1(d\omega)/\mathsf{P}(d\omega)$ *is such that*

$$\frac{\mathsf{P}_1(d\omega)}{\mathsf{P}(d\omega)} = \frac{\mathsf{P}_1(d\omega)}{\mathsf{P}_1^0(d\omega)} \frac{\mathsf{P}_1^0(d\omega)}{\mathsf{P}(d\omega)}. \tag{2.11}$$

(We recall that we consider here the case $\mathsf{P} = \mathsf{P}^0$.)

II.2.2 Conditions of Equivalence Associated with Hilbert Spaces $L_T(F)$ and $L_T(F_1)$

Let P and P_1 be Gaussian measures on the σ-algebra $\mathfrak{A}(T)$ generated by variables $\eta(\varphi)$ of the type given by (1.20) where the function parameter $\varphi(\lambda)$ runs over the space L_T^0. By Theorem 2 one may consider two cases. In case one the correlation functions $B(\varphi, \psi) = \langle\varphi, \psi\rangle_F$ and $B_1(\varphi, \psi) = \langle\varphi, \psi\rangle_{F_1}$ coincide:

$$\langle\varphi, \psi\rangle_F = \langle\varphi, \psi\rangle_{F_1} \tag{2.12}$$

for all $\varphi, \psi \in L_T^0$. In case two the mean $\mathsf{M}_1\eta(\varphi)$ is equal to zero. Let us consider case one where the Gaussian measures P and P_1 differ only in the mean:

$$\mathsf{M}\eta(\varphi) = 0, \qquad \mathsf{M}_1\eta(\varphi) = a(\varphi), \qquad \varphi \in L_T^0.$$

Theorem 3. *Under the condition given by* (2.12) *the Gaussian measures* P *and* P_1 *are equivalent if and only if the mean value* $a(\varphi)$ *is a linear continuous functional on the Hilbert space* $L_T(F)$:

$$a(\varphi) = \langle\varphi, \psi\rangle_F, \qquad \varphi \in L_T^0, \tag{2.13}$$

for some $\psi(\lambda)$ *in* $L_T(F)$.

Proof. The continuity of the linear functional $a(\varphi)$ on the Hilbert space $L_T(F)$ is equivalent to the boundedness of this functional, and has already been proved (Subsection III.1.3).

Let the mean value $a(\varphi)$ be a continuous linear functional. The mean $a(\varphi)$, $\varphi \in L_T^0$, can be described, as any linear continuous functional can be, by formula (2.13) where $\psi(\lambda) \in L_T(F)$ is defined uniquely since the subspace L_T^0 is dense in $L_T(F)$. Let $\{\varphi_i\} \in L_T^0$ be a complete orthonormal system in $L_T(F)$. As seen from (2.3), the entropy distance between the Gaussian measures P and P_1 on the σ-algebra $\mathfrak{A}_n$ generated by the variables $\eta_k = \eta(\varphi_k)$, $k = 1, \ldots, n$, is

$$r_n = 2 \sum_{k=1}^{n} a_k^2$$

(where $a_k = a(\varphi_k)$) and

$$\lim_{n\to\infty} r_n = 2 \sum_{k=1}^{\infty} a_k^2 = 2 \sum_{k=1}^{\infty} \langle\varphi_k, \psi\rangle_F^2 = 2\|\psi\|_F^2 < \infty.$$

Therefore, (2.10) is satisfied and thus (see Theorem 1), the Gaussian measures P and P_1 are equivalent on the σ-algebra $\mathfrak{A} = \lim \mathfrak{A}_n$. By Lemma 1 this fact is equivalent to the equivalence of P and P_1 on the σ-algebra $\mathfrak{A}(T)$. The theorem is proved. □

We consider now the Gaussian measures P and P_1 (on the σ-algebra $\mathfrak{A}(T)$) with identical means equal to zero.

Let us define the operator A on the Hilbert space $L_T(F)$ into the Hilbert space $L_T(F_1)$ by

$$A\varphi(\lambda) = \varphi(\lambda) \tag{2.14}$$

for all $\varphi(\lambda) \in L_T^0$. As before, it will be assumed that (1.27) is satisfied, since if this condition is violated, the measures P and P_1 will be orthogonal. This condition is equivalent to the fact that the operator A is bounded and has a bounded inverse. This can be expressed as

$$A^*A \asymp E, \tag{2.15}$$

where A^* is the adjoint operator of A, and E is the identity operator, (2.15) implies that

$$\|(A^*A)\varphi\|_F \asymp \|\varphi\|_F, \qquad \varphi \in L_T(F).$$

Note that

$$\langle A^*A\varphi, \psi\rangle_F = \langle A\varphi, A\psi\rangle_{F_1} = \langle \varphi, \psi\rangle_{F_1} \tag{2.16}$$

for any φ, ψ.

We consider the difference

$$\Delta = E - A^*A. \tag{2.17}$$

Lemma 4. *If the operator Δ is completely continuous,* (2.15) *as well as* (1.27) *will be satisfied if and only if the operator Δ has no eigenvalue equal to one.*

Proof. It is obvious that the condition given by (2.15) is equivalent to the fact that the operator A^*A is bounded and has a bounded inverse $(A^*A)^{-1}$.

Further, since the operator A^*A is positive, the difference $\Delta = E - A^*A$ is such that

$$\delta = \sup_{\|\varphi\|_F = 1} \langle \Delta\varphi, \varphi\rangle \leq 1.$$

We have

$$\langle \varphi, \varphi\rangle_F - \langle A^*A\varphi, \varphi\rangle_F \leq \delta\langle \varphi, \varphi\rangle_F,$$

$$\langle A^*A\varphi, \varphi\rangle_F \geq (1 - \delta)\langle \varphi, \varphi\rangle_F.$$

Therefore, the bounded operator $(A^*A)^{-1}$ exists for $\delta \neq 1$. On the other hand, if 1 is the eigenvalue of the operator $E - A^*A$, 0 will be the eigenvalue of the operator A^*A, and therefore the inverse operator $(A^*A)^{-1}$ does not exist. □

Theorem 4. *Under the condition given by* (1.27) *the Gaussian measures* P *and* P_1 *are equivalent if and only if* $\Delta = E - A^*A$ *is Hilbert–Schmidt.**

Proof. Let us consider the spectral representation of the bounded symmetric operator Δ:

$$\Delta = \int \mu E(d\mu),$$

where $E(d\mu)$ is the spectral family of projection operators (*unitary decomposition*). It is seen that

$$A^*A = \int (1 - \mu)E(d\mu).$$

We assume that the spectrum of the operator Δ is not purely discrete. Then outside of a neighborhood $(-\varepsilon, \varepsilon)$ there is an infinite number of spectral points, and therefore an infinite number of nonoverlapping intervals $[\mu_k, \mu_{k+1}]$, $k = 1, 2, \ldots$, such that all invariant orthogonal subspaces of the form

$$E[\mu_k, \mu_{k+1}]L_T(F)$$

are different from zero. Let us choose an element φ_k, $\|\varphi_k\|_F = 1$, from each subspace mentioned above for which

$$\langle A^*A\varphi_k, \varphi_j\rangle_F = \langle \varphi_k, \varphi_j\rangle_{F_1} = \begin{cases} \sigma_k^2 & \text{for } j = k \\ 0 & \text{for } j \neq k, \end{cases}$$

and

$$\mu_k \leqslant 1 - \sigma_k^2 \leqslant \mu_{k+1}, \qquad (1 - \sigma_k^2)^2 \geqslant \varepsilon^2.$$

The entropy distance r_n between the Gaussian measures P and P_1 on the σ-algebra $\mathfrak{A}_n$ generated by the variables $\eta_k = \eta(\varphi_k)$, $k = 1, \ldots, n$ is such that (see (2.9))

$$r_n \asymp \sum_{k=1}^{n} (1 - \sigma_k^2)^2 \geqslant \varepsilon^2 n.$$

It is seen that $r_n \to \infty$ as $n \to \infty$, and by Theorem 1 the Gaussian measures P and P_1 are orthogonal on the σ-algebra $\mathfrak{A} = \lim_{n\to\infty} \mathfrak{A}_n$.

Thus, the spectrum of the operator Δ (therefore of the operator A^*A as well) is purely discrete for the equivalent measures P and P_1. If $\varphi_1, \varphi_2, \ldots,$ is a complete orthonormal system of functions with eigenvalues $\mu_1, \mu_2, \ldots,$ the condition

$$\sum_k \mu_k^2 < \infty \tag{2.18}$$

* We recall here that by a Hilbert–Schmidt operator we mean a compact operator, the complete system of its eigenvalues $\mu_1, \mu_2, \ldots$ satisfying the condition $\sum_k \mu_k^2 < \infty$.

Theorem 4 is due to J. Feldman, "Equivalence and perpendicularity of Gaussian processes," *Pacific J. Math.* **8** (1958), 699–708; **9** (1959), 1295–1296. We deduce it easily from Theorem 1, whereas the proof suggested by Feldman is rather complicated.

is equivalent to the fact that

$$\lim_{n\to\infty} r_n \asymp \sum_{k=1}^{\infty} (1-\sigma_k^2)^2 < \infty,$$

where $\sigma_k^2 = 1 - \mu_k$, $k = 1, 2, \ldots$, is the complete system of eigenvalues of the operator A^*A and r_n is the entropy distance between P and P_1 on the σ-algebra $\mathfrak{A}_n$ generated by the variables $\eta_k = \eta(\varphi_k)$, $k = 1, \ldots, n$. Therefore, (2.18) (as well as (1.27)) is a necessary and sufficient condition for the Gaussian measures P and P_1 to be equivalent, as was to be proved. □

Let us note that (2.18) can be written as

$$\sum_{k,j} \langle \Delta\varphi_k, \varphi_j \rangle_F^2 < \infty$$

and that for any orthonormal system $\psi_1, \psi_2, \ldots \in L_T(F)$

$$\begin{aligned}\sum_{k,j} \langle \Delta\psi_k, \psi_j \rangle_F^2 &= \sum_k \left[\sum_j \langle \Delta\psi_k, \psi_j \rangle_F\right]^2 \\ &= \sum_k \|\Delta\psi_k\|_F^2 \\ &= \sum_k \left[\sum_j \langle \Delta\psi_k, \varphi_j \rangle_F^2\right] \\ &= \sum_j \left[\sum_k \langle \psi_k, \Delta\varphi_j \rangle_F^2\right] \leqslant \sum_j \|\Delta\varphi_j\|_F^2 = \sum_{k,j} \langle \Delta\varphi_k, \varphi_j \rangle_F^2.\end{aligned}$$

In the above relations the pertinent inequalities become equalities.

It is readily seen that the operator Δ is a Hilbert–Schmidt operator (we have here a discrete spectrum and the complete system of eigenvalues satisfies (2.18)) if and only if

$$\sum_{k,j} \langle \Delta\varphi_k, \varphi_j \rangle_F^2 < \infty \tag{2.19}$$

for any complete orthonormal system $\varphi_1, \varphi_2, \ldots$.

(2.19) can immediately be expressed in terms of correlation functionals of the distributions P and P_1, since

$$\begin{aligned}\langle \Delta\varphi, \psi \rangle_F &= \langle \varphi, \psi \rangle_F - \langle A^*A\varphi, \psi \rangle_F \\ &= \langle \varphi, \psi \rangle_F - \langle \varphi, \psi \rangle_{F_1} = B(\varphi, \psi) - B_1(\varphi, \psi)\end{aligned}$$

for any φ, $\psi \in L_T(F)$. Therefore, under the condition given by (1.27) a necessary and sufficient condition for equivalence of the Gaussian measures P and P_1 is that for any complete orthonormal system $\varphi_1, \varphi_2, \ldots \in L_T(F)$

$$\sum_{k,j} b(\varphi_k, \varphi_j)^2 < \infty, \tag{2.20}$$

where

$$b(\varphi, \psi) = B(\varphi, \psi) - B_1(\varphi, \psi), \qquad \varphi, \psi \in L_T(F).$$

Example. Let $T = (-\infty, \infty)$ be the whole real line. In this case $L_T(F)$ is the space of functions $\varphi(\lambda)$ such that $\int_{-\infty}^{\infty} |\varphi(\lambda)|^2 F(d\lambda) < \infty$. The condition given by (1.27) (implying, in particular, that $\int_\Delta F(d\lambda) \asymp \int_\Delta F_1(d\lambda)$) is equivalent to the fact that the spectral measures F and F_1 are equivalent and that the density $f(\lambda) = F_1(d\lambda)/F(d\lambda)$ of these measures satisfies the condition $f(\lambda) \asymp 1$ (for almost all λ). The operator A^*A can easily be found from the equality

$$\langle A^*A\varphi, \psi\rangle_F = \langle \varphi, \psi\rangle_{F_1} = \langle f\varphi, \psi\rangle_F, \qquad \varphi, \psi \in L_T(F).$$

In fact, since $f\varphi \in L_T(F)$, the operator A^*A is the operator of multiplication by the density $f(\lambda)$. Therefore, the operator $\Delta = E - A^*A$ is the operator of multiplication by the function $1 - f(\lambda)$. It is seen that the function $\varphi(\lambda)$ is an eigenfunction for the operator Δ if and only if $\varphi(\lambda) = 0$ almost everywhere outside of the set $\{\lambda\colon 1 - f(\lambda) - \mu\}$, where μ is the corresponding eigenvalue. It can easily be seen that Δ is a Hilbert–Schmidt operator if and only if $1 - f(\lambda) = 0$ almost everywhere with respect to the continuous segment of the spectral measure $F(d\lambda)$ (i.e., continuous segments of the spectral measures F and F_1 coincide). For the discrete segment of $F(d\lambda)$ concentrated at the points $\lambda_1, \lambda_2, \ldots$, the condition

$$\sum_k \left[1 - \frac{F_1(\lambda_k)}{F(\lambda_k)}\right]^2 < \infty$$

is satisfied, where

$$\mu_k = 1 - f(\lambda_k) = 1 - \frac{F_1(\lambda_k)}{F(\lambda_k)}, \qquad k = 1, 2, \ldots,$$

is the complete system of nonzero eigenvalues corresponding to eigenfunctions of the form

$$\varphi_k(\lambda) = \begin{Bmatrix} 0 \text{ for } \lambda \neq \lambda_k \\ 1 \text{ for } \lambda = \lambda_k \end{Bmatrix}, \qquad k = 1, 2, \ldots$$

(see (2.18)).

We discuss in detail the condition given by (2.20) and the corollaries following from this condition.

Let us consider a Hilbert space $L_{T\times T}(F \times F)$, analogous to the space $L_T(F)$ introduced before. Let us consider actually the linear space $L_{T\times T}^0$ of functions of the form

$$\varphi(\lambda, \mu) = \sum_{k,j} c_{kj} e^{i(\lambda s_k - \mu t_j)} \tag{2.21}$$

(where $s_k, t_j \in T$ and where c_{kj} are real coefficients). We define $L_{T\times T}(F \times F)$ as a Hilbert space obtained by means of the closure of $L_{T\times T}^0$ over the scalar product

$$\langle \varphi, \psi \rangle_{F \times F} = \iint \varphi(\lambda, \mu)\overline{\psi(\lambda, \mu)}F(d\lambda)F(d\mu). \tag{2.22}$$

It is obvious that if $\varphi', \varphi'' \in L_T(F)$, then

$$\varphi(\lambda, \mu) = \varphi'(\lambda)\overline{\varphi''(\mu)} \tag{2.23}$$

enters into the space $L_{T \times T}(F \times F)$. In this case the system of functions of the type given by (2.23) is complete in $L_{T \times T}(F \times F)$. This fact follows from the definition of the Hilbert spaces $L_T(F)$ and $L_{T \times T}(F \times F)$ considered if we note that the functions $\varphi(\lambda, \mu)$ of the type given by (2.23) enter into $L_{T \times T}(F \times F)$ if φ' and φ'' are functions of the type given by (1.21), and

$$\begin{aligned}
&\|\varphi(\lambda, \mu) - \psi(\lambda, \mu)\|_{F \times F}^2 \\
&= \iint |\varphi'(\lambda)\overline{\varphi''(\mu)} - \psi'(\lambda)\overline{\psi''(\mu)}|^2 F(d\lambda)F(d\mu) \\
&\leqslant 2 \iint [|\varphi''(\mu)|^2|\varphi'(\lambda) - \psi'(\lambda)|^2 + |\psi'(\lambda)|^2|\varphi''(\mu) - \psi''(\mu)|^2] \times F(d\lambda)F(d\mu) \\
&- 2[\|\varphi''\|_F^2\|\varphi' - \psi'\|_F^2 + \|\psi'\|_F^2\|\varphi'' - \psi''\|_F^2]
\end{aligned}$$

for any functions φ', φ'' and ψ', ψ'' of the type mentioned above.

Let $\varphi_1, \varphi_2, \ldots$ be an orthonormal system in a Hilbert space $L_T(F)$. Then, obviously, the functions

$$\varphi_{kj}(\lambda, \mu) = \varphi_k(\lambda)\overline{\varphi_j(\mu)}, \qquad k, j = 1, 2, \ldots,$$

comprise a complete orthonormal system in the Hilbert space $L_{T \times T}(F \times F)$. Under the condition given by (2.20) let

$$b_{kj} = b(\varphi_k, \varphi_j), \qquad k, j = 1, 2, \ldots,$$

and let $\psi_0(\lambda, \mu) \in L_{T \times T}(F \times F)$ be defined as

$$\psi_0(\lambda, \mu) = \sum_{k, j} b_{kj}\varphi_{kj}(\lambda, \mu). \tag{2.24}$$

(2.24) yields the decomposition of $\psi_0(\lambda, \mu)$ over the orthonormal system $\varphi_{kj}(\lambda, \mu)$ so that

$$b(\varphi_k, \varphi_j) = \langle \varphi_k\overline{\varphi}_j, \psi_0 \rangle_{F \times F}$$

This relation obviously extends to any linear combinations

$$\varphi'(\lambda) = \sum_k c'_k\varphi_k(\lambda), \qquad \varphi''(\mu) = \sum_j c''_j\varphi_j(\mu),$$

so that

$$b(\varphi', \varphi'') = \langle \varphi'\overline{\varphi''}, \psi_0 \rangle_{F \times F}. \tag{2.25}$$

In (2.25) the limiting transition leads us to arbitrary functions $\varphi', \varphi'' \in L_T(F)$.

On the other hand, if the difference between the correlation functionals $b(\varphi', \varphi'')$, $\varphi', \varphi'' \in L_T(F)$ permits a representation of the type given by (2.25), for any orthonormal system $\varphi_1, \varphi_2, \ldots$, we have

$$\sum_{k,j} b(\varphi_k, \varphi_j) \leqslant \|\psi_0\|^2_{F\times F},$$

that is, (2.20) is satisfied.

(2.25) implies that the bilinear functional $b(\varphi', \varphi'')$ on $L_T(F)$ regarded as a functional on the class of functions $\varphi'(\lambda)\overline{\varphi''(\mu)} \in L_{T\times T}(F \times F)$ extends to a linear continuous functional on the Hilbert space $L_{T\times T}(F \times F)$.

We have thus arrived at the following result, similar to a certain extent to Theorem 3.

Theorem 5. *Under the condition given by (1.27) a necessary and sufficient condition for equivalence of the Gaussian measures* P *and* P_1 *is that the difference*

$$b(\varphi', \varphi'') = B(\varphi', \varphi'') - B_1(\varphi', \varphi''),$$

being a functional on the class of functions $\varphi'(\lambda)\overline{\varphi''(\mu)} \in L_{T\times T}(F \times F)$, *can be extended as to be linear continuous on the Hilbert space* $L_{T\times T}(F \times F)$.

Let us take now the space $L_{T\times T}(F \times F_1)$, defined similarly to $L_{T\times T}(F \times F)$, with the scalar product

$$\langle\varphi, \psi\rangle_{F\times F_1} = \iint \varphi(\lambda, \mu)\overline{\psi(\lambda, \mu)}F(d\lambda)F_1(d\mu)$$

(compare with (2.22)). It is easily seen* that under the condition given by (1.27)

$$\|\varphi\|^2_{F\times F} \asymp \|\varphi\|^2_{F\times F_1} \tag{2.26}$$

* It is useful to note that $L_{T\times T}(F \times F)$ is a function model for a Hilbert space of variables $\eta(\omega', \omega'')$ on the product $\Omega \times \Omega$ that is the closure of variables of the form

$$\eta(\omega', \omega'') = \sum_{k,j} c_{kj}\xi(\omega', s_k)\xi(\omega'', t_j)$$

with a scalar product of the form

$$\iint \eta_1(\omega', \omega'')\eta_2(\omega', \omega'')\mathsf{P}(d\omega') \times \mathsf{P}(d\omega'').$$

In fact, the variables $\eta(\omega', \omega'')$ are associated with functions $\varphi(\lambda, \mu)$ of the type given by (2.21) and

$$\iint \eta_1(\omega', \omega'')\eta_2(\omega', \omega'')\mathsf{P}(d\omega') \times \mathsf{P}(d\omega'') = (\varphi_1, \varphi_2)_{F\times F}.$$

Similarly,

$$\iint \eta_1(\omega', \omega'')\eta_2(\omega', \omega'')\mathsf{P}(d\omega') \times \mathsf{P}_1(d\omega'') = (\varphi_1, \varphi_2)_{F\times F_1}.$$

(1.27) implies that

$$\int [\eta(\omega)]^2\mathsf{P}(d\omega) \asymp \int [\eta(\omega)]^2\mathsf{P}_1(d\omega)$$

for any variables of the form $\eta(\omega) = \sum_k c_k\xi(\omega, t_k)$. Hence, under the condition given by (1.27) we obtain

$$\|\varphi\|^2_{F\times F} = \int \mathsf{P}(d\omega'') \int \left|\sum_k \left[\sum_j c_{kj}\xi(\omega'', t_j)\right]\xi(\omega', s_k)\right|^2 \mathsf{P}(d\omega')$$

$$\asymp \int \mathsf{P}(d\omega'') \int \left|\sum_k \left[\sum_j c_{kj}\xi(\omega'', t_j)\right]\xi(\omega', s_k)\right|^2 \mathsf{P}_1(d\omega') = \|\varphi\|^2_{F\times F_1}.$$

for $\varphi \in L^0_{T\times T}$. This relation shows that the Hilbert spaces $L_{T\times T}(F \times F)$ and $L_{T\times T}(F \times F_1)$ are equivalent:

$$L_{T\times T}(F \times F) = L_{T\times T}(F \times F_1),$$

(2.26) holding true for all

$$\varphi \in L_{T\times T}(F \times F) \qquad (\varphi \in L_{T\times T}(F \times F_1)).$$

Under the condition given by (2.26) the functional $b(\varphi', \varphi'')$ of the type given by (2.25) extends into the Hilbert space $L_{T\times T}(F \times F_1)$ and is representable as

$$b(\varphi', \varphi'') = \langle \varphi'\overline{\varphi''}, \psi \rangle_{F\times F_1}, \tag{2.27}$$

where $\psi(\lambda, \mu)$ is a fixed element in $L_{T\times T}(F \times F_1)$.

Similarly, under the condition given by (2.26) it follows from (2.27) that the functional $b(\varphi', \varphi'')$ is representable as (2.25). In fact, (1.27) as well as (2.26) follow from (2.27). Since for any $\varphi \in L^0_T$

$$b(\varphi, \varphi) = \|\varphi\|^2_F - \|\varphi\|^2_{F_1} = \langle \varphi(\lambda)\varphi(\mu), \psi \rangle,$$

we have

$$\left|\|\varphi\|^2_F - \|\varphi\|^2_{F_1}\right| \leqslant \|\varphi(\lambda)\overline{\varphi(\mu)}\|_{F\times F_1}\|\psi\|_{F\times F_1} = \|\varphi\|_F\|\varphi\|_{F_1}\|\psi\|_{F\times F_1}.$$

It is seen that if $\|\varphi\|_F \to 0$, then $\|\varphi\|_{F_1} \to 0$, and vice versa. Therefore, the operator A introduced above from $L_T(F)$ into $L_T(F_1)$ is continuous and has a continuous inverse A^{-1}. But this fact (see (2.15)) is equivalent to (1.27).

Therefore, if in Theorem 5 the space $L_{T\times T}(F \times F)$ is replaced by $L_{T\times T}(F \times F_1)$, the additional condition given by (1.27) is satisfied automatically. More precisely, the following holds:*

Theorem 6. *A necessary and sufficient condition for equivalence of the Gaussian measures* P *and* P_1 *is that the difference*

$$b(\varphi', \varphi'') = B(\varphi', \varphi'') - B_1(\varphi', \varphi''),$$

being a functional on the class of functions $\varphi'(\lambda)\overline{\varphi''(\mu)} \in L^0_{T\times T}$, *can be extended as to be linear continuous on the Hilbert space* $L_{T\times T}(F \times F_1)$.

III.3 General Conditions for Equivalence and Formulas for Density of Equivalent Distributions

Let P and P_1 be Gaussian measures on the σ-algebra $\mathfrak{A}(T)$ generated by all values of $\xi(t)$, $t \in T$, of the stationary (with respect to the measure P) random process $\xi(t)$; T is an arbitrary set on the real line.

* Theorem 6 and a simpler Theorem 3 are suggested by Parzen (see, for instance, E. Parzen, "Probability density functionals and reproducing kernel Hilbert spaces," *Time Series Analysis* (ed. M. Rosenblatt), 1963.

Let us consider first the case where these measures differ only in the mean

$$a(t) = \mathsf{M}_1\xi(t)$$

(more precisely, where $\mathsf{P}_1^0 = \mathsf{P}$—see Theorem 2).

Theorem 7. *The Gaussian measures* P *and* P_1 *are equivalent on the σ-algebra* $\mathfrak{A}(T)$ *if and only if the mean value* $a(t)$ *permits a spectral representation of the form*

$$a(t) = \int e^{-t\lambda t}\psi(\lambda)F(d\lambda) \tag{3.1}$$

for $t \in T$, *where* $\psi(\lambda)$ *is a function satisfying the condition* $\int|\psi(\lambda)|^2F(d\lambda) < \infty$. *For equivalent measures the integral equation given by* (3.1) *has a solution* $\psi(\lambda) \in L_T(F)$ *and the density* $p(\omega) = \mathsf{P}_1(d\omega)/\mathsf{P}(d\omega)$ *can be expressed as*

$$p(\omega) = D\exp\left\{\int\psi(\lambda)\Phi(d\lambda)\right\}, \tag{3.2}$$

where $\psi(\lambda)$ *is the solution of* (3.1) *from the space* $L_T(F)$ *and* D *is a normalizing multiplier,* $D = e^{-\|\varphi\|_F^2/2}$.

(Here, as above, $F(d\lambda)$ is the spectral measure and $\Phi(d\lambda)$ is the stochastic spectral measure of the stationary process $\xi(t)$).

Proof. By Theorem 3 a necessary and sufficient condition for P and P_1 to be equivalent is that (3.1) hold true for some function $\psi(\lambda)$ in $L_T(F)$ since this is equivalent to (2.13) because the system of functions $\varphi(\lambda) = e^{i\lambda t}$, $t \in T$, is complete. Further, if (3.1) holds for a function $\psi(\lambda)$ in the Hilbert space $L_{(-\infty,\infty)}(F)$ of all square-integrable functions, it will hold as well for a function $\psi_0(\lambda)$ in $L_T(F)$, i.e., the projection of $\psi(\lambda)$ onto the subspace $L_T(F) \subseteq L_{(-\infty,\infty)}(F)$, since

$$\langle e^{i\lambda t}, \psi(\lambda)\rangle_F = \langle e^{i\lambda t}, \psi_0(\lambda)\rangle_F$$

for all $t \in T$ (when $e^{i\lambda t} \in L_T(F)$).

We consider now the equivalent measures P and P_1 and the density $p(\omega) = \mathsf{P}_1(d\omega)/\mathsf{P}(d\omega)$ of these measures. Let us choose the complete orthonormal system of functions $\varphi_1(\lambda), \varphi_2(\lambda), \ldots \in L_T(F)$. Let us also consider the densities $p_n(\omega) = \mathsf{P}_1(d\omega)/\mathsf{P}(d\omega)$ on the σ-algebras $\mathfrak{A}_n$, each of which is generated by the variables $\eta_k = \eta(\varphi_k)$, $k = 1, \ldots, n$, of the type given by (1.20). As mentioned before (see Subsection III.1.3),

$$p(\omega) = \lim_{n\to\infty} p_n(\omega).$$

It is obvious (see (2.2)) that

$$\begin{aligned} p_n(\omega) &= \exp\left\{\sum_{k=1}^n a_k\eta_k(\omega) - \frac{1}{2}\sum_{k=1}^n a_k^2\right\} \\ &= \exp\left\{\int\psi_n(\lambda)\Phi(d\lambda) \quad \frac{1}{2}\|\psi_n\|_F^2\right\}, \end{aligned}$$

where

$$a_k = \mathsf{M}_1\eta_k = \langle\varphi_k, \psi\rangle_F, \qquad k = 1, 2, \ldots,$$

$\psi(\lambda) \in L_T(F)$ is the function in (3.1), and

$$\psi_n(\lambda) = \sum_{k=1}^{n} a_k\varphi_k(\lambda).$$

We have

$$\lim_{n\to\infty} \psi_n(\lambda) = \lim \sum_{k=1}^{n} \langle\psi, \varphi_k\rangle_F\varphi_k(\lambda) = \psi(\lambda)$$

and

$$\lim_{n\to\infty} \|\psi_n(\lambda)\|_n^2 = \lim \sum_{k=1}^{n} a_k^2 = \|\psi(\lambda)\|_F^2.$$

Therefore,

$$p(\omega) = \lim_{n\to\infty} p_n(\omega) = \exp\left\{\int \psi(\lambda)\Phi(d\lambda) - \frac{1}{2}\|\psi(\lambda)\|_F^2\right\},$$

which was to be proved. □

We take next the Gaussian measures P and P_1 on the σ-algebra $\mathfrak{A}(T)$ with zero mean values and arbitrary correlation functions.

We consider the space $H_2(T)$ as well as the spaces $H_n(T)$. The space $H_2(T)$ introduced in Section I.5 is the closed linear hull of the variables $\xi(s)\xi(t) - B(s, t)$, $s, t \in T$.

Let us consider the linear space everywhere dense in $H_2(T)$ of all variables represented in the symmetric form

$$\eta = \sum_{k,j} c_{kj}[\xi(t_k)\xi(t_j) - B(t_k, t_j)] \tag{3.3}$$

(with symmetric real coefficients $c_{kj} = c_{jk}$, $k, j = 1, 2, \ldots$), which is the linear hull of the variables $\xi(s)\xi(t) - B(s, t)$. Noting the formula for moments (see Chapter I, (5.10))

$$\begin{aligned}&\mathsf{M}\xi(t_1)\xi(t_2)\xi(t_3)\xi(t_4)\\ &= B(t_1, t_2)B(t_3, t_4) + B(t_1, t_3)B(t_2, t_4) + B(t_1, t_4)B(t_2, t_3),\end{aligned}$$

it is easy to compute that

$$\begin{aligned}\mathsf{M}\eta'\eta'' &= \sum_{k,j}\sum_{m,n} c'_{kj}c''_{mn}B(t_k, t_m)B(t_j, t_n) + \sum_{k,j}\sum_{m,n} c'_{kj}c''_{mn}B(t_k, t_n)B(t_j, t_m)\\ &= 2\sum_{k,j}\sum_{m,n} c'_{kj}c''_{mn}B(t_k, t_m)B(t_j, t_n)\end{aligned} \tag{3.4}$$

for any variables

$$\eta' = \sum_{k,j} c'_{kj}[\xi(t_k)\xi(t_j) - B(t_k, t_j)]$$

and

$$\eta'' = \sum_{k,j} c''_{kj}[\xi(t_k)\xi(t_j) - B(t_k, t_j)]$$

of the given type. If we introduce the stochastic measure $\Psi(d\lambda, d\mu)$, setting

$$\Psi(\Delta_1 \times \Delta_2) = \Phi(\Delta_1)\overline{\Phi(\Delta_2)} - F(\Delta_1 \cap \Delta_2), \tag{3.5}$$

we can easily see that each variable of the type given by (3.3) can be expressed as

$$\eta = \iint \varphi(\lambda, \mu)\Psi(d\lambda, d\mu), \tag{3.6}$$

where

$$\varphi(\lambda, \mu) = \sum_{k, j} c_{kj} e^{i(\lambda t_k - \mu t_j)}. \tag{3.7}$$

We have from (3.4) that

$$\mathsf{M}\eta'\eta'' = 2\iint \varphi'(\lambda, \mu)\overline{\varphi''(\lambda, \mu)}F(d\lambda)F(d\mu) = 2\langle \varphi', \varphi'' \rangle_{F \times F}, \tag{3.8}$$

where

$$\varphi'(\lambda, \mu) = \sum_{k, j} c'_{kj} e^{i(\lambda t_k - \mu t_j)}$$

and

$$\varphi''(\lambda, \mu) = \sum_{k, j} c''_{kj} e^{i(\lambda t_k - \mu t_j)}.$$

It is seen from (3.6)–(3.8) that the convergent sequence $\eta_n \in H_2$ is associated with a sequence of functions $\varphi_n \in L_{T \times T}(F \times F)$. Therefore, any variable $\eta \in H_2$ as the limit of a sequence of η_n of the type given by (3.3) can be represented by (3.6), where the function $\varphi(\lambda, \mu) \in L_{T \times T}(F \times F)$ is the limit of the corresponding functions φ_n of the type given by (3.7); given any function $\varphi(\lambda, \mu)$, (3.6) defines a certain variable $\eta \in H_2(T)$.

Theorem 8. *The Gaussian measures* P *and* P_1 *(with zero mean values) are equivalent on the σ-algebra* $\mathfrak{A}(T)$ *if and only if the difference between the correlation functions*

$$b(s, t) = B(s, t) - B_1(s, t)$$

is representable as

$$b(s, t) = \iint e^{-i(\lambda s - \mu t)}\psi(\lambda, \mu)F(d\lambda)F_1(d\mu) \tag{3.9}$$

for $s, t \in T$, *where the function* $\psi(\lambda, \mu)$ *is such that*

$$\iint |\psi(\lambda, \mu)|^2 F(d\lambda)F_1(d\mu) < \infty.$$

For the equivalent measures P *and* P_1 *the integral equation given by* (3.9) *has a solution* $\psi(\lambda, \mu) \in L_{T \times T}(F \times F_1)$. *The density* $p(\omega) = \mathsf{P}_1(d\omega)/\mathsf{P}(d\omega)$ *of the equivalent measures can be expressed as*

$$p(\omega) = D\exp\left\{-\frac{1}{2}\iint \psi(\lambda, \mu)\Psi(d\lambda, d\mu)\right\}, \tag{3.10}$$

where $\psi(\lambda, \mu)$ *is the solution of* (3.9) *from the space* $L_{T\times T}(F \times F_1)$; D *is a normalizing multiplier.*

Proof. Since the functions $\varphi(\lambda, \mu) = e^{i(\lambda s - \mu t)}$, $s, t \in T$, form a complete system in the Hilbert space $L_{T\times T}(F \times F_1)$, the relation given by (3.9) is equivalent, for $\psi(\lambda, \mu) \in L_{T\times T}(F \times F_1)$, to the relation given by (2.27), with the latter implying the equivalence of the Gaussian measures P and P_1 according to Theorem 6. If (3.9) holds true for a function $\psi(\lambda, \mu)$ in the Hilbert space of all square-integrable (with respect to $F(d\lambda) \times F_1(d\mu)$) functions, it will also hold for the function $\psi_0(\lambda, \mu)$, the projection of $\psi(\lambda, \mu)$ onto the subspace $L_{T\times T}(F \times F_1)$.

We consider next the equivalent measures P and P_1 and choose the sequence of points $t_1, t_2, \ldots$ everywhere dense in the set T. It is seen that the variables $\eta_k = \xi(t_k)$, $k = 1, 2, \ldots$, form a complete system in the Hilbert space $L_T(F)$. Let us consider the densities $p_n(\omega) = \mathsf{P}_1(d\omega)/\mathsf{P}(d\omega)$ on the σ-algebras $\mathfrak{A}_n$, each of which is generated by variables η_k, $k = 1, \ldots, n$. By virtue of (2.1),

$$\log p_n(\omega) = \mathsf{M} \log p_n - \frac{1}{2} \sum_{k,\, j=1} c_{kj}[\xi(t_k)\xi(t_j) - B(t_k, t_j)], \tag{3.11}$$

where $\{c_{kj}\}$ is the difference between the matrices inverse to the correlation matrices $\{B_1(t_k, t_j)\}$ and $\{B(t_k, t_j)\}$. The corresponding variables

$$\eta_n(\omega) = \sum_{k,\, j=1}^{n} c_{kj}[\xi(t_k)\xi(t_j) - B(t_k, t_j)]$$

appearing in (3.11) belong to the space $H_2(T)$ and are representable by (3.6):

$$\eta_n = \iint \psi_n(\lambda, \mu)\Psi(d\lambda, d\mu), \tag{3.12}$$

where

$$\psi_n(\lambda, \mu) = \sum_{k,\, j=1}^{n} c_{kj} e^{i(\lambda t_k - \mu t_j)}. \tag{3.13}$$

It is easy to verify that each function $\psi_n(\lambda, \mu)$ satisfies an equation of the type given by (3.9),

$$\iint e^{-i(\lambda s - \mu t)}\psi_n(\lambda, \mu)F(d\lambda)F_1(d\mu) = b(s, t), \tag{3.14}$$

for $s, t = t_1, \ldots, t_n$. In fact, this equality can be rewritten in matrix form

$$\{B(t_k, t_j)\}\{c_{kj}\}\{B_1(t_k, t_j)\} = \{b(t_k, t_j)\},$$

where

$$\{c_{kj}\} = \{B_1(t_k, t_j)\}^{-1} - \{B(t_k, t_j)\}^{-1},$$

and it follows immediately that

$$\{B(t_k, t_j)\}\{c_{kj}\} = \{B(t_k, t_j)\}\{B_1(t_k, t_j)\}^{-1} - E,$$
$$\{B(t_k, t_j)\}\{c_{kj}\}\{B_1(t_k, t_j)\} = \{B(t_k, t_j)\} - \{B_1(t_k, t_j)\} = \{b(t_k, t_j)\}.$$

We can rewrite (3.14) also as

$$\langle \varphi, \psi_n \rangle_{F \times F_1} = b(s, t), \qquad s, t \in T_n,$$

where $T_n = \{t_1, \ldots, t_n\}$ and $\varphi(\lambda, \mu) = e^{i(\lambda s - \mu t)}$, $s, t \in T_n$. It is clear that for $m \leqslant n$ the function $\psi_m(\lambda, \mu)$ coincides with the projection of the element $\psi_n(\lambda, \mu) \in L_{T_n \times T_n}(F \times F_1)$ onto the subspace $L_{T_m \times T_m}(F \times F_1)$ so that

$$\|\psi_n - \psi_m\|^2_{F \times F_1} = \|\psi_m\|^2_{F \times F_1} - \|\psi_n\|^2_{F \times F_1} \to 0$$

as $m, n \to \infty$, since the sequence $\|\psi_n\|^2_{F \times F_1}$, $n = 1, 2, \ldots$, turns out to be monotone decreasing (and bounded from below by zero) and $\lim \|\psi_n\|^2$ exists. It is also seen that since the Hilbert space $L_{T \times T}(F \times F_1)$ coincides with the closure of extending spaces $L_{T_n \times T_n}(F \times F_1)$, $n = 1, 2, \ldots$, with each function $\psi_n(\lambda, \mu)$ in (3.14) being the projection of $\psi(\lambda, \mu) \in L_{T \times T}(F \times F_1)$ by (3.9), $\lim_{n \to \infty} \psi_n(\lambda, \mu) \in L_{T \times T}(F \times F_1)$ has the property that

$$\psi(\lambda, \mu) = \lim_{n \to \infty} \psi_n(\lambda, \mu).$$

This fact implies that the variables η_n of the type given by (3.12) appearing in (3.11) converge in the mean to the variable

$$\eta = \iint \psi(\lambda, \mu) \Psi(d\lambda, d\mu) \in H_2(T).$$

It was shown before (see (1.33)) that the limit

$$\lim_{n \to \infty} \mathsf{M} \log p_n$$

exists for equivalent measures. Therefore, the density $p(\omega) = \mathsf{P}_1(d\omega)/\mathsf{P}(d\omega)$ on the σ-algebra $\mathfrak{A}(T)$ can be determined by the limit relation given by (1.34),

$$\begin{aligned}
\log p(\omega) &= \lim_{n \to \infty} \mathsf{M} \log p_n + \lim_{n \to \infty} [\log p_n - \mathsf{M} \log p_n] \\
&= \lim_{n \to \infty} \mathsf{M} \log p_n - \frac{1}{2} \lim_{n \to \infty} \eta_n(\omega) \\
&= \lim_{n \to \infty} \mathsf{M} \log p_n - \frac{1}{2} \lim_{n \to \infty} \iint \psi_n(\lambda, \mu) \Psi(d\lambda, d\mu) \\
&= \log D - \frac{1}{2} \iint \psi(\lambda, \mu) \Psi(d\lambda, d\mu),
\end{aligned}$$

which yields (3.10). □

III.4 Further Investigation of Equivalence Conditions

III.4.1 Gaussian Measures with Different Mean Values

Let us examine the conditions for equivalence of Gaussian measures P and P_1 on the σ-algebra $\mathfrak{A}(T)$ proved in Theorem 7. We consider the case where the spectral measure $F(d\lambda)$ is absolutely continuous and has the bounded density

$$f(\lambda) = \frac{F(d\lambda)}{d\lambda}.$$

A necessary and sufficient condition for Gaussian measures P and P_1 differing only in the mean

$$a(t) = M_1\xi(t), \qquad t \in T, \tag{4.1}$$

to be equivalent is that the function $a(t)$ permit a representation

$$a(t) = \int e^{-i\lambda t}\psi(\lambda)f(\lambda)\,d\lambda, \qquad t \in T, \tag{4.2}$$

where $\psi(\lambda)$ is a function satisfying the condition

$$\int |\psi(\lambda)|^2 f(\lambda)\,d\lambda < \infty$$

(see Theorem 7). Let

$$\varphi(\lambda) = \psi(\lambda)f(\lambda). \tag{4.3}$$

The right-hand side in (4.2) defines a function

$$a(t) = \int e^{-i\lambda t}\varphi(\lambda)\,d\lambda, \qquad -\infty < t < \infty, \tag{4.4}$$

which coincides with the mean value given by (4.1) for $t \in T$, the function $\varphi(\lambda)$ of the type given by (4.3) being square-integrable,

$$\int |\varphi(\lambda)|^2\,d\lambda \leqslant C \int |\psi(\lambda)|^2 f(\lambda)\,d\lambda < \infty,$$

for $f(\lambda) \leqslant C$.

This means that the function $a(t)$, $-\infty < t < \infty$, defined by (4.4) is square-integrable, and $\varphi(\lambda)$ is the Fourier transform:

$$\varphi(\lambda) = \frac{1}{2\pi}\int e^{i\lambda t}a(t)\,dt. \tag{4.5}$$

As can be seen from (4.3), the function $\varphi(\lambda)$ is square-integrable and also it satisfies the condition

$$\int \frac{|\varphi(\lambda)|^2}{f(\lambda)}\,d\lambda < \infty. \tag{4.6}$$

Next, let us assume that the function $a(t)$, $t \in T$, can be extended to the entire real line $-\infty < t < \infty$ as to be a square-integrable function $a(t)$, $-\infty < t < \infty$, whose Fourier transform $\varphi(\lambda)$ satisfies (4.6). Then for all t

$$a(t) = \int e^{-i\lambda t}\varphi(\lambda)\,d\lambda = \int e^{-i\lambda t}\psi(\lambda)f(\lambda)\,d\lambda,$$

where the function $\psi(\lambda) = \varphi(\lambda)/f(\lambda)$ satisfies the condition $\int|\psi(\lambda)|^2 f(\lambda)\,d\lambda < \infty$. By Theorem 7 the above implies that the Gaussian measures P and P_1 are equivalent.

Thus, we have proved the following result:

Theorem 9. *A necessary and sufficient condition for the Gaussian measures* P *and* P_1 *to be equivalent is that the function* $a(t)$, $t \in T$ *(the mean) be extendable to the square-integrable function* $a(t)$, $-\infty < t < \infty$, *whose Fourier transform satisfies the condition given by* (4.6).

We shall consider next the case where the set T is a finite interval of the real line (say, $T = [0, \tau]$).

Let us consider the spectral densities $f(\lambda)$ satisfying either the condition

$$f(\lambda) \leqslant K(1 + \lambda^2)^{-n} \tag{4.7}$$

or the condition

$$f(\lambda) \geqslant k(1 + \lambda^2)^{-n}, \tag{4.8}$$

where K and k are positive constants. The results obtained below will be extended further on (see Subsection III.4.3) to the case where the spectral densities satisfy either the condition

$$\overline{\lim_{\lambda \to \infty}}\, f(\lambda)|\lambda|^{2n} < \infty \tag{4.9}$$

or the condition

$$\underline{\lim_{\lambda \to \infty}}\, f(\lambda)|\lambda|^{2n} > 0. \tag{4.10}$$

Theorem 10. *For the spectral density* $f(\lambda)$ *of the type given by* (4.7) *a necessary and sufficient condition for the Gaussian measures* P *and* P_1 *to be equivalent is that the function* $a(t)$ *(the mean value) have on the interval* $T = [0, \tau]$ *considered an absolutely continuous* $(n-1)$*th derivative* $a^{(n-1)}(t)$ *such that*

$$\int_T [a^{(n)}(t)]^2\, dt < \infty. \tag{4.11}$$

For the spectral density $f(\lambda)$ *of the type given by* (4.8) *the condition indicated above is sufficient for equivalence of the measures* P *and* P_1.

Proof. It is seen that under the condition given by (4.7) the function $\varphi(\lambda)$ satisfying (4.6) is such that

$$\int (1 + |\lambda|^n)^2 |\varphi(\lambda)|^2\, d\lambda < \infty, \tag{4.12}$$

and, for any $m = 0, 1, \ldots, n - 1$,

$$\int |\lambda|^m |\varphi(\lambda)|\, d\lambda \leqslant \left[\int \left(\frac{|\lambda|^m}{1 + |\lambda|^n}\right)^2 d\lambda\right]^{1/2} \left[\int (1 + |\lambda|^n)^2 |\varphi(\lambda)|^2\, d\lambda\right]^{1/2} < \infty.$$

Therefore, the function

$$a(t) = \int e^{-i\lambda t} \varphi(\lambda)\, d\lambda$$

has $n - 1$ derivatives. Let us show that the $(n - 1)$th derivative

$$a^{(n-1)}(t) = \int (-i\lambda)^{n-1} e^{-i\lambda t} \varphi(\lambda)\, d\lambda$$

is absolutely continuous, the condition given by (4.11) being satisfied.

For any nonoverlapping intervals (t_k, t_{k+1}), $k = 1, 2, \ldots$, we have

$$\begin{aligned}\sum_k [a^{(n-1)}(t_{k+1}) - a^{(n-1)}(t_k)] &= \int \left[\sum_k (e^{-i\lambda t_{k+1}} - e^{-i\lambda t_k}) \right] (-i\lambda)^{n-1} \varphi(\lambda)\, d\lambda \\ &= \int \left[\sum_k \int_{t_k}^{t_{k+1}} (-i\lambda) e^{-i\lambda t}\, dt \right] (-i\lambda)^{n-1} \varphi(\lambda)\, d\lambda \\ &= \int \varphi_\Delta(\lambda)(-i\lambda)^n \varphi(\lambda)\, d\lambda, \end{aligned} \tag{4.13}$$

where

$$\varphi_\Delta(\lambda) = \int e^{-i\lambda t} \chi_\Delta(t)\, dt$$

is the Fourier transform of the function $\chi_\Delta(t)$, which is the indicator of the set $\Delta = \bigcup_k (t_k, t_{k+1})$. By virtue of (4.12) it is seen that

$$\begin{aligned}\left| \sum_k [a^{(n-1)}(t_{k+1}) - a^{(n-1)}(t_k)] \right| &\leqslant \left[\int |\chi_\Delta(\lambda)|^2\, d\lambda \right]^{1/2} \left[\int |\lambda|^{2n} |\varphi(\lambda)|^2\, d\lambda \right]^{1/2} \\ &\leqslant C \left[\int |\chi_\Delta(t)|^2\, dt \right]^{1/2} = C\delta^{1/2}, \end{aligned}$$

where C is a constant and

$$\delta = \sum_k (t_{k+1} - t_k),$$

which fact proves the absolute continuity of the function $a^{(n-1)}(t)$. It is also seen that its derivative

$$a^{(n)}(t) = \int (-i\lambda)^n c^{-i\lambda t} \varphi(\lambda)\, d\lambda$$

(defined for almost all t) coincides with the Fourier transform of the square-integrable function $(-i\lambda)^n \varphi(\lambda)$, and hence satisfies (4.11).

Next, let the spectral density $f(\lambda)$ satisfy (4.8) and let (4.11) be satisfied. It is seen that the function $a(t)$ can be extended from the interval $T = [0, \tau]$ to the entire real line so as to have the $(n - 1)$th absolute continuous derivative $a^{(n-1)}(t)$ and vanish outside of an interval $T' \supseteq T$, with

$$\int [a^{(n)}(t)]^2\, dt < \infty.$$

Let

$$\psi(\lambda) = \frac{1}{2\pi} \int e^{i\lambda t} a^{(n)}(t)\, dt.$$

By integrating n-times by parts we have

$$\varphi(\lambda) = \frac{1}{2\pi} \int e^{i\lambda t} a(t)\, dt = (i\lambda)^{-n} \psi(\lambda)$$

and

$$\int |\lambda|^{2n} |\varphi(\lambda)|^2 \, d\lambda = \int |\psi(\lambda)|^2 \, d\lambda = \frac{1}{2\pi} \int [a^{(n)}(t)]^2 \, dt < \infty.$$

Therefore, the condition given by (4.6) is satisfied and, by Theorem 9, the Gaussian measures P and P_1 are equivalent, as was to be proved. □

III.4.2 Gaussian Measures with Different Correlation Functions

Let us discuss now the conditions for equivalence of the Gaussian measures P and P_1 on the σ-algebra $\mathfrak{A}(T)$ proved in Theorem 8. We consider the case where the spectral measures $F(d\lambda)$ and $F_1(d\lambda)$ are absolutely continuous and have the bounded densities

$$f(\lambda) = F(d\lambda)/d\lambda \quad \text{and} \quad f_1(\lambda) = F_1(d\lambda)/d\lambda.$$

Let

$$b(s, t) = B(s, t) - B_1(s, t), \qquad s, t \in T,$$

be the difference between the corresponding correlation functions

$$B(s, t) = \int e^{i\lambda(s-t)} f(\lambda)\, d\lambda$$

and

$$B_1(s, t) = \int e^{i\lambda(s-t)} f_1(\lambda)\, d\lambda.$$

Theorem 11. *A necessary and sufficient condition for equivalence of the Gaussian measures P and P_1 is that the difference between their correlation functions $b(s, t)$, $s, t \in T$, be extendable to a square-integrable function $b(s, t)$ in the entire plane $-\infty < s, t < \infty$, whose Fourier transform*

$$\varphi(\lambda, \mu) = \frac{1}{4\pi^2} \iint e^{i(\lambda s - \mu t)} b(s, t)\, ds\, dt$$

satisfies the condition

$$\iint \frac{|\varphi(\lambda, \mu)|^2}{f(\lambda) f_1(\mu)} \, d\lambda\, d\mu < \infty. \tag{4.14}$$

Proof. By Theorem 8 a necessary and sufficient condition for equivalence of P and P_1 is that the function $b(s, t)$, $s, t \in T$, permit the representation

$$b(s, t) = \int e^{-i(\lambda s - \mu t)} \psi(\lambda, \mu) f(\lambda) f_1(\mu)\, d\lambda\, d\mu,$$

where $\psi(\lambda, \mu)$ is a function satisfying the condition

$$\iint |\psi(\lambda, \mu)|^2 f(\lambda) f_1(\mu)\, d\lambda\, d\mu < \infty.$$

Let

$$\varphi(\lambda, \mu) = \psi(\lambda, \mu) f(\lambda) f_1(\mu).$$

For bounded densities $f(\lambda)$ and $f_1(\mu)$ the function $\varphi(\lambda, \mu)$ satisfying (4.14) is square-integrable. The Fourier transform $b(s, t)$ of this function coincides on the set $T \times T$ with the difference of the correlation functions considered; therefore the conditions of Theorem 11 are necessary. On the other hand, if these conditions are satisfied, the function $b(s, t)$, $s, t \in T$, can be expressed as

$$b(s, t) = \int e^{-i(\lambda s - \mu t)} \varphi(\lambda, \mu)\, d\lambda\, d\mu$$

(where $\varphi(\lambda, \mu)$ satisfies (4.14)), which fact implies, as indicated above, the equivalence of the Gaussian measures P and P_1. The theorem is proved. $\square$

Let us note that the function $b(s, t)$ is given in the whole plane $-\infty < s$, $t < \infty$ by

$$b(s, t) = \iint e^{-i\lambda(s-t)} [f(\lambda) - f_1(\lambda)]\, d\lambda \tag{4.15}$$

when the set T is the entire real line.

Obviously, if the difference

$$g(\lambda) = f(\lambda) - f_1(\lambda), \qquad -\infty < \lambda < \infty,$$

is different from zero on a set of a positive measure, then

$$\iint |b(s, t)|^2\, ds\, dt = \infty,$$

and, by Theorem 11, the measures P and P_1 will be orthogonal. Therefore, the Gaussian measures P and P_1 are equivalent on the σ-algebra $\mathfrak{A}(T)$ with $T = (-\infty, \infty)$ if and only if they coincide (if the spectral densities $f(\lambda)$ and $f_1(\lambda)$ coincide), which agrees with the example given on page 82.

From now on we shall consider only the case where T is a finite interval on the real line (say, $T = [0, \tau]$), and where the spectral density $f(\lambda)$ satisfies (4.8) for some n. The results obtained further on will be extended to the spectral density $f(\lambda)$ of the type given by (4.10) in Subsection III.4.3.

We note that if the spectral density $f(\lambda)$ decreases faster than any power $|\lambda|^{-2n}$ (as $\lambda \to \infty$), more precisely, if the stationary process $\xi(t)$ is "analytic" so that

$$\mathsf{M}\left[\xi(t) - \sum_{k=1}^{n} \frac{\xi^{(k)}(0)}{k!} t^k\right]^2 = \int \left| e^{i\lambda t} - \sum_{k=1}^{n} \frac{(i\lambda)^k}{k!} t^k \right|^2 f(\lambda)\, d\lambda$$
$$\leqslant \left(\frac{|t|^{n+1}}{(n+1)!}\right)^2 \int \lambda^{2(n+1)} f(\lambda)\, d\lambda \to 0 \qquad \text{as } n \to \infty, \tag{4.16}$$

the completion of the σ-algebra $\mathfrak{A}(T)$ contains the σ-algebra $\mathfrak{A}(-\infty, \infty)$ and, as mentioned above, the measures P and P_1 are equivalent if and only if P and P_1 simply coincide.

As was shown before (see Subsection II.1.1), the space $L_T(F)$ consists of integral analytic functions under the condition given by (4.8).

Returning to the investigation of the equivalence conditions given by Theorem 4, let us consider the pertinent operator $\Delta = E - A^*A$ on the Hilbert space $L_T(F)$.

Lemma 5. *Let the space $L_T(F)$ consist of entire analytic functions, and let $\Delta = E - A^*A$ be a Hilbert–Schmidt operator. Then* (1.27) *is equivalent to the condition that the operator Δ have no eigenvalue equal to one.*

Proof. Because of the boundedness of the operator $A^*A = E - \Delta$ we have

$$\langle A^*A\varphi, \varphi\rangle_F = \|\varphi\|_{F_1}^2 \leqslant C\|\varphi\|_F^2$$

for all $\varphi(\lambda) \in L_T(F)$ (C is a constant). Assume there is an eigenfunction $\varphi(\lambda) \in L_T(F)$ with the eigenvalue equal to one. Let $\varphi_n(\lambda)$, $n = 1, 2, \ldots$, be a sequence of functions in L_T^0 convergent to $\varphi(\lambda)$ in the space $L_T(F)$. This sequence vanishes in the space $L_T(F_1)$:

$$\|\varphi_n\|_{F_1}^2 = \langle\varphi_n, \varphi_n\rangle_F - \langle[E - A^*A]\varphi_n, \varphi_n\rangle_F \to 0$$

as $n \to \infty$. Therefore, the limit function $\varphi(\lambda)$ is zero for almost all λ for which $f_1(\lambda) > 0$. Then the analytic function $\varphi(\lambda)$ is identical to zero, which contradicts the fact that it is an eigenfunction. The lemma is proved. □

This result enables us to strengthen Theorems 4 and 5 by omitting the additional condition given by (1.27) (more precisely, this condition is satisfied automatically if $\Delta = E - A^*A$ is a Hilbert–Schmidt operator).

This, in turn, enables us to obtain the following result (almost the same as Theorem 8):

A necessary and sufficient condition for equivalence of the Gaussian measures P *and* P_1 *on the algebra* $\mathfrak{U}(T)$ *is that the difference* $b(s, t) = B(s, t) - B_1(s, t)$ *between the correlation functions permit a representation of the form (compare with* (3.9))

$$b(s, t) = \iint e^{-i(\lambda s - \mu t)}\varphi(\lambda, \mu)F(d\lambda)F(d\mu) \tag{4.17}$$

for $s, t \in T$, *where the function* $\varphi(\lambda, \mu)$ *is such that*

$$\iint |\varphi(\lambda, \mu)|^2 F(d\lambda)F(d\mu) < \infty. \tag{4.18}$$

From this it is easy to deduce (in the same way as Theorem 11 is deduced) the following result.

Theorem 12. *Under the condition given by* (4.8) *a necessary and sufficient condition for equivalence of the Gaussian measures* P *and* P_1 *on the* σ*-algebra*

$\mathfrak{A}(T)$ is that the difference between the correlation functions, $b(s, t)$, s, $t \in T$ (where $T = [0, \tau]$), *be extendable to a square-integrable function $b(s, t)$, $-\infty < s, t < +\infty$, whose Fourier transform satisfies the condition**

$$\iint \frac{|\varphi(\lambda, \mu)|^2}{f(\lambda) f(\mu)} \, d\lambda \, d\mu < \infty. \tag{4.19}$$

Next, using (4.19) let us consider the function

$$\varphi(\lambda, \mu) = \frac{1}{4\pi^2} \iint e^{i(\lambda s - \mu t)} b(s, t) \, ds \, dt, \tag{4.20}$$

where $b(s, t)$ is the corresponding extension of the difference between the correlation functions given on the square $T \times T$.

Let the spectral density $f(\lambda)$ satisfy (4.7). Then it follows from (4.19) that

$$\iint (1 + |\lambda|^n)^2 (1 + |\mu|^n)^2 |\varphi(\lambda, \mu)|^2 \, d\lambda \, d\mu < \infty \tag{4.21}$$

and for any $k, m = 0, 1, \ldots, n - 1$

$$\begin{aligned}
&\iint |\lambda|^k |\mu|^m |\varphi(\lambda, \mu)| \, d\lambda \, d\mu \\
&\leqslant \left[\int \left(\frac{|\lambda|^k}{1 + |\lambda|^n} \right)^2 d\lambda \int \left(\frac{|\mu|^m}{1 + |\mu|^n} \right)^2 d\mu \right]^{1/2} \\
&\times \left[\iint (1 + |\lambda|^n)^2 (1 + |\mu|^n)^2 |\varphi(\lambda, \mu)|^2 \, d\lambda \, d\mu \right]^{1/2} < \infty.
\end{aligned}$$

Therefore, the function

$$b(s, t) = \iint e^{-i(\lambda s - \mu t)} \varphi(\lambda, \mu) \, d\lambda \, d\mu$$

has all derivatives up to the order $n - 1$ in each variable:

$$\frac{\partial^{k+m} b(s, t)}{\partial s^k \, \partial t^m} = \iint (-i\lambda)^k (i\mu)^m e^{-i(\lambda s - \mu t)} \varphi(\lambda, \mu) \, ds \, dt, \qquad k, m = 0, \ldots, n - 1.$$

We shall show that the function $C(s, t) = \partial^{2(n-1)} b(s, t) / \partial s^{n-1} \, \partial t^{n-1}$ is absolutely continuous. This implies that $C(s, t)$ is a *distribution function* of some absolutely continuous Borel measure† $m(\Delta)$ on the square $T \times T$, that is,

$$m(\Delta) = C(s'', t'') - C(s'', t') - C(s', t'') + C(s', t')$$

for any rectangle $\Delta = (s', s''] \times (t', t'']$.

* It is seen that Theorem 12 still holds if the space $L_T(F)$ consists of analytic functions (see Chapter II).

† In other words, the additive function of bounded variation $m(\Delta)$ is absolutely continuous with respect to the Lebesgue measure.

It can easily be seen that for any set $\Delta = \bigcup_k \Delta_k$ that is the union of a finite number of nonoverlapping rectangles $\Delta_k = (s_k', s_k''] \times (t_k', t_k'']$ we have the equality

$$m(\Delta) = \iint \varphi_\Delta(\lambda, \mu)(-i\lambda)^n(i\mu)^n\varphi(\lambda, \mu)\, d\lambda\, d\mu,$$

where

$$\varphi_\Delta(\lambda, \mu) = \iint e^{-i(\lambda s - \mu t)}\chi_\Delta(s, t)\, ds\, dt$$

is the Fourier transform of the function $\chi_\Delta(s, t)$, the indicator of the set Δ (compare with (4.13)). By virtue of (4.21)

$$\begin{aligned}|m(\Delta)| &\leqslant \left[\iint |\varphi_\Delta(\lambda, \mu)|^2\, d\lambda\, d\mu\right]^{1/2} \left[\iint |\lambda|^{2n}|\mu|^{2n}|\varphi(\lambda, \mu)|^2\, d\lambda\, d\mu\right]^{1/2} \\ &\leqslant C\left[\iint |\chi_\Delta(s, t)|^2\, ds\, dt\right]^{1/2} Cl(\Delta)^{1/2},\end{aligned}$$

where $l(\Delta)$ is the Lebesgue measure of the set Δ. It is seen that the additive function $m(\Delta)$ is actually absolutely continuous with respect to Lebesgue measure.

It is clear that the density $c(s, t) = m(ds\, dt)/(ds\, dt)$ coincides with the derivative $\partial^{2n}b(s, t)/(\partial s^n\, \partial t^n)$ (existing for almost all $s, t \in T \times T$). It is seen that under the condition (4.21) the derivative $c(s, t) = \partial^{2n}b(s, t)/(\partial s^n\, \partial t^n)$ is the Fourier transform of the square-integrable function $\psi(\lambda, \mu) = (-i\lambda)^n(i\mu)^n\varphi(\lambda, \mu)$. In fact,

$$\frac{\partial^{2n}b(s, t)}{\partial s^n\, \partial t^n} = \iint (-i\lambda)^n(i\mu)^n e^{-i(\lambda s - \mu t)}\varphi(\lambda, \mu)\, d\lambda\, d\mu,$$

and therefore

$$\iint_{T \times T} \left[\frac{\partial^{2n}b(s, t)}{\partial s^n\, \partial t^n}\right]^2 ds\, dt < \infty. \tag{4.22}$$

Let the spectral density $f(\lambda)$ satisfy (4.8) and let the difference of the correlation functions $b(s, t)$ have the absolutely continuous derivative $\partial^{2(n-1)}b(s, t)/(\partial s^{n-1}\, \partial t^{n-1})$, with (4.22) being satisfied.

Obviously, the function $b(s, t)$ can be extended to the whole plane $-\infty < s$, $t < \infty$ so that it has the derivative of the type indicated above and so that it vanishes outside of a finite square $T' \times T' \supseteq T \times T$. By iterated integration and $2n$-multiple integration by parts, we obtain from the formula

$$\psi(\lambda, \mu) = \frac{1}{4\pi^2} \iint_{T' \times T'} e^{i(\lambda s - \mu t)} \frac{\partial^{2n}}{\partial s^n\, \partial t^n} b(s, t)\, ds\, dt$$

that

$$\varphi(\lambda, \mu) = \frac{1}{4\pi^2} \iint_{T' \times T'} e^{i(\lambda s - \mu t)} b(s, t)\, ds\, dt = \frac{1}{(i\lambda)^n(-i\mu)^n} \psi(\lambda, \mu).$$

Therefore

$$\iint_{-\infty}^{\infty} |\lambda|^{2n}|\mu|^{2n}|\varphi(\lambda, \mu)|^2 \, d\lambda \, d\mu = \iint_{-\infty}^{\infty} |\psi(\lambda, \mu)|^2 \, d\lambda \, d\mu < \infty,$$

from which (4.19) of Theorem 12 follows for the spectral density $f(\lambda)$ of the type given by (4.8).

Therefore, for a density $f(\lambda)$ of the type givenby (4.7), the condition given by (4.22) follows from (4.19); conversely, for a density of the type given by (4.8), the condition given by (4.19) follows from (4.22), implying that the operator Δ in the Hilbert space $L_T(F)$ defined by the corresponding function

$$b(s, t) = \langle \Delta e^{i\lambda s}, e^{i\lambda t} \rangle_F, \qquad s, t \in T,$$

is a Hilbert–Schmidt operator.

We recall now that the function $b(s, t)$ depends only on the difference $s - t$ (see 4.15)).

It is seen that the absolute continuity of the function $\partial^{2(n-1)}b(s, t)/(\partial s^{n-1}\, \partial t^{n-1})$ (of the pair of variables s, t) is equivalent to the existence of $2n - 1$ absolutely continuous derivative $b^{(2n-1)}(t)$, $-\tau \leqslant t \leqslant \tau$, in the difference

$$b(t) = B(t) = B_1(t), \qquad -\tau < t < \tau,$$

of the correlation functions $B(t)$ and $B_1(t)$ considered, (4.22) implying that

$$\iint_{T \times T} [b^{(2n)}(s - t)]^2 \, ds \, dt < \infty. \tag{4.23}$$

We have arrived at the following result.

Theorem 13. *For the Gaussian measures* P *and* P_1 *to be equivalent on the* σ-*algebra* $\mathfrak{A}(T)$ *with a spectral density* $f(\lambda)$ *of the type given by* (4.7) *it is necessary that the difference of the correlation functions* $b(t) = B(t) - B_1(t)$ *have derivatives up to the order* $2n - 1$ *on the interval* $(-\tau, \tau)$; *the* $(2n - 1)$*th derivative* $b^{(2n-1)}(t)$ *has to be absolutely continuous and the* (*for almost all* t) *derivative* $b^{(2n)}(t)$ *has to satisfy* (4.23). *For the spectral density* $f(\lambda)$ *of the type given by* (4.8) *these conditions are sufficient for* P *and* P_1 *to be equivalent.*

Let us note that, according to Theorem 13, for a spectral density of the type given by

$$f(\lambda) \asymp (1 + \lambda^2)^{-n} \tag{4.24}$$

the condition given by (4.23) is necessary and sufficient for the Gaussian measures P and P_1* to be equivalent.

* For the condition given by (4.23), compare G. Feldman, "Some classes of equivalent Gaussian processes on an interval," *Pacific J. Math.* **10** (1960), 1200–1220.

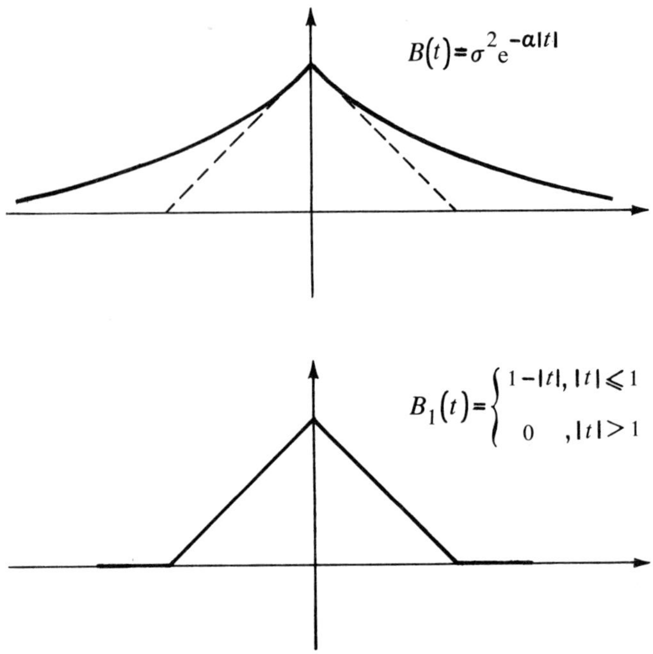

Figure 1

Example. Let

$$f(\lambda)=\frac{\sigma^2}{\pi}\frac{\alpha}{(\lambda^2+\alpha^2)}$$

where σ^2 and α are positive parameters. The corresponding correlation function is

$$B(t)=\sigma^2 e^{-\alpha|t|}.$$

Let the correlation function $B_1(t)$ be as follows:

$$B_1(t)=\begin{cases}1-|t| & \text{for } |t|\leqslant 1,\\ 0 & \text{for } |t|>1\end{cases}$$

(see Figure 1). The corresponding spectral density is

$$f_1(\lambda)=\frac{1}{\pi}\frac{1-\cos\lambda}{\lambda^2}.$$

Obviously, (4.23) is satisfied (for $n=1$) if the parameters σ^2, α, and τ are such that $\sigma^2\alpha=1$, $\tau\leqslant 1$. In this case the Gaussian measures P and P_1 are equivalent on the σ-algebra $\mathfrak{A}(T)$, $T=[0,\tau]$. For other values of σ^2, α, and τ these measures will be orthogonal.

The nonintersecting carriers of the orthogonal measures P and P_1 were described in Section III.1.2.

III.4.3 Some Spectral Conditions for Equivalence

The condition given by (4.23) obtained above shows that the equivalence of the Gaussian measures P and P_1 on the σ-algebra $\mathfrak{A}(T)$ (where T is any finite interval) depends only on the behavior of spectral densities "at infinity," more precisely, for the spectral density $f(\lambda)$ of the type given by (4.8) we have the following simple fact.

Theorem 14. *Arbitrary variation of $f(\lambda)$ on any finite interval (such that it yields a spectral density $f_1(\lambda)$) leads to the Gaussian measure P_1 equivalent to the primary measure P.*

Proof. In fact, in this case the difference of the correlation functions is the Fourier transform of the finite function $g(\lambda) = f(\lambda) - f_1(\lambda)$,

$$b(t) = \int e^{i\lambda t} g(\lambda)\, d\lambda, \tag{4.25}$$

and it has derivatives of all orders; therefore, in particular, the equivalence condition given by (4.23) is satisfied for any finite interval T. □

This fact enables us to extend the results obtained before to the case where the spectral density $f(\lambda)$ satisfies (4.10), which is weaker than (4.8), viz:

$$\varliminf_{\lambda\to\infty} f(\lambda)|\lambda|^{2n} > 0.$$

For example, Theorem 9 can be made stronger as follows.

Theorem 15. *For the spectral density $f(\lambda)$ of the type given by* (4.10) *for equivalence of the Gaussian measures P and P_1 that differ only in the mean $a(t)$, $t \in T$ (where T is a certain interval), it is necessary and sufficient that the function $a(t)$, $t \in T$, be extendable to a square-integrable function $a(t)$, $-\infty < t < \infty$, whose Fourier transform satisfies the condition*

$$\int_{|\lambda|>R} \frac{|\varphi(\lambda)|^2}{f(\lambda)}\, d\lambda < \infty \tag{4.26}$$

for any $R < \infty$.

Proof. Let $\tilde{f}(\lambda)$ satisfy (4.8) and let $\tilde{f}(\lambda)$ coincide with $f(\lambda)$ for $|\lambda| > R$. Let $\tilde{\mathsf{P}}$ be a Gaussian measure with the zero mean and the spectral density $\tilde{f}(\lambda)$. As mentioned above, the measures P and $\tilde{\mathsf{P}}$ are equivalent; therefore, the equivalence of P_1 and P is the same as the equivalence of P_1 and $\tilde{\mathsf{P}}$. By Theorem 9, however, (4.6) is a necessary and sufficient condition for P_1 and $\tilde{\mathsf{P}}$ to be equivalent:

$$\int_{-\infty}^{\infty} \frac{|\varphi(\lambda)|^2}{\tilde{f}(\lambda)}\, d\lambda < \infty.$$

Due to the relation

$$\tilde{f}(\lambda) \asymp 1 \qquad \text{for } |\lambda| \leqslant R$$

this is equivalent to the condition

$$\int_{|\lambda|>R} \frac{|\varphi(\lambda)|^2}{\tilde{f}(\lambda)} \, d\lambda < \infty,$$

which yields (4.26) since $\tilde{f}(\lambda) = f(\lambda)$ for $|\lambda| > R$. The theorem is proved. □

Let us consider the spectral densities $f(\lambda)$ satisfying (4.9), which is weaker than (4.7):

$$\overline{\lim_{\lambda\to\infty}} f(\lambda)|\lambda|^{2n} < \infty.$$

Each density $f(\lambda)$ of this type coincides with a density $\tilde{f}(\lambda)$ of the type given by (4.7) for sufficiently large $|\lambda|$:

$$f(\lambda) = \tilde{f}(\lambda) \qquad \text{for } |\lambda| > R.$$

Introducing the pertinent Gaussian measure $\tilde{\mathsf{P}}$ in a similar way as in proving Theorem 15, we can generalize Theorem 10 as follows:

The condition given by (4.11) is necessary for equivalence of the Gaussian measures P and P_1 if the spectral density $f(\lambda)$ satisfies (4.9).

Similarly, for the spectral density $f(\lambda)$ of the type given by (4.10) a density $\tilde{f}(\lambda)$ can be chosen as to satisfy (4.8) and coincide with $f(\lambda)$ for $|\lambda| > R$. It is seen that the condition given by (4.11) is sufficient for equivalence of the Gaussian measures P and P_1 if the spectral density $f(\lambda)$ satisfies (4.10).

Let us consider next Gaussian measures P and P_1 with zero mean values and spectral densities $f(\lambda)$ and $f_1(\lambda)$, with $f(\lambda)$ being bounded and of the type given by (4.10).

As before, let us consider the Gaussian measure $\tilde{\mathsf{P}}$ with a bounded spectral density $\tilde{f}(\lambda)$ of the type given by (4.8) coinciding with $f(\lambda)$ outside of a certain interval $|\lambda| \leqslant R$. The Gaussian measures P and $\tilde{\mathsf{P}}$ are equivalent; therefore, the primary measures P_1 and P will be equivalent if and only if the measures P_1 and $\tilde{\mathsf{P}}$ have this property. For equivalence of the Gaussian measures P_1 and $\tilde{\mathsf{P}}$ the condition given by (4.19) is necessary and sufficient:

$$\iint_{-\infty}^{\infty} \frac{|\tilde{\varphi}(\lambda, \mu)|^2}{\tilde{f}(\lambda)\tilde{f}(\mu)} \, d\lambda \, d\mu < \infty,$$

where $\tilde{\varphi}(\lambda, \mu)$ is the Fourier transform of some extension of the function

$$\tilde{b}(s, t) = \int e^{i\lambda(s-t)}[\tilde{f}(\lambda) - f_1(\lambda)] \, d\lambda, \qquad t \in T.$$

Since $\tilde{f}(\lambda) \asymp 1$ for $|\lambda| \leqslant R$ and $\tilde{f}(\lambda) = f(\lambda)$ for $|\lambda| > R$, this condition is equivalent to the fact that

$$\int_{|\lambda|>R} \int_{|\mu|>R} \frac{|\tilde{\varphi}(\lambda, \mu)|^2}{\tilde{f}(\lambda)\tilde{f}(\mu)} \, d\lambda \, d\mu < \infty. \tag{4.27}$$

But the function $\tilde{b}(s, t)$ for $s, t \in T$ differs from the function

$$b(s, t) = \int e^{i\lambda(s-t)}[f(\lambda) - f_1(\lambda)]\, d\lambda$$

only by a term of the form

$$c(s, t) = \int e^{i\lambda(s-t)}[\tilde{f}(\lambda) - f(\lambda)]\, d\lambda.$$

Since the difference $f(\lambda) - \tilde{f}(\lambda)$ vanishes for $|\lambda| > R$, the function $c(s, t)$ is infinitely differentiable and, obviously, permits an extension from the square $T \times T$ for which the Fourier transform $\psi(\lambda, \mu)$ satisfies the condition

$$\iint |\lambda|^{2n}|\mu|^{2n}|\psi(\lambda, \mu)|^2\, d\lambda\, d\mu < \infty,$$

where n is the same as that in (4.10). It is seen that if the function $\tilde{\varphi}(\lambda, \mu)$ satisfies (4.27), the function

$$\varphi(\lambda, \mu) = \tilde{\varphi}(\lambda, \mu) + \psi(\lambda, \mu),$$

being the Fourier transform of the appropriate extension of the function

$$b(s, t) = \tilde{b}(s, t) + c(s, t), \qquad s, t \in T,$$

will satisfy the condition

$$\int_{|\lambda| > R} \int_{|\mu| > R} \frac{|\varphi(\lambda, \mu)|^2}{f(\lambda) f(\mu)}\, d\lambda\, d\mu < \infty. \tag{4.28}$$

Conversely, if some extension of the function $b(s, t)$ has the Fourier transform $\varphi(\lambda, \mu)$ satisfying (4.28), the function $\tilde{b}(s, t)$ will have the same property, that is, the Gaussian measures P_1 and $\tilde{\mathsf{P}}$ will be equivalent.

We have obtained the following generalization of Theorem 12.

Theorem 16. *Under the condition given by* (4.10) *a necessary and sufficient condition for equivalence of the Gaussian measures* P *and* P_1 *on the* σ-*algebra* $\mathfrak{A}(T)$ *(*T *is any finite interval) is that the difference between their correlation functions* $b(s, t)$, $s, t \in T$, *be extendable to a square-integrable function* $b(s, t)$, $-\infty < s, t < \infty$, *whose Fourier transform* $\varphi(\lambda, \mu)$ *satisfies* (4.28) *for any* R.

From this it readily follows that Theorem 13 holds true for spectral densities $f(\lambda)$ satisfying the corresponding conditions given by (4.7) and (4.8) as well as densities of a more general type satisfying the corresponding conditions given by (4.9) and (4.10). In particular, for the spectral density $f(\lambda)$ of the type

$$0 < \varliminf_{\lambda \to \infty} f(\lambda)|\lambda|^{2n} \leqslant \varlimsup_{\lambda \to \infty} f(\lambda)|\lambda|^{2n} < \infty \tag{4.29}$$

the condition given by (4.23) is necessary and sufficient for equivalence of the Gaussian measures P and P_1 on the σ-algebra $\mathfrak{A}(T)$.

Each spectral density $f(\lambda)$ of the type given by (4.29) is such that

$$f(\lambda) \asymp |\varphi(\lambda)|^2 \tag{4.30}$$

for sufficiently large $|\lambda|$, where $\varphi(\lambda)$ is the Fourier transform of some square-integrable finite function:

$$\varphi(\lambda) = \int e^{i\lambda t} c(t)\, dt$$

($c(t) = 0$ outside of a finite interval). Moreover, we shall show next that a density $f(\lambda)$ such that

$$0 < \underline{\lim}_{\lambda \to \infty} f(\lambda)|\lambda|^r \leqslant \overline{\lim}_{\lambda \to \infty} f(\lambda)|\lambda|^r < \infty, \tag{4.31}$$

for some $r > 1$, satisfies (4.30) (we recall that $f(\lambda)$ is an integrable function).

In fact, the product of any functions of the type given by (4.30) belongs to this type, since

$$f_1(\lambda) f_2(\lambda) \asymp |\varphi_1(\lambda)\varphi_2(\lambda)|^2,$$

where

$$\varphi_1(\lambda)\varphi_2(\lambda) = \int e^{i\lambda t}[c_1(t) * c_2(t)]\, dt$$

is the Fourier transform of the finite function

$$c_1(t) * c_2(t) = \int c_1(t - s) c_2(s)\, ds,$$

which is the convolution of the corresponding finite functions $c_1(t)$ and $c_2(t)$. Hence it suffices to show that all functions $f(\lambda)$ satisfying (4.31) at $0 < r < 1$ belong to the type given by (4.30). Let

$$\varphi(\lambda) = \int_{-\delta}^{\delta} e^{i\lambda t} |t|^{r-1}\, dt = 2 \int_0^{\delta} \cos \lambda t \cdot t^{r-1}\, dt.$$

Substituting $\lambda t = u$, we have

$$\varphi(\lambda) = 2\lambda^{-r} \int_0^{\lambda\delta} \cos u \cdot u^{r-1}\, du \sim k\lambda^{-r}$$

(as $\lambda \to \infty$), since the limit

$$k = \lim_{t \to \infty} \int_0^t \cos u \cdot u^{r-1}\, du \qquad (0 < r < 1)$$

exists.

It is seen that (4.30) holds true for the density $f(\lambda)$ of the type given by (4.31).

Theorem 17. *For spectral densities of the type given by* (4.30) *a necessary and sufficient condition for equivalence of the Gaussian measures* P *and* P_1 *on the σ-algebra* $\mathfrak{A}(T)$ *(where T is any finite interval) is that the function*

$$h(\lambda) = \frac{f(\lambda) - f_1(\lambda)}{f(\lambda)}$$

*satisfy the condition**

$$\int_{|\lambda|>R} |h(\lambda)|^2 \, d\lambda < \infty \tag{4.32}$$

for any $R < \infty$.

Proof. We consider first the case where the function

$$h(\lambda) = \frac{f(\lambda) - f_1(\lambda)}{f(\lambda)}$$

is square-integrable and

$$f(\lambda) = |\varphi(\lambda)|^2,$$

where $\varphi(\lambda)$ is the Fourier transform of a finite function $c(t)$. Let us find an extension of the difference between the correlation functions

$$b(s, t) = \int e^{i\lambda(s-t)} h(\lambda) f(\lambda) \, d\lambda$$

such that (4.19) from Theorem 12 is satisfied for this extension.

Using the fact that the Fourier transform of the product coincides with the convolution of inverse transforms, we have

$$b(s, t) = \iint a(u - v)c(s - u)c(t - v) \, du \, dv,$$

where $a(t)$ is the Fourier transform of a function $h(\lambda)$. In this representation the functions $b(s, t)$, $s, t \in T$, and the finite functions $c(s - u)$ and $c(t - v)$ vanish if the variables u and v lie outside of a finite interval T', so that

$$b(s, t) = \iint_{T' \times T'} a(u - v)c(s - u)c(t - v) \, du \, dv, \qquad s, t \in T.$$

Let us choose an extension $a(u, v)$, $-\infty < u, v < \infty$, of the function $a(u - v)$, $u, v \in T'$, such that the function $a(u, v)$ is square-integrable in the whole plane, and let us denote by $\psi(\lambda, \mu)$ the Fourier transform of the function $a(u, v)$. For all s, t we define the function $b(s, t)$, $-\infty < s, t < \infty$, setting

$$b(s, t) = \iint a(u, v)c(s - u)c(t - v) \, du \, dv, \qquad -\infty < s, t < \infty.$$

It is seen that this function coincides with the difference between the correlation functions for $s, t \in T$, its Fourier transform

$$\varphi(\lambda, \mu) = \psi(\lambda, \mu)|\varphi(\lambda)|^2$$

satisfying (4.19). Therefore, in this case the Gaussian measures P and P_1 are equivalent.

Let us consider an arbitrary density $f(\lambda)$ of the type given by (4.30), assuming temporarily that

$$f_1(\lambda) \geqslant f(\lambda).$$

* We note that if $h(\lambda) \to 0$ as $\lambda \to \infty$, the spectral density $f_1(\lambda)$ as well as the spectral density $f(\lambda)$ belong to the type given by (4.30), i.e., $f_1(\lambda) \asymp f(\lambda)$, for sufficiently large λ.

It is clear that for sufficiently large λ (say, $|\lambda| > R$), the spectral density $f(\lambda)$ is such that

$$f(\lambda) \asymp \tilde{f}(\lambda) \quad \text{and} \quad f(\lambda) \geqslant \tilde{f}(\lambda),$$

where $\tilde{f}(\lambda)$ is the spectral density of the above type, i.e., $\tilde{f}(\lambda) = |\varphi(\lambda)|^2$, and where $\varphi(\lambda)$ is the Fourier transform of a finite function. We can consider without loss of generality that $f(\lambda) = f_1(\lambda)$ for $|\lambda| \leqslant R$, since each Gaussian measure with a spectral density different from $f(\lambda)$ only on a finite interval is equivalent to the primary Gaussian measure P. For this choice of $f(\lambda)$ the function $h(\lambda)$ satisfying (4.32) will be square-integrable, since $h(\lambda) = 0$ for $|\lambda| \leqslant R$. Let

$$\tilde{f}_1(\lambda) = \tilde{f}(\lambda) + [f_1(\lambda) - f(\lambda)].$$

It can be seen that

$$\tilde{h}(\lambda) = \frac{\tilde{f}(\lambda) - \tilde{f}_1(\lambda)}{\tilde{f}(\lambda)} = \frac{f(\lambda) - f_1(\lambda)}{\tilde{f}(\lambda)} = h(\lambda)\frac{f(\lambda)}{\tilde{f}(\lambda)} \asymp h(\lambda);$$

therefore, the function $\tilde{h}(\lambda)$ is square-integrable. As proved above, the Gaussian measures $\tilde{\mathsf{P}}$ and $\tilde{\mathsf{P}}_1$ corresponding to the spectral densities $\tilde{f}(\lambda)$ and $\tilde{f}_1(\lambda)$ will be equivalent. Hence for some extension of the difference between the correlation functions

$$\begin{aligned} b(s, t) &= \int e^{i\lambda(s-t)}[\tilde{f}(\lambda) - \tilde{f}_1(\lambda)]\,d\lambda \\ &= \int e^{i\lambda(s-t)}[f(\lambda) - f_1(\lambda)]\,d\lambda, \qquad s, t \in T, \end{aligned}$$

so that the corresponding Fourier transform $\varphi(\lambda, \mu)$ satisfies the condition given by (4.28), namely

$$\int_{|\lambda|>R}\int_{|\mu|>R} \frac{|\varphi(\lambda, \mu)|^2}{\tilde{f}(\lambda)\tilde{f}(\mu)}\,d\lambda\,d\mu < \infty,$$

and *a fortiori* the condition

$$\int_{|\lambda|>R}\int_{|\mu|>R} \frac{|\varphi(\lambda, \mu)|^2}{f(\lambda)f(\mu)}\,d\lambda\,d\mu < \infty,$$

since $f(\lambda) \geqslant \tilde{f}(\lambda)$ for $|\lambda| > R$. It is seen that the equivalence condition given by (4.23) from Theorem 16 is satisfied by the Gaussian measures P and P_1.

To complete the proof of Theorem 17 we need to get rid of the temporary limitation according to which $f_1(\lambda) \geqslant f(\lambda)$. This can be done if we consider the Gaussian measure P_2 with spectral density $f_2(\lambda) = f(\lambda) + \max[0, f_1(\lambda) - f(\lambda)]$. It is seen that $f_2(\lambda) \geqslant f(\lambda)$, $f_2(\lambda) \geqslant f_1(\lambda)$, and that the conditions of the type given by (4.32),

$$\int_{|\lambda|>R}\left(\frac{f_2(\lambda) - f(\lambda)}{f(\lambda)}\right)^2 d\lambda < \infty, \qquad \int_{|\lambda|>R}\left(\frac{f_2(\lambda) - f_1(\lambda)}{f_1(\lambda)}\right)^2 d\lambda < \infty,$$

are satisfied for some $R < \infty$, (since $f(\lambda) \asymp f_1(\lambda)$ for sufficiently large λ). Hence the Gaussian measures P and P_2 as well as P_1 and P_2 are equivalent.

Therefore, the Gaussian measures P and P_1 will also be equivalent. The theorem is proved. □

In conclusion, we note that the condition given by (4.32) is close to being a necessary condition for equivalence of the Gaussian measures P and P_1. In fact, as was actually shown in Subsection III.1.2, for spectral densities of the general type given by (4.31) the Gaussian measures P and P_1 will be orthogonal if the corresponding function

$$h(\lambda) = \frac{f(\lambda) - f_1(\lambda)}{f(\lambda)}$$

is such that

$$\lim_{\lambda \to \infty} h(\lambda)\lambda^{1/2} = \infty \tag{4.33}$$

(which follows from relations of the type given by (7.10), Chapter I).*

* For Theorem 17 and the condition given by (4.33), compare these two papers: D. S. Apokorin, "Gaussian measures corresponding to generalized stationary processes," *Teoria Verojatn. i Primen.* **12** No 4 (1967), 638–646 (English translation); V. G. Alekseev, "On conditions for the perpendicularity of Gaussian measures corresponding to two stationary processes," *Teoria Verojatn. i Primen.* **7** No 3 (1963), 286–289 (English translation).

CHAPTER IV

Conditions for Regularity of Stationary Random Processes

IV.1 Preliminaries

Let us consider a stationary narrow-sense random process $\xi(t)$ with continuous or discrete time t. We denote, as before, by $\mathfrak{A}(T)$ the σ-algebra of events generated by the process on the set T, that is, $\mathfrak{A}(T)$ is the minimal σ-algebra containing events such as

$$\{\xi(t_1) \in E_1, \ldots, \xi(t_s) \in E_s\}, \qquad t_1, \ldots, t_s \in T,$$

the E_j being Borel sets on the real line.* Algebras of the form $\mathfrak{A}(-\infty, t)$ determine the past of the process (before time t), algebras of the form $\mathfrak{A}(t, \infty)$ determine the future of the process (after time t).

If for any $\tau > 0$ the σ-algebras $\mathfrak{A}(-\infty, t)$, and $\mathfrak{A}(t + \tau, \infty)$ are independent, then, for all $A \in \mathfrak{A}(-\infty, t)$ and $B \in \mathfrak{A}(t + \tau, \infty)$,

$$\mathsf{P}(AB) - \mathsf{P}(A)P(B) = 0. \tag{1.1}$$

In the general case the left side of this equality (or similar variables) can be taken as the basis for defining the dependence between the σ-algebras $\mathfrak{A}(-\infty, t)$ and $\mathfrak{A}(t + \tau, \infty)$. Conditions of various kinds to decrease the dependence between $\mathfrak{A}(-\infty, t)$ and $\mathfrak{A}(t + \tau, \infty)$ as τ (the regularity condition) increases are widely used in extending the theorems associated with sequences of independent random variables to stationary processes.†

* We restrict ourselves here to real processes for the sake of simplicity. The theorems which follow hold true for complex processes as well. One should remember that for complex Gaussian processes, $\mathsf{M}\xi(t)\xi(s) = 0$ for all t, s (see [6]). However, in investigating the Hilbert space $H(-\infty, \infty)$ there will be no difference whether the process is complex or real. Hence it will be more convenient to consider this space over the field of complex numbers.

† See, for instance, [16] and [24].

1. A stationary process $\xi(t)$ is said to be *regular* if the σ-algebra

$$\mathfrak{A}(-\infty, -\infty) = \mathfrak{A}(-\infty) \stackrel{\text{def}}{=} \bigcap_t \mathfrak{A}(-\infty, t)$$

is trivial, that is, contains only events of probability zero or one.

Let us note that this condition can be expressed in terms of the differences $\mathsf{P}(AB) - \mathsf{P}(A)\mathsf{P}(B)$: for all $B \in \mathfrak{A}(-\infty, \infty)$

$$\sup_{A \in \mathfrak{A}(-\infty, t)} |\mathsf{P}(AB) - \mathsf{P}(A)\mathsf{P}(B)| \xrightarrow[t \to -\infty]{} 0. \tag{1.2}$$

In fact, let the process $\xi(t)$ be regular. We denote by χ_A the indicator of the event A, i.e.,

$$\chi_A = \chi_A(\omega) = \begin{cases} 1, & \omega \in A, \\ 0, & \omega \notin A. \end{cases}$$

Further, let the random variables η_1, η_2 be defined by the equalities $\eta_1 = \chi_A - \mathsf{P}\{A\}$, $\eta_2 = \chi_B - \mathsf{P}\{B\}$. Then

$$\mathsf{M}\eta_1\eta_2 = \mathsf{P}(AB) - \mathsf{P}(A)\mathsf{P}(B).$$

The random variable η_1 is measurable with respect to $\mathfrak{A}(-\infty, t)$, and therefore,

$$\begin{aligned} \mathsf{M}\eta_1\eta_2 &= \mathsf{M}\{\eta_1\}\mathsf{M}\{\eta_2|\mathfrak{A}(-\infty, t)\} \\ &\leqslant (\mathsf{M}\eta_1^2)^{1/2}[\mathsf{M}(\mathsf{M}(\eta_2|\mathfrak{A}(-\infty, t))^2]^{1/2} \\ &\leqslant \mathsf{M}(\mathsf{M}(\eta_2|\mathfrak{A}(-\infty, t)))^2 \xrightarrow[t \to -\infty]{} \mathsf{M}(\mathsf{M}(\eta_2|\mathfrak{A}(-\infty)))^2 = 0. \end{aligned}$$

Otherwise, let the process $\xi(t)$ be unregular. Then the σ-algebra $\mathfrak{A}(-\infty)$ is nontrivial and it contains at least one event A for which $0 < \mathsf{P}\{A\} < 1$. For all t, $A \in \mathfrak{A}(-\infty, t)$ so that

$$\lim_{B \in \mathfrak{A}(-\infty, t)} |\mathsf{P}(AB) - \mathsf{P}(A)\mathsf{P}(B)| \geqslant \mathsf{P}\{A\} - \mathsf{P}^2\{A\} \not\to 0.$$

If we take in (1.1) the supremum over $B \in \mathfrak{A}(t + \tau, \infty)$, we shall arrive at the following.

2. The stationary process $\xi(t)$ satisfies the *strong mixing condition** if

$$\alpha(\tau) = \sup_{A \in \mathfrak{A}(-\infty, t),\, B \in \mathfrak{A}(t+\tau, \infty)} |\mathsf{P}(AB) - \mathsf{P}(A)\mathsf{P}(B)| \xrightarrow[\tau \to \infty]{} 0. \tag{1.3}$$

The value $\alpha(\tau)$ characterizes the mixing rate and is referred to as a *mixing coefficient*.

3. We call the process $\xi(t)$ *absolutely regular* if

$$\beta(\tau) = \mathsf{M} \sup_{A \in \mathfrak{A}(t+\tau, \infty)} |\mathsf{P}\{A|\mathfrak{A}(-\infty, t)\} - \mathsf{P}\{A\}| \xrightarrow[\tau \to \infty]{} 0. \tag{1.4}$$

* The strong mixing condition became well known due to the fruitful application of this condition to limit theorems for weakly dependent variables (see, for example, [16], [24]); it was used first for this purpose, apparently, in Rosenblatt, "A central limit theorem and strong mixing condition," *Proc. Nat. Acad. Sci. USA* **42** (1956).

The value $\beta(\tau)$ is referred to as an (*absolute*) *regularity coefficient*.

It is easily seen that $\alpha(\tau) \leqslant \beta(\tau)$.

In fact, for all $A \in \mathfrak{A}(t+\tau, \infty)$ and $B \in \mathfrak{A}(-\infty, t)$ we have the inequality

$$|\mathsf{P}(AB) - \mathsf{P}(A)\mathsf{P}(B)| = \left|\int_B (\mathsf{P}\{A|\mathfrak{A}(-\infty, t)\} - \mathsf{P}\{A\})\, d\mathsf{P}\right| \leqslant \beta(\tau).$$

Therefore, the absolute regularity condition imposes constraints on the process $\xi(t)$ more stringent than the strong mixing conditions. Taking the supremum in (1.4) instead of averaging over the whole space of elementary events Ω, we arrive at even more stringent conditions.

4. The process $\xi(t)$ satisfies the uniform strong mixing condition if

$$\varphi(\tau) = \operatorname*{vrai\,sup}_{\omega \in \Omega} \sup_{A \in \mathfrak{A}(t+\tau, \infty)} |\mathsf{P}\{A|\mathfrak{A}(-\infty, t)\} - \mathsf{P}\{A\}| \xrightarrow[\tau \to \infty]{} 0. \tag{1.5}$$

It can easily be verified that

$$\varphi(\tau) = \sup_{A \in \mathfrak{A}(t+\tau, \infty),\, B \in \mathfrak{A}(-\infty, t)} \frac{|\mathsf{P}(AB) - \mathsf{P}(A)\mathsf{P}(B)|}{\mathsf{P}(A)}$$

Another definition of regularity is related to the concept of amount of information.

Let $\{\xi_1(t), t \in T\}$ and $\{\xi_2(s), s \in S\}$ be two arbitrary random processes (two families of random variables). We denote by $\mathfrak{A}_1$ and $\mathfrak{A}_2$ the minimal σ-algebras of events generated by the families $\{\xi_1(t), t \in T\}$ and $\{\xi_2(s), s \in S\}$, respectively. By the *amount of information of the random process* $\{\xi_1(t), t \in T\}$ *given by the process* $\{\xi_2(s), s \in S\}$ we mean

$$I(\xi_1, \xi_2) = \sup \sum \mathsf{P}(A_i B_j) \ln \frac{\mathsf{P}(A_i B_j)}{\mathsf{P}(A_i)\mathsf{P}(B_j)}, \tag{1.6}$$

where the supremum is taken over all possible finite decompositions of the space of elementary outcomes Ω onto the nonintersecting events $(A_1, A_2, \ldots, A_n)$, $(B_1, B_2, \ldots, B_m)$, where $A_i \in \mathfrak{A}_1$, $B_j \in \mathfrak{A}_2$. It is easy to see that $I(\xi_1, \xi_2) = I(\xi_2, \xi_1)$. Applying Jensen's inequality to the convex function $x \ln x$, $x > 0$, we can show that $I(\xi_1, \xi_2) \geqslant 0$, with $I(\xi_1, \xi_2) = 0$ if and only if the σ-algebras $\mathfrak{A}_1$ and $\mathfrak{A}_2$ are independent.*

Computing the amount of information contained in the past $\{\xi(s), s \leqslant t\}$ of our process with respect to the future $\{\xi(s), s \geqslant t+\tau\}$, we arrive at the following definition.

5. The stationary process $\xi(t)$ is referred to as *information regular* if

$$I(\tau) = I\{(\xi(s), s \leqslant t), (\xi(s), s \geqslant t+\tau)\} \xrightarrow[\tau \to \infty]{} 0. \tag{1.7}$$

The value $I(\tau)$ is sometimes referred to as the *information coefficient of regularity*.

* See, for instance, [21].

Next, the left side of (1.1) can be written as $\mathsf{M}\eta_1\eta_2$ where $\eta_1 = \chi_A - \mathsf{P}\{A\}$, $\eta_2 = \chi_B - \mathsf{P}\{B\}$. If we consider arbitrary random variables η_1, η_2, we arrive at the following notion, provided only that the mathematical expectation $\mathsf{M}\eta_1\eta_2$ can be defined.

By the *maximal correlation coefficient between the systems of random variables* $\{\xi_1(t), t \in T\}$ *and* $\{\xi_2(s), s \in S\}$ we mean

$$r(\xi_1, \xi_2) = \sup \mathsf{M}\eta_1\eta_2,$$

where the supremum is taken over all η_1 and η_2 measurable with respect to the σ-algebras $\mathfrak{A}_1$ and $\mathfrak{A}_2$ and such that

$$\mathsf{M}\eta_1 = \mathsf{M}\eta_2 = 0, \qquad \mathsf{M}|\eta_1|^2 = \mathsf{M}|\eta_2|^2 = 1.$$

This notion has a simple geometric meaning. Let us denote by $\boldsymbol{H}$ the totality of all random variables η with a finite expectation $\mathsf{M}|\eta|^2$; $\boldsymbol{H}$ is a Hilbert space with the scalar product $(\eta_1, \eta_2) = \mathsf{M}\eta_1\eta_2$. If $\boldsymbol{H}_1$ and $\boldsymbol{H}_2$ are subspaces of $\boldsymbol{H}$ consisting of random variables measurable with respect to $\mathfrak{A}_1$ and $\mathfrak{A}_2$, respectively, then $r(\xi_1, \xi_2)$ is the cosine of the minimal angle between $\boldsymbol{H}_1$ and $\boldsymbol{H}_2$ (compare with 2′ below).

The notion of the maximal coefficient of correlation leads us to the following *complete regularity condition* for the stationary process $\xi(t)$.

6.

$$r(\tau) = r((\xi(s), s \leqslant t), (\xi(s), s \geqslant t + \tau)) \xrightarrow[\tau \to \infty]{} 0. \tag{1.8}$$

If (1.8) is satisfied, we call the random process $\xi(t)$ *completely regular.*

Obviously, $r(\tau) \geqslant \alpha(\tau)$ always. However, for Gaussian stationary processes, as is well known,*

$$\alpha(\tau) \leqslant r(\tau) \leqslant 2\pi\alpha(\tau), \tag{1.9}$$

and therefore the Gaussian stationary process $\xi(t)$ satisfies the strong mixing condition if and only if it is completely regular.

Restricting ourselves to Gaussian processes, the main object of investigation in this book, we naturally attempt to express the conditions given by (1)–(6) in spectral terms, since the spectral or correlation functions define a Gaussian process completely. It is clear that in this case these conditions have to be formulated in terms of the spaces $H(T)$, and, naturally, this equivalent formulation will continue to have meaning for arbitrary stationary wide-sense processes. Hence, it is appropriate to give some natural analogs of these conditions for stationary wide-sense processes. In such processes, orthogonality replaces independence, the spaces $H(T)$ are to be considered instead of the σ-algebras $\mathfrak{A}(T)$, and finally, the operator of taking the conditional expectation $\mathsf{M}\{\cdot|\mathfrak{A}(T)\}$ is to be replaced by the operator of projecting $\hat{\mathsf{M}}\{\cdot|H(t)\}$ (in a Hilbert space H) onto a subspace $H(T)$.

Let $\xi(t)$, $\mathsf{M}\xi(t) \equiv 0$, be a random stationary wide-sense process.

* See [24], p. 249.

(1′) The process $\xi(t)$ is called (*linearly*) *regular* if the space

$$H(-\infty, -\infty) = H(-\infty) \stackrel{\text{def}}{=} \bigcap_t H(-\infty, t)$$

is trivial, that is, it consists of random variables equal to zero with probability one. As with (1.1), the process $\xi(t)$ is linearly regular if and only if

$$\|\hat{\mathsf{M}}\{\eta|H(-\infty, -\tau)\}\| = \mathsf{M}^{1/2}\{(\hat{\mathsf{M}}\{\eta|H(-\infty, -\tau)\})^2\} \xrightarrow[\tau\to\infty]{} 0. \quad (1.10)$$

It is seen that if $\xi(t)$ is regular in the sense of the definition given by (1), it will be linearly regular as well.

The linear regularity condition plays an essential role in the prediction theory of stationary random processes. One of the main results of this theory implies that the stationary wide-sense process $\xi(t)$ is linearly regular if and only if this process permits the Wold representation*

$$\xi(t) = \sum_{-\infty}^{t} c(t-s)\zeta(s) \qquad \left(\sum_0^\infty |c(t)|^2 < \infty\right) \quad (1.11)$$

for discrete t, and

$$\xi(t) = \int_{-\infty}^{t} c(t-s)\zeta(ds) \qquad \left(\int_0^\infty |c(t)|^2\, dt < \infty\right) \quad (1.12)$$

for continuous t, where $\zeta(ds)$ is the stochastic orthogonal measure on the time axis: $\mathsf{M}\zeta(\Delta_1)\zeta(\Delta_2) = |\Delta_1 \cap \Delta_2|$ ($|\Delta|$ denotes the length of the time interval Δ). From (1.11) and (1.12) it follows that the σ-algebra $\mathfrak{A}(-\infty, t)$ is contained in the σ-algebra $\mathfrak{B}(-\infty, t)$ generated by values of $\zeta(\Delta)$ on the time axis $(-\infty, t)$ and

$$\bigcap_t \mathfrak{A}(-\infty, t) \subseteq \bigcap_t \mathfrak{B}(-\infty, t).$$

For a Gaussian process, orthogonality is equivalent to independence and $\zeta(ds)$ is the stochastic measure with independent values (in particular, in the case of discrete time t we deal with a sequence of independent Gaussian variables $\zeta(s)$, $s = t, t-1, \ldots$). According to the zero–one law the intersection $\bigcap_t \mathfrak{B}(-\infty, t)$ is trivial, and therefore the regularity of the Gaussian stationary process $\xi(t)$ (in the sense of the definition given by (1)) follows from the linear regularity of this process. Thus we have arrived at the following statement:

The Gaussian stationary process $\xi(t)$ is regular if and only if this process is linearly regular.

Linearly regular stationary processes are readily described. A stationary process $\xi(t)$ is linearly regular if and only if the spectral measure of this process is absolutely continuous and if the spectral density $f(\lambda)$ satisfies the condition

$$\int_{-\pi}^{\pi} \ln f(\lambda)\, d\lambda > -\infty \quad (1.13)$$

* See, for instance, [24], pp. 77, 162.

for discrete t, and the condition

$$\int_{-\infty}^{\infty} \frac{\ln f(\lambda)}{1+\lambda^2}\, d\lambda > -\infty \tag{1.14}$$

for continuous t (for this, see [24], pp. 85 and 161, or see Section II.2).

Further, in a similar manner as we did in passing from (1.1) to (1.2), we can take in (1.10) the supremum over η, thus arriving at the following:

(2′) The process $\xi(t)$ is called *completely (linearly) regular if*

$$\rho(\tau) = \sup_{\eta \in H(0,\infty),\ \|\eta\|=1} \|\hat{\mathsf{M}}\{\eta | H(-\infty, -\tau)\}\| \xrightarrow[t\to\infty]{} 0. \tag{1.15}$$

We call the value $\rho(\tau)$ the *coefficient of complete (linear) regularity.*

Obviously, $\rho(\tau) \leqslant r(\tau)$ always. But for Gaussian processes we have the following fact, which is fundamental for our further investigation:*

$$\rho(\tau) = r(\tau). \tag{1.16}$$

This implies that the Gaussian stationary process $\xi(t)$ is completely regular (in the sense of the definition given by (6)) if and only if this process is completely linearly regular.

It is easily seen that

$$\rho(\tau) = \sup|\mathsf{M}\eta_1\bar{\eta}_2| = \sup|(\eta_1, \eta_2)|,$$

where the supremum is taken over all $\eta_1 \in H(-\infty, 0)$ and $\eta_2 \in H(\tau, \infty)$, with $\mathsf{M}|\eta_1|^2 = \mathsf{M}|\eta_2|^2 = 1$; therefore $\rho(\tau)$ is the cosine of the minimal angle between the subspaces $H(-\infty, 0)$ and $H(\tau, \infty)$. We note that the condition $\rho(\tau) < 1$ is stronger than the (linear) regularity condition.

Next, in addition to the minimal angle between the subspaces $H(-\infty, t)$ and $H(t+\tau, \infty)$, other similar indices of these subspaces can be developed by starting from the following fact.

Let us denote by $\mathscr{P}_\tau^-(=\mathscr{P}_\tau)$ the operator projecting H onto $H(-\infty, -\tau)$, and let us denote by $\mathscr{P}_\tau^+$ the projector onto $H(\tau, \infty)$. Let us also introduce the operators

$$B_\tau^- = B_\tau = \mathscr{P}_\tau^- \mathscr{P}_0^+ \mathscr{P}_\tau^-, \qquad B_\tau^+ = \mathscr{P}_0^+ \mathscr{P}_\tau^- \mathscr{P}_0^+, \qquad \tau \geqslant 0.$$

The subspaces $H(-\infty, 0)$ and $H(\tau, \infty)$ are orthogonal if and only if $B_\tau^+ = 0$. This leads to the conjecture that the regularity conditions should be equivalent to the convergence of the operators B_τ to zero, as $\tau \to \infty$, in any (uniform, Hilbert–Schmidt, kernel) topology; we can see above that this fact makes it possible to reformulate all regularity conditions described above.

Regularity conditions expressed in terms of Hilbert spaces $H(T)$ permit an analytic formulation since there exists an isometric correspondence $\xi(t) \leftrightarrow e^{i\lambda t}$ with the subspaces $L_T(F)$ described in Chapter II. Furthermore, for Gaussian processes other conditions are available as well (since, for

* See [24], p. 249.

example, $\rho(\tau) \asymp r(\tau) \asymp \alpha(\tau)$). The present chapter, Chapter V, and Chapter VI deal essentially with the translation of the regularity conditions described above into the analytic language, as well as with the solution of analytic problems thus obtained, in order to derive criteria for regularity expressed in spectral terms.

IV.2 Regularity Conditions and Operators B_τ

Let $\xi(t)$ be a stationary Gaussian process. We consider the operators* $B_\tau^-(=B_\tau)$ and B_τ^+ introduced in Section IV.1. These are positive self-adjoint operators. We shall show that all the regularity conditions formulated in Section IV.1 for the Gaussian process $\xi(t)$ can be expressed in terms of convergence $B_\tau \to 0$ as $\tau \to \infty$ (in any sense).

Theorem 1. *The stationary Gaussian process $\xi(t)$ is regular if and only if the operators B_τ vanish weakly as $\tau \to \infty$, that is, if for any random variable $\eta \in H(-\infty, \infty)$*

$$\|B_\tau \eta\| = \mathsf{M}^{1/2}|B_\tau \eta|^2 \xrightarrow[\tau \to \infty]{} 0.$$

According to Section IV.1 we need not distinguish between regularity and linear regularity. Hence Theorem 1 follows immediately from the definition of linear regularity and the inequality $\|B_\tau\| \leqslant \|\mathscr{P}_\tau \eta\|$.

Theorem 2. *A necessary and sufficient condition for a stationary Gaussian process $\xi(t)$ to be completely regular is that the operators B_τ converge to zero uniformly; furthermore, the regularity coefficient is given by $\rho(\tau) = \|B_\tau\|$.*

Naturally, Theorem 2 holds for arbitrary stationary processes if we consider linearly completely regular processes. For Gaussian processes, this theorem yields a strong mixing criterion as well, due to (1.9) and (1.16).

Theorem 2 follows immediately from the definition of the regularity coefficient $\rho(\tau)$. In fact, by definition

$$\begin{aligned}
\rho(\tau) &= \sup_{\eta \in H(0, \infty), \|\eta\| = 1} \|\mathscr{P}_\tau \eta\| \\
&= \sup(\mathscr{P}_\tau \mathscr{P}_0^+ \eta, \mathscr{P}_0^+ \eta)^{1/2} \\
&= \sup(B_\tau^+ \eta, \mathscr{P}_0^+ \eta) \\
&= \sup_{\eta \in H(-\infty, \infty), \|\eta\| = 1} (B_\tau^- \eta, \eta) = \|B_\tau^-\| = \|B_\tau\|.
\end{aligned}$$

* The operators B_τ are introduced into the random processes theory by I. M. Gelfand and A. M. Yaglom, "On the computation of precision of information about a random function contained in another similar function," *UMN* **12** (1957). The relationship between the operators B_τ and the regularity conditions is noted in A. M. Yaglom, "Stationary Gaussian processes satisfying the strong mixing condition and best predictable functionals," *Bernoulli–Bayes–Laplace Anniversary Volume*, Springer-Verlag, 1965.

A more interesting version of Theorem 2 can be proved for processes with discrete time.

Theorem 3. *The stationary process $\xi(t)$ with discrete time is completely regular if and only if it is regular and if the operator B_1 is compact.*

Lemma 1. *Let H_1 and H_2 be subspaces of a separable Hilbert space H. We denote by $\mathscr{P}_1$ and $\mathscr{P}_2$ the operators of projection on H into H_1 and H_2, respectively. Next, let $B_1 = \mathscr{P}_1\mathscr{P}_2\mathscr{P}_1$, $B_2 = \mathscr{P}_2\mathscr{P}_1\mathscr{P}_2$. If the operators B_1, B_2 are compact, an orthogonal basis $\{\{e_{1j}\} \oplus \{e'_{1k}\}\}$ can be chosen in H_1, and an orthogonal basis $\{\{e_{2j}\} \oplus \{e'_{2k}\}\}$ can be chosen in H_2, as to have the following properties:*

1. *all vectors e'_{1k} are orthogonal to H_2, all vectors e'_{2k} are orthogonal to H_1;*
2. *the scalar products (e_{1i}, e_{2j}) are equal to zero if $i \neq j$;*
3. *the e_{1j} are eigenvectors of the operators B_1, and the e_{2j} are eigenvectors of the operator B_2.*

Proof of Lemma 1. It can easily be deduced from elementary properties of projection operators $\mathscr{P}_i$, namely, $\mathscr{P}_i^2 = \mathscr{P}_i$ and $\mathscr{P}_i^* = \mathscr{P}_i$, that B_1 and B_2 are positive self-ajoint operators that are either completely continuous or not completely continuous together. The completely continuous self-ajoint operator B_1 possesses a sequence (complete in its range) of orthonormal eigenvectors $e_{11}, e_{12}, \ldots$, corresponding to nonzero eigenvalues of $\lambda_1, \lambda_2, \ldots$ (completeness implies here that any vector $\varphi = B_1 h$ orthogonal to the vectors e_{1j}, is equal to zero).*

Set

$$e_{2j} = \frac{\mathscr{P}_2 e_{1j}}{\|\mathscr{P}_2 e_{1j}\|} = \frac{1}{\sqrt{\lambda_1}} \mathscr{P}_2 e_{1j}.$$

It is easily seen that the e_{2j} are eigenvectors of the operator B_2 corresponding to eigenvalues λ_j. Indeed,

$$B_2 e_{2j} = \frac{1}{\sqrt{\lambda_1}} \mathscr{P}_2\mathscr{P}_1\mathscr{P}_2 e_{1j} = \frac{1}{\sqrt{\lambda_1}} \mathscr{P}_2 B_1 e_{1j} = \lambda_1 e_{2j}.$$

By considering the projections $\mathscr{P}_1 e_{2j}$ we can convince ourselves that the vectors $\{e_{2j}\}$ constitute all eigenvectors of the operator B_2. Further, let the orthonormal vectors $\{e'_{1k}\}$ ($\{e'_{2k}\}$) complete the system of $\{e_{1j}\}$ ($\{e_{2j}\}$) so as to form a basis in H_1 (H_2). Let us prove that the basis $\{\{e_{1j}\} \oplus \{e'_{1k}\}\}$ and the basis $\{\{e_{2j}\} \oplus \{e'_{2k}\}\}$ have the necessary properties.

We prove first that vectors e'_{1k} are orthogonal to H_2, that is, $\mathscr{P}_2 e'_{1k} = 0$ for each k. In fact,

$$\|\mathscr{P}_2 e'_{1k}\|^2 = (e'_{1k}, \mathscr{P}_2 e'_{1k}) = (\mathscr{P}_1 e'_{1k}, \mathscr{P}_2\mathscr{P}_1 e'_{1k}) = (e'_{1k}, B_1 e'_{1k}) = 0$$

* See, for instance, [2], p. 189.

since the vector $B_1 e'_{1k}$, orthogonal to each vector e_{1j}, is equal to zero because of completeness. It can be proved in a similar way that $e'_{2k} \perp H_1$ for each k.

Computing scalar products of (e_{1i}, e_{2j}), we have

$$\sqrt{\lambda_1}(e_{1i}, e_{2j}) = (\mathscr{P}_1 e_{1i}, \mathscr{P}_2 e_{1j}) = (\mathscr{P}_2 \mathscr{P}_1 e_{1i}, e_{1j})$$
$$= (B_1 e_{1i}, e_{1j}) = \lambda_1(e_{1i}, e_{1j}) = \lambda_1 \delta_{ij},$$

where δ_{ij} is Kronecker's delta. □

Proof of Theorem 3. Let the operator B_1 be completely continuous. Let $e_1, e_2, \ldots$ be normalized eigenvectors of this operator, and let $\lambda_1 \geqslant \lambda_2 \geqslant \ldots$ be the corresponding eigenvalues. By Lemma 1 the complement of $e_1, e_2, \ldots$ up to a basis in $H(-\infty, -1)$ is orthogonal to $H(0, \infty)$; hence we can consider without loss of generality that the vectors $e_1, e_2, \ldots$ form a basis in $H(-\infty, -1)$. We can express an arbitrary element $\eta \in H(-\infty, -\tau)$ as

$$\eta = \sum_1^\infty a_j e_j,$$

where

$$|a_j| = |(\eta, e_j)| = |(\eta, \mathscr{P}_\tau e_j)| \leqslant \|\mathscr{P}_\tau e_j\|.$$

We have $\|\mathscr{P}_\tau e_j\| \xrightarrow[\tau\to\infty]{} 0$ for any fixed j because of the regularity. Therefore,

$$\begin{aligned}
\|B_\tau\| &= \sup_{\|\eta\|=1} \|\mathscr{P}_0^+ \eta\| \\
&= \sup_{\|\eta\|=1} \Big(\sum_{i,j} a_i \bar{a}_j (\mathscr{P}_0^+ e_i, \mathscr{P}_0^+ e_j)\Big)^{1/2} \\
&= \sup_{\|\eta\|=1} \Big(\sum_{i,j} a_i a_j (B_1 e_i, e_j)\Big)^{1/2} \\
&= \sup_{\|\eta\|=1} \Big(\sum_1^\infty |a_j|^2 \lambda_j\Big)^{1/2} \\
&\leqslant \lambda_1 s \max_{1\leqslant j\leqslant s} \|\mathscr{P}_\tau e_j\|^2 + \lambda_{s+1} = \lambda_{s+1} + o(1)
\end{aligned}$$

and is arbitrarily small if $\lim_s \lambda_s = 0$.

To prove sufficiency, we note first that it suffices to prove that the operator B_1 is completely continuous on $H(-\infty, -1)$. We write the space $H(-\infty, -1)$ as $H(-\infty, -\tau) \oplus R_\tau$ where the orthogonal complement R_τ of the space $H(-\infty, -\tau)$ in the space $H(-\infty, -1)$ is finite-dimensional. If Q_τ denotes the projector on $H(-\infty, -1)$ onto R_τ, $\mathscr{P}_\tau + Q_\tau$ is the identity operator on $H(-\infty, -1)$. Therefore,

$$B_1 = (\mathscr{P}_\tau + Q_\tau) B_1 (\mathscr{P}_\tau + Q_\tau) = \mathscr{P}_\tau B_1 \mathscr{P}_\tau + K_\tau = B_\tau + K_\tau,$$

where the operator K_τ is finite-dimensional. Then

$$\|B_1 - K_\tau\| = \|B_\tau\| = \rho(\tau) \xrightarrow[\tau\to\infty]{} 0,$$

implying that the operator B_1 permits arbitrarily close approximation by finite-dimensional operators; therefore the operator B_1 is compact.* □

Note. In proving sufficiency for Theorem 3 we have proved a more general result, namely:

Let $\xi(t)$ be a stationary Gaussian process with discrete or continuous time. Let also the operator B_τ be compact for any $\tau > 0$. Then the process $\xi(t)$ will be completely regular.

We shall next prove the following.

Theorem 4. *A necessary and sufficient condition for a stationary Gaussian process to be absolutely regular is that this process be regular for any τ_0 and that the operator B_{τ_0} be a compact operator with a finite trace.*† *In this case each operator B_τ, $\tau \geqslant \tau_0$, has a finite trace, and the (absolute) regularity coefficient is given by*

$$\beta(\tau) \asymp \sqrt{\operatorname{Sp} B_\tau}, \qquad \tau \to 0, \tag{2.1}$$

where $\operatorname{Sp} B_\tau$ *denotes the trace of the operator B_τ. More precisely, if $\beta(\tau) \to 0$, then*

$$\frac{1}{2\pi} \leqslant \varliminf_{\tau} \frac{\beta(\tau)}{\sqrt{\operatorname{Sp} B_\tau}} \leqslant \varlimsup_{\tau} \frac{\beta(\tau)}{\sqrt{\operatorname{Sp} B_\tau}} \leqslant \frac{1}{2\sqrt{2\pi}}.$$

We shall establish first some general properties of the coefficient $\beta(\tau)$.‡ Let $\{\eta_1(u), u \in U; \eta_2(v), v \in V\}$ be an arbitrary system of Gaussian random variables. Let us denote by $\mathfrak{A}_1$, $\mathfrak{A}_2$, and $\mathfrak{A}$ the σ-algebras of events generated by $\{\eta_1(u), u \in U\}$, $\{\eta_2(v), v \in V\}$, and $\{\eta_1(u), \eta_2(v); u \in U, v \in V\}$, respectively. Next, let H, H_1, and H_2 be linear spaces spanned by $\{\eta_1, \eta_2\}$, $\{\eta_1\}$, and $\{\eta_2\}$, respectively. Let us denote by $\hat{\mathscr{P}}_1$, $\hat{\mathscr{P}}_2$ the projectors on H into H_1, H_2. Finally, let

$$\hat{B}_1 = \hat{\mathscr{P}}_1\hat{\mathscr{P}}_2\hat{\mathscr{P}}_1, \qquad \hat{B}_2 = \hat{\mathscr{P}}_2\hat{\mathscr{P}}_1\hat{\mathscr{P}}_2,$$

and

$$\beta = \beta(\{\eta_1(u), \eta_2(v)\}) = \mathsf{M} \sup_{A \in \mathfrak{A}_2} |\mathsf{P}\{A|\mathfrak{A}_1\} - \mathsf{P}\{A\}|.$$

We shall denote by Q_1, Q_2, and Q probability measures generated on the algebras $\mathfrak{A}_1$, $\mathfrak{A}_2$, and $\mathfrak{A}$ by the random variables $\{\eta_1(u), u \in U; \eta_2(v), v \in V\}$. We shall denote by $\tilde{Q}$ the probability measure on $\mathfrak{A}$ coinciding with the

* See, for instance, [7], p. 47.

† That is a kernel trace.

‡ In proving Theorem 4 we use some results from V. A. Volkonsky and Yu. A. Rozanov, "Some limit theorems for random functions: II," *Teoria Verojatn. i Primen.* **6** (1961), 186–198 (English translation).

measure Q_1 on $\mathfrak{A}_1$, with the measure Q_2 on $\mathfrak{A}_2$ and being such that the algebras $\mathfrak{A}_1$ and $\mathfrak{A}_2$ are independent with respect to the measure Q (assume $\tilde{Q} = Q_1 \times Q_2$).

Lemma 2. *We have the equality*

$$\beta = \frac{1}{2} \operatorname{Var}(Q - \tilde{Q}).$$

Proof of Lemma 2. We prove first that

$$\frac{1}{2} \operatorname{Var}(Q - \tilde{Q}) \leqslant \beta. \tag{2.2}$$

It suffices to show that we have inequality

$$|Q(C) - \tilde{Q}(C)| \leqslant \beta$$

for all events C of the form $C = \bigcup_i A_i B_i$ where $A_i \in \mathfrak{A}_1$, $B_i \in \mathfrak{A}_2$, and each pair of $A_i B_i$ is disjoint.

If C is an event of the type indicated above, it can always be expressed as $C = \bigcup_i A_i B_i'$, where any two events B_i' and B_j' are either joint or disjoint (and all events $A_i B_i'$ are disjoint). Let us write B_i instead of B_i'. Setting $Q(C) > \tilde{Q}(C)$ we obtain

$$\begin{aligned} Q(C) - \tilde{Q}(C) &= \sum_i [\mathsf{P}\{A_i B_i\} - \mathsf{P}\{A_i\}\mathsf{P}\{B_i\}] \\ &= \sum_i \int_{B_i} [\mathsf{P}\{A_i | \mathfrak{A}_2\} - \mathsf{P}\{A_i\}]\, d\mathsf{P}. \end{aligned} \tag{2.3}$$

If $B_{i_1}, \ldots, B_{i_l}$ are some joint events from the sequence $B_1, B_2, \ldots$, then the corresponding events A_{i_s} do not intersect. Setting $\tilde{A}_i = \bigcup_{s=1}^{l} A_{i_s}$, we find that

$$\sum_s \int_{B_{i_s}} [\mathsf{P}\{A_{i_s} | \mathfrak{A}_2\} - \mathsf{P}\{A_{i_s}\}]\, d\mathsf{P} = \int_{B_{i_1}} [\mathsf{P}\{\tilde{A}_i | \mathfrak{A}_2\} - \mathsf{P}\{\tilde{A}_i\}]\, d\mathsf{P}. \tag{2.4}$$

From this and (2.3) it follows that

$$Q(C) - \tilde{Q}(C) = \sum_i \int_{\tilde{B}_i} [\mathsf{P}\{\tilde{A}_i | \mathfrak{A}_2\} - \mathsf{P}\{\tilde{A}_i\}]\, d\mathsf{P},$$

where $\tilde{B}_1, \tilde{B}_2, \ldots$ is a sequence of disjoint events from $\mathfrak{A}$. Therefore,

$$\begin{aligned} |Q(C) - \tilde{Q}(C)| &\leqslant \sum_i \int_{\tilde{B}_i} \sup_{A \in \mathfrak{A}_1} |\mathsf{P}\{A | \mathfrak{A}_2\} - \mathsf{P}\{A\}|\, d\mathsf{P} \\ &= \int_{\cup \tilde{B}_i} \sup_{A \in \mathfrak{A}_1} |\mathsf{P}\{A | \mathfrak{A}_2\} - \mathsf{P}\{A\}|\, d\mathsf{P} \leqslant \beta. \end{aligned}$$

We have proved (2.2).

Further, we shall prove the reverse inequality. Denote by $\tilde{\mathfrak{A}}_1$ a subset of $\mathfrak{A}_1$ consisting of events of the type $\{(\eta_1(u_1), \ldots, \eta_1(u_s)) \in \Lambda\}$ where Λ is the union of 2s-dimensional cubes with rational vertices. The number of events $\tilde{\mathfrak{A}}_1$ is countable. We can assume that in defining β the supremum is to be taken only over $A \in \tilde{\mathfrak{A}}_1$ (as the case may require, the reader may include this requirement in the definition of β). Each elementary event $\omega \in \Omega$ is associated with an event $A_j \in \tilde{\mathfrak{A}}$ such that

$$\mathsf{P}\{A_j|\mathfrak{A}_2\} - \mathsf{P}\{A_j\} \geqslant \sup_{A \in \mathfrak{A}_1} |\mathsf{P}\{A|\mathfrak{A}_2\} - \mathsf{P}\{A\}| - \varepsilon,$$

where $\varepsilon > 0$. Let us denote by B_j the events consisting of all ω corresponding to one A_j. Obviously, $B_j \in \mathfrak{A}_2$. We can consider that all events B_j are disjoint (otherwise, we could consider the events $\tilde{B}_1 = B_1$, $\tilde{B}_2 = B_2 \backslash B_1$, $\tilde{B}_3 \backslash (B_2 \cup B_1), \ldots$). Therefore, the events $A_j B_j$ are disjoint too. Hence

$$\begin{aligned}
\beta &= \int_\Omega \sup_{A \in \mathfrak{A}_1} |\mathsf{P}\{A|\mathfrak{A}_2\} - \mathsf{P}\{A\}| \, d\mathsf{P} \\
&\leqslant \sum_j \int_{B_j} [\mathsf{P}\{A_j|\mathfrak{A}_2\} - \mathsf{P}\{A_j\}] \, d\mathsf{P} + \varepsilon \\
&= \sum_j [\mathsf{P}\{A_j B_j\} - \mathsf{P}\{A_j\}\mathsf{P}\{B_j\}] + \varepsilon \\
&= [Q(\textstyle\bigcup A_j B_j) - \tilde{Q}(\bigcup A_j B_j)] + \varepsilon \leqslant \frac{1}{2} \operatorname{Var}(Q - \tilde{Q}) + \varepsilon,
\end{aligned}$$

thus proving the lemma. □

Lemma 3. *If the sets U and V are finite, we have the inequalities*

$$\begin{aligned}
&\frac{1}{2\pi}\sqrt{\operatorname{Sp} B_1} - \frac{1}{2}\left(2 \operatorname{Sp} B_1 + \frac{1}{2} \frac{3 \operatorname{Sp} B_1 + \frac{1}{2}(\operatorname{Sp} B_1)^2}{(1 - \|B_1\|)^2}\right) \\
&\leqslant \beta \\
&\leqslant \frac{1}{2\sqrt{2\pi}}\sqrt{\operatorname{Sp} B_1} + \frac{1}{2}\left(2 \operatorname{Sp} B_1 + \frac{1}{2} \frac{3 \operatorname{Sp} B_1 + \frac{1}{2}(\operatorname{Sp} B_1)^2}{(1 - \|B_1\|)^2}\right).
\end{aligned} \tag{2.5}$$

The constants $1/(2\sqrt{2\pi}$ and $1/(2\pi)$ for $\sqrt{\operatorname{Sp} B_1}$ cannot be improved.

Proof. By Lemma 1 it can be assumed that $\{\eta_1(u), u \in U\} = \{\eta_{11}, \ldots, \eta_{1n}\}$, $\{\eta_2(v), v \in V\} = \{\eta_{21}, \ldots, \eta_{2n}\}$, that the vectors $\{\eta_{11}, \ldots, \eta_{1n}\}$, $\{\eta_{21}, \ldots, \eta_{2n}\}$ consist of independent normal variables with zero mean and dispersion equal to one, and that

$$\mathsf{M}\eta_{1i}\eta_{2i} = \delta_{ij}\rho_i, \qquad 1 \geqslant \rho_1 \geqslant \ldots \geqslant \rho_n > 0.$$

Also,

$$\operatorname{Sp} B_1 = \operatorname{Sp} B_2 = \sum \rho_i^2, \qquad \|B_1\| = \|B_2\| = \rho_1^2.$$

The assertion of the lemma is trivial in the case $\rho_1 = 1$; hence we take $\rho_1 < 1$. In the assumptions made above, denoting by $p_1(x)$, $p_2(y)$, and $p_{12}(x, y)$ (where $x = (x_1, \ldots, x_n)$, $y = (y_1, \ldots, y_n)$) the probability densities of the Gaussian vectors $(\eta_{11}, \ldots, \eta_{1n})$, $(\eta_{21}, \ldots, \eta_{2n})$, and $(\eta_{11}, \ldots, \eta_{1n}, \eta_{21}, \ldots, \eta_{2n})$, we have

$$\operatorname{Var}[Q - \tilde{Q}] = \int_{R^n} \int_{R^n} |p_{12}(x, y) - p_1(x)p_2(y)|\, dx\, dy,$$

$$p_1(x) = (2\pi)^{-n/2} e^{-(1/2)\sum x_i^2}, \qquad p_2(y) = (2\pi)^{-n/2} e^{-(1/2)\sum y_i^2}, \tag{2.6}$$

$$p_{12}(x, y) = (2\pi)^{-n} \prod \frac{1}{\sqrt{1 - \rho_i}} \exp\left\{ -\frac{x_i^2 + y_i^2 - 2\rho_i x_i y_i}{2(1 - \rho_i^2)} \right\}.$$

We rewrite the right-hand side of (2.6) as

$$\int_{R^n} \int_{R^n} \left| \frac{p_{12}(x, y)}{p_1(x)p_2(y)} - 1 \right| p_1(x)p_2(y)\, dx\, dy = \tilde{\mathsf{M}}|e^{-\zeta} - 1|,$$

where

$$\zeta = -\ln \frac{p_{12}(\eta_1, \eta_2)}{p_1(\eta_1)p_2(\eta_2)} = \sum \frac{\rho_i^2 \eta_{1i}^2 + \rho_i^2 \eta_{2i}^2 - 2\rho_i \eta_{1i} \eta_{2i}}{2(1 - \rho_i^2)} + \frac{1}{2} \ln(1 - \rho_i^2),$$

and where the mathematical expectation $\tilde{\mathsf{M}}$ is computed with respect to the measure $\tilde{Q}$, that is, assuming all variables $(\eta_{11}, \ldots, \eta_{2n})$ to be independent.

We shall expand $e^{-\zeta}$ in a Taylor series with the remainder term as a Lagrange value:

$$e^{-\zeta} = 1 - \zeta + \frac{\zeta^2}{2} e^{-\theta\zeta}, \quad 0 < \theta < 1.$$

Then

$$\begin{aligned}
\mathsf{M}|\zeta^{-\zeta} - 1| &= \int_{\{\zeta \geqslant 0\}} \left| \zeta - \frac{\zeta^2}{2} e^{-\theta\zeta} \right| d\tilde{Q} + \int_{\{\zeta < 0\}} \left| \zeta - \frac{\zeta^2}{2} e^{-\theta\zeta} \right| d\tilde{Q} \\
&\geqslant \int_{\{\zeta \geqslant 0\}} \left(|\zeta| - \frac{\zeta^2}{2} \right) d\tilde{Q} + \int_{\{\zeta < 0\}} \left(|\zeta| - \frac{\zeta^2}{2} e^{-\zeta} \right) d\tilde{Q} \\
&\geqslant \tilde{\mathsf{M}}|\zeta| - \frac{1}{2} \tilde{\mathsf{M}}\zeta^2 - \frac{1}{2} \tilde{\mathsf{M}}\zeta^2 e^{-\zeta}.
\end{aligned}$$

The inequality

$$\tilde{\mathsf{M}}|e^{-\zeta} - 1| \leqslant \tilde{\mathsf{M}}|\zeta| - \frac{1}{2} \tilde{\mathsf{M}}\zeta^2 - \frac{1}{2} \tilde{\mathsf{M}}\zeta^2 e^{-\zeta} \tag{2.7}$$

can be deduced in a similar manner.

Noting that $\tilde{\mathsf{M}}\zeta^2 e^{-\zeta} = \tilde{\mathsf{M}}\zeta^2$, where M denotes the mathematical expectation with respect to Q, we obtain

$$\tilde{\mathsf{M}}|\zeta| - \frac{1}{2}(\tilde{\mathsf{M}}\zeta^2 + \mathsf{P}\zeta^2) \leqslant 2\beta \leqslant \tilde{\mathsf{M}}|\zeta| + \frac{1}{2}(\tilde{\mathsf{M}}\zeta^2 + \mathsf{M}\zeta^2). \tag{2.8}$$

We introduce new random variables u_i, v_i, letting

$$u_i = \frac{1}{\sqrt{2}}(\eta_{1i} - \eta_{2i}), \qquad v_i = \frac{1}{\sqrt{2}}(\eta_{1i} + \eta_{2i}).$$

It is seen that $\tilde{\mathsf{M}} u_i v_i = \mathsf{M} u_i v_i = 0$; therefore the Gaussian random variables $(u_1, \ldots, u_n, v_1, \ldots, v_n)$ are independent with respect to the distributions $\tilde{Q}$ and Q. Furthermore,

$$\tilde{\mathsf{M}} u_i = \tilde{\mathsf{M}} v_i = 0, \qquad \tilde{\mathsf{M}} u_i^2 = \tilde{\mathsf{M}} v_i^2 = 1,$$

$$\mathsf{M} u_i = \mathsf{M} v_i = 0, \qquad \mathsf{M} u_i^2 = \left(1 - \frac{\rho_i}{2}\right), \quad \mathsf{M} v_i^2 = \left(1 + \frac{\rho_i}{2}\right).$$

In the new variables we have

$$\zeta = \frac{1}{2} \sum \left(\frac{\rho_i(1 + \rho_i)u_i^2 - \rho_i(1 - \rho_i)v_i^2}{1 - \rho_i^2} + \ln(1 - \rho_i^2) \right),$$

from which it easily follows that

$$\tilde{\mathsf{M}}\zeta^2 - \tilde{\mathsf{D}}\zeta + (\tilde{\mathsf{M}}\zeta)^2 \leqslant \frac{1}{4}\left[2 \operatorname{Sp} B_1 + \frac{2 \operatorname{Sp} B_1}{(1 - \|B_1\|)^2} + \left(\operatorname{Sp} B_1 \frac{\|B_1\|^2}{1 - \|B_1\|} \right)^2 \right],$$

$$\mathsf{M}\zeta^2 \leqslant \frac{1}{4}\left[2 \operatorname{Sp} B_1 + \frac{2 \operatorname{Sp} B_1}{(1 - \|B_1\|)^2} + \left(\frac{\operatorname{Sp} B_1}{1 - \|B_1\|} \right)^2 \right].$$

It is more difficult to obtain an estimate of $\tilde{\mathsf{M}}|\zeta|$. First, we replace ζ by a simpler random variable,

$$\zeta_1 = \frac{1}{2} \sum \rho_i (u_i^2 - v_i^2),$$

using in this case the obvious inequalities

$$\begin{gathered} \tilde{\mathsf{M}}|\zeta_1| - \tilde{\mathsf{M}}|\zeta - \zeta_1| \leqslant \tilde{\mathsf{M}}|\zeta| \leqslant \tilde{\mathsf{M}}|\zeta_1| + \tilde{\mathsf{M}}|\zeta - \zeta_1|, \\ \tilde{\mathsf{M}}|\zeta - \zeta_1| \leqslant \operatorname{Sp} B_1 \left(1 - \frac{\|B_1\|^{1/2}}{1 - \|B_1\|} \right). \end{gathered} \tag{2.9}$$

The estimate $\tilde{\mathsf{M}}|\zeta_1|$ follows from the following: □

Lemma 4. *We have inequalities*

$$\frac{1}{\sqrt{2\pi}} \sqrt{\operatorname{Sp} B_1} \geqslant \tilde{\mathsf{M}}|\zeta_1| \geqslant \frac{1}{\pi} \sqrt{\operatorname{Sp} B_1}. \tag{2.10}$$

Proof of Lemma 4 and Lemma 3. First, we compute the characteristic function $a(\theta)$ of the random variable ζ_1. The characteristic function of each random variable u_i^2, v_i^2 is given by

$$\frac{1}{\sqrt{2\pi}} \int_{-\infty}^{\infty} e^{i\theta x^2} e^{-(x^2/2)}\, dx = \frac{1}{\sqrt{1 - 2i\theta}};$$

therefore

$$a(\theta) = \prod_i (1 + \rho_i^2\theta^2)^{-1/2}.$$

Next, from the equality

$$\int_{-\infty}^{\infty} \frac{\sin \alpha\theta}{\theta}\, d\theta = \pi \operatorname{sign} \alpha$$

we obtain

$$\begin{aligned} \mathsf{M}|\zeta_1| &= \mathsf{M}\zeta_1 \operatorname{sign} \zeta_1 \\ &= \frac{1}{2\pi i} \mathsf{M}\left(\zeta_1 \int_{-\infty}^{\infty} \frac{e^{i\zeta_1\theta} - e^{-i\zeta_1\theta}}{\theta}\, d\theta\right) \\ &= \frac{1}{2\pi} \int_{-\infty}^{\infty} \frac{a'(\theta) - a'(-\theta)}{\theta}\, d\theta \\ &= \frac{1}{2\pi} \int_{-\infty}^{\infty} \prod (1 + \rho_i^2\theta^2)^{-1/2} \left(\sum_i \frac{\rho_i^2}{1 + \rho_i^2\theta^2}\right) d\theta. \end{aligned}$$

Expressing $\rho_i^2/\operatorname{Sp} B_1$ in terms of λ_i, it follows from the last equality that

$$\tilde{\mathsf{M}}|\zeta_1| = \frac{\sqrt{\operatorname{Sp} B_1}}{2\pi} \int_{-\infty}^{\infty} \prod (1 + \lambda_1\theta^2)^{-1/2} \left(\sum_i \frac{\lambda_i}{1 + \lambda_i\theta^2}\right) d\theta, \qquad \sum_i \lambda_i = 1. \tag{2.11}$$

Let us find the upper and lower bounds of the functions

$$\Phi(\lambda_1, \ldots, \lambda_n) = \int_{-\infty}^{\infty} \prod (1 + \lambda_i\theta^2)^{-1/2} \left(\sum_i \frac{\lambda_i}{1 + \lambda_i\theta^2}\right) d\theta$$

in the simplexes $\sum \lambda_i = 1$, $\lambda_i \geqslant 0$. A standard method for finding extrema (for example, the method using Lagrange multipliers) demonstrates via simple computations that the function Φ attains a minimum (maximum) if the coordinates are equal, $\lambda_1 = \lambda_2 = \ldots$. It occurs also at the vertices of the simplex $\sum \lambda_i = 1$, $\lambda_i \geqslant 0$. Therefore, the integral in the right-hand side of (2.11) has the upper bound expressed as

$$\begin{aligned} \sup_s \int_{-\infty}^{\infty} \left(1 + \frac{\theta^2}{s}\right)^{-(s+2)/2} d\theta &= \sup_s 4\pi\sqrt{s}\, \frac{\Gamma(s-1)}{2^s \left[\Gamma\left(\frac{s}{2}\right)\right]^2} \\ &= \lim_{s\to\infty} \sqrt{2\pi}\, \frac{\sqrt{s-2}(s-2)^{s-2}\sqrt{s}e^2}{s^{s-1}} = \sqrt{2\pi}. \end{aligned} \tag{2.12}$$

Here $\Gamma(s)$ denotes the Euler Γ-function, to which we have applied Stirling's formula. Similarly, the integral in the right-hand side of (2.11) is not less than

$$\int_{-\infty}^{\infty} (1 + \theta^2)^{-3/2}\, d\theta = 2. \tag{2.13}$$

The estimates given by (2.12) and (2.13) and the equality given by (2.11) enable us to write

$$\frac{1}{\sqrt{2\pi}}\sqrt{\operatorname{Sp} B_1} \geqslant \tilde{\mathsf{M}}|\zeta_1| \geqslant \frac{1}{\pi}\sqrt{\operatorname{Sp} B_1}.$$

It is seen from the proof that (2.12) and (2.13) cannot be improved; Lemma 4 as well as Lemma 3 are proved. □

Lemma 5. *Under the conditions of Lemma 3*

$$\int_{-\infty}^{\infty}\left|\frac{p_{12}(x, y)}{p_1(x)p_2(y)} - 1\right|^2 p_1(x)p_2(y)\,dx\,dy = \prod_i (1-\rho_i^2)^{-1} - 1. \qquad (2.14)$$

Lemma 5 can be proved by means of direct computation of the left-hand side of (2.14).

Proof of Theorem 4. Let Q_τ be the measure generated by the given random process $\xi(t)$ on the union of σ-algebras $\mathfrak{A}(-\infty, -\tau)$ and $\mathfrak{A}(0, \infty)$. We denote by $\tilde{Q}_\tau$ the measure coinciding with Q on $\mathfrak{A}(-\infty, -\tau)$ and $\mathfrak{A}(0, \infty)$ and being such that the σ-algebras $\mathfrak{A}(-\infty, -\tau)$, $\mathfrak{A}(0, \infty)$ are $\tilde{Q}_\tau$-independent. By Lemma 2

$$\beta(\tau) = \frac{1}{2}\operatorname{Var}(Q_\tau - \tilde{Q}_\tau). \qquad (2.15)$$

Let the process $\xi(t)$ be absolutely regular, that is, $\lim_{\tau\to\infty}\beta(\tau) = 0$. Using (2.15) we can find a number τ_0 such that

$$\frac{1}{2}\operatorname{Var}(Q_\tau - \tilde{Q}_\tau) = \beta(\tau) < \frac{1}{2}$$

for all $\tau \geqslant \tau_0$. Therefore, the measures Q_τ and $\tilde{Q}_\tau$ are nonsingular for $\tau \geqslant \tau_0$; these measures are Gaussian. According to the results obtained in Section III.2 these measures are mutually absolutely continuous as well.

Let us choose a basis $\eta_{11}, \eta_{12}, \ldots,$ in the space $H(-\infty, -\tau)$ and a basis $\eta_{21}, \eta_{22}, \ldots,$ in the space $H(0, \infty)$. We consider the vector $(\eta_{11}, \ldots, \eta_{1n}, \eta_{21}, \ldots, \eta_{2n})$. With the measures Q and $\tilde{Q}$ generated by this vector in the way described above (see the paragraph following the statement of Theorem 4), the operator B_1 and eigenvalues of the operator B_1 will be written as Q_n, $\tilde{Q}_n$, B_{1n}, and ρ_{in}^2, respectively. Let $p_n = dQn/d\tilde{Q}n$. It is easy to show that

$$\mathsf{M}\ln p_n = -\frac{1}{2}\sum \ln(1-\rho_{in}^2). \qquad (2.16)$$

It follows from the fact that the measures Q_τ and $\tilde{Q}_\tau$ are absolute continuous (see Section III.2) that

$$\sup_n |\mathsf{M}\ln p_n| < \infty. \qquad (2.17)$$

From this and (2.16) the inequality

$$\sup_n \operatorname{Sp} B_{1n} < \infty$$

follows. The operators B_{1n} converge weakly to the operator B_τ as $n \to \infty$, that is, $\lim_n (B_{1n}\eta, \eta) = (B_\tau \eta, \eta)$ (the operators B_{1n} are defined in an obvious manner everywhere in $H(-\infty, \infty)$). Therefore, the operator B_τ has finite trace.

Next, let q_τ be the density of Q_τ in $\tilde{Q}_\tau$. The results of the end of Section III.1 imply that for $\tau > \tau_0$ we have

$$q_\tau = \tilde{\mathsf{M}}(q_{\tau 0} | \mathfrak{A}_\tau),$$

where $\mathfrak{A}_\tau$ denotes the union of σ-algebras $\mathfrak{A}(-\infty, -\tau)$ and $\mathfrak{A}(0, \infty)$. By virtue of (2.14),

$$\tilde{\mathsf{M}}|q_\tau - 1|^2 < \infty.$$

By familiar theorems from the theory of conditional mathematical expectations, it follows from the above that the random variables q_τ converge in the mean to q_∞ as $\tau \to \infty$. The absolutely regular process $\xi(t)$ is regular, that is, the intersection $\bigcap_\tau \mathfrak{A}(-\infty, -\tau)$ is trivial. Therefore, it is necessary that $q_\infty = 1$.

Indeed, the random variable $q_\infty = \tilde{\mathsf{M}}(q_{\tau 0} | \mathfrak{A}_\infty)$ is measurable with respect to $\mathfrak{A}_\infty = \mathfrak{A}(0, \infty)$ and by definition of the conditional mathematical expectation we have

$$\int_A q_\infty \, d\tilde{Q} = \int_A q_{\tau_0} \, d\tilde{Q} = \int_A dQ = Q(A) = \tilde{Q}(A) = \int_A 1 \cdot d\tilde{Q}.$$

for each $A \in \mathfrak{A}_\infty$.

These equalities imply that q_∞ is equal to one with probability one. By virtue of (2.14),

$$\operatorname{Sp} B_\tau \leqslant \tilde{\mathsf{M}}|q_\tau - 1|^2 = \tilde{\mathsf{M}}|q_\tau - q_\infty|^2 \xrightarrow[\tau \to \infty]{} 0;$$

applying Lemma 3 we arrive at the limit relations given by (2.1).

Conversely, for any τ_0 let the operator B_{τ_0} be a completely continuous operator with finite trace. Then (see the note to Theorem 3) the process $\xi(t)$ is completely regular and also $\|B_\tau\| \to 0$. We assume that $\|B_{\tau_0}\| < 1$. It is also seen that all operators B_τ together with operators B_{τ_0}, $\tau > \tau_0$, have finite trace and that

$$\operatorname{Sp} \mathrm{B}_\tau \leqslant \operatorname{Sp} \mathrm{B}_{\tau_0}.$$

Using the compactness of the operators B_τ we choose now a basis η_{11}, $\eta_{12}, \ldots, \eta_{21}, \eta_{22} \ldots$ in the way indicated in Lemma 1. Then, by virtue of (2.16), we infer that

$$\sup_n |\tilde{\mathsf{M}} \ln p_n| \leqslant \operatorname{Sp} B_\tau \left(1 + \frac{\|B_\tau\|}{1 - \|B_\tau\|}\right).$$

Returning to Section III.2 we can see that the measures Q_τ and $\tilde{Q}_\tau$ are absolutely continuous. Furthermore, as noted before, the process $\xi(t)$ is completely regular, so that, *a fortiori*, it is regular. As we have proved above, in this case

$$\tilde{\mathsf{M}}|q_\tau - 1|^2 \xrightarrow[\tau\to\infty]{} 0,$$

and therefore $\operatorname{Sp} B_\tau \xrightarrow[\tau\to\infty]{} 0$ due to (2.14). Invoking Lemma 3 we prove the theorem. □

Note. In the case for processes with discrete time the theorem can be reformulated as follows:

A necessary and sufficient condition for the stationary Gaussian process $\xi(t)$ with discrete time to be absolutely regular is that the operator B_1 be a compact operator with finite trace. In fact, as was noted in the proof of Theorem 3, in the processes with discrete time the operators B_τ differ from the operator B_1 by a finite-dimensional operator only.

Theorem 5. *The stationary Gaussian process $\xi(t)$ satisfies the regularity condition given by* (1.5) *if and only if $B_\tau = 0$ for all $\tau > \tau_0$.*

The proof of this simple theorem can be found in [16]. We note here only that the equality $B_\tau = 0$ implies that the spaces $H(-\infty, -\tau)$ and $H(0, \infty)$ are orthogonal and, therefore, the σ-algebras $\mathfrak{A}(-\infty, -\tau)$ and $\mathfrak{A}(0, \infty)$ are independent.

We shall prove in the next section that the conditions for absolute regularity and those for information regularity are equivalent.

IV.3 Conditions for Information Regularity

Theorem 6. *The stationary Gaussian process $\xi(t)$ is regular with respect to information if and only if the operator B_{τ_0} is a compact operator with finite trace for any τ_0. In this case each operator B_τ, $\tau > \tau_0$, has finite trace, and the information regularity coefficient is given by*

$$I(\tau) = -\frac{1}{2}\sum \ln(1 - \rho_i^2) \sim \frac{1}{2}\operatorname{Sp} B_\tau, \tag{3.1}$$

where ρ_i^2 is the ith eigenvalue of the operator B_τ.

We shall deduce Theorem 6 from a more general result (Theorem 7, to follow). Let $\{\xi(u),\ u \in U;\ \eta(v),\ v \in V\}$ be a system of Gaussian random variables. Next, let

$$I_{\xi\eta} = I\{\xi(u),\ u \in U,\ \eta(v),\ v \in V\}$$

be the information contained in the random variables $\{\xi(u),\ u \in U\}$ about $\{\eta(v),\ v \in V\}$. As usual, let H_1, H_2 be linear closed hulls of the pertinent variables $\{\xi(u),\ u \in U\}$, $\{\eta(v),\ v \in V\}$; let $\mathfrak{A}_1$, $\mathfrak{A}_2$ denote the σ-algebras generated by variables $\{\xi(u),\ u \in U\}$, $\{\eta(v),\ v \in V\}$; finally, let B_1, B_2 be nonnegative self-adjoint operators defined on the spaces H_1 and H_2 as shown above. Then the following theorem holds.*

Theorem 7. *The information $I_{\xi\eta}$ is finite if and only if the operator B_1 (and therefore, B_2) is a compact operator with finite trace, and $\|B_1\| < 1$. In this case*

$$I_{\xi\eta} = -\frac{1}{2}\sum_i \ln(1-\rho_i^2), \tag{3.2}$$

where the $\rho_1^2 \geqslant \rho_2^2 \geqslant \ldots$ are the eigenvalues of the operator B_1.

Proof of Theorem 7. (1) First, we compute the information contained in one Gaussian random variable ξ with respect to another variable η related in a Gaussian manner. We can obviously assume

$$\mathsf{M}\xi = \mathsf{M}\eta = 0, \qquad \mathsf{M}\xi^2 = \mathsf{M}\eta^2 = 1.$$

Furthermore, let $\mathsf{M}\xi\eta = \rho$; therefore, the mutual density of the variables ξ and η is given by

$$p(x, y) = \frac{1}{\sqrt{2\pi(1-\rho^2)}}\exp\left\{-\frac{x^2+y^2-2\rho xy}{2(1-\rho^2)}\right\}.$$

We have

$$I_{\xi\eta} = \sup_{A_i, B_j}\sum \ln\frac{\mathsf{P}\{A_iB_j\}}{\mathsf{P}\{A_i\}\mathsf{P}\{B_j\}}\mathsf{P}\{A_iB_j\}, \tag{3.3}$$

where the upper bound is taken over events of the type $A_i = \{\xi \in E_i\}$, $B_j = \{\eta \in E'_j\}$, where the E_i, E'_j are linear Borel sets. Therefore,

$$I_{\xi\eta} = \sup_{E_i, E'_j}\sum \ln\frac{Q_{12}\{E_i \times E'_j\}}{Q_1\{E_i\}Q_2\{E'_j\}}Q_{12}(E_i \times E'_j), \tag{3.4}$$

where Q_{12}, Q_1, Q_2 are measures in the plane and on the line induced by the distributions of the random variables (ξ, η) and ξ, η:

$$Q_{12}(E \times F) = \mathsf{P}\{\xi \in E, \eta \in F\},$$
$$Q_1(E) = \mathsf{P}\{\xi \in E\}, \qquad Q_2(F) = \mathsf{P}\{\eta \in F\}.$$

First, let $\rho < 1$. The measure Q_{12} is absolutely continuous with respect to Lebesgue measure, and the measures $Q_{12}, Q_1 \times Q_2$ are mutually absolutely continuous with the density

$$\frac{Q_{12}(dx\,dy)}{Q_1(dx)Q_2(dy)} = \frac{p(x, y)}{p(x)p(y)},$$

where $p(x) = (1/\sqrt{2\pi})e^{-x^2/2}$ denotes the density of the variables ξ, η. Hence the sum in the right-hand side of (3.4) is the integral sum of the Lebesgue

* I. M. Gelfand and A. M. Yaglom, "Computation of information about a random function contained in another random function," *UMN* **12** (1957).

integral of the function $\ln \rho(x, y)/(p(xp(y))\rho(x, y)$. Therefore

$$I_{\xi\eta} = \int_{-\infty}^{\infty} \int_{-\infty}^{\infty} \ln \frac{p(x, y)}{p(x)p(y)} p(x, y)\, dx\, dy$$

$$= \mathsf{M} \ln \frac{p(\xi, \eta)}{p(\xi)p(\eta)} = -\frac{1}{2} \ln(1 - \rho^2). \tag{3.5}$$

If $\rho = 1$, then $I_{\xi\eta} = \infty$. To make sure this is the case it suffices to take in (3.3) $A_1 = B_1, \ldots, A_n = B_n$ with probabilities $\mathsf{P}\{A_i\} = 1/n$. For all n we have

$$I_{\xi\eta} > \ln n.$$

Since for $\rho = 1$ the right-hand side of (3.5) goes into infinity, (3.5) holds true in this case as well.

(2) Let the sets U, V be finite. The information $I_{\xi\eta}$ is invariant with respect to nondegenerate transforms of the spaces H_1 and H_2. This follows immediately from the definition of the information given by (1.6) and from the obvious fact that the σ-algebras $\mathfrak{A}_1$, $\mathfrak{A}_2$ are invariant with respect to the transforms mentioned. Hence, using the above as well as Lemma 1 we can assume that

$$\{\xi(u), u \in U\} = (\xi_1, \ldots, \xi_n), \qquad \{\eta(v), v \in V\} = (\eta_1, \ldots, \eta_n),$$

where each pair of $(\xi_i, \eta_i\}$ is independent and

$$\mathsf{M}\xi_i = \mathsf{M}\eta_i = 0, \qquad \mathsf{D}\xi_i = \mathsf{D}\eta_i = 1, \qquad \mathsf{M}\xi_i\eta_i = \rho_i.$$

It suffices to take in (1.6) the supremum over the events A_i, B_j:

$$A_i = \{\xi_1 \in E_{i1}, \ldots, \xi_n \in E_{in}\}, \qquad B_j = \{\eta_1 \in E'_{j1}, \ldots, \eta_n \in E'_{jn}\}.$$

Then it follows from the independence of the pair of (ξ_i, η_i) that

$$I_{\xi\eta} = \sum_{i=1}^{n} I_{\xi_i\eta_i},$$

and therefore, by virtue of (3.5)

$$I_{\xi\eta} = -\frac{1}{2} \sum_{i=1}^{n} \ln(1 - \rho_i^2);$$

$I_{\xi\eta} < \infty$ if and only if $\rho_1 = \sqrt{\|B_1\|} < 1$.

(3) Next we shall consider the general case. It was noted in (1) of the proof of Theorem 7 that $I_{\xi\eta} = \infty$ if $\|B_1\| = 1$. Hence it suffices to take the case with $\|B_1\| < \infty$.

Suppose B_1 is not an operator with finite trace. We choose n random variables $\xi_1, \ldots, \xi_n$ from among the family $\{\xi(u), u \in U\}$ and we choose n random variables $\eta_1, \ldots, \eta_n$ from among the family $\{\eta(v), v \in V\}$. We denote by $B_1^{(n)}$ the operator associated with these variables. Naturally, the operators $B_1^{(n)}$ can be considered in the whole space H_{12} spanned by $\{\xi(u), \eta(v)\}$. Let us denote by $\lambda_1^{(n)} \geqslant \lambda_2^{(n)} \geqslant \ldots \geqslant \lambda_n^{(n)}$ the eigenvalues of the operator $B_1^{(n)}$ (it is finite-dimensional). It is seen that $\lambda_1^{(n)} = \|B_1^{(n)}\| \leqslant \|B_1\| < 1$.

The random variables $(\xi_1, \ldots, \xi_n)$, $(\eta_1, \ldots, \eta_n)$ can be chosen so that the operators $B_1^{(n)}$ converge weakly to B_1. Then

$$\lim_n \operatorname{Sp} B_1^{(n)} = \lim_n \sum_i \lambda_i^{(n)} = \infty.$$

It follows from the definition of the information given by (1.6) that

$$I_{\xi\eta} \geqslant I((\xi_1, \ldots, \xi_n), (\eta_1, \ldots, \eta_n)).$$

As we have established in Subsection IV.3.2, we have

$$I((\xi_1, \ldots, \xi_n), (\eta_1, \ldots, \eta_n)) = -\frac{1}{2}\sum_i \ln(1 - \lambda_i^{(n)})$$
$$= \frac{1}{2}\sum_{k=1}^{\infty} \frac{1}{k} \sum_i (\lambda_i^{(n)})^k \geqslant \frac{1}{2} \operatorname{Sp} B_1^{(n)}.$$

Therefore, if $\operatorname{Sp} B_1 = \infty$, then $I_{\xi\eta} = \infty$.

Let B_1 be a completely continuous operator with finite trace. Using Lemma 1 we take in the spaces H_1, H_2 the basis $(\xi_1, \xi_2, \ldots)$, $(\eta_1, \eta_2, \ldots)$ consisting of eigenvalues of the operators B_1, B_2. For such choice of the variables $\mathsf{M}\xi_i\eta_i = \rho_i$, where the ρ_i^2 are eigenvalues of the operators B_1, B_2.

Choosing n sufficiently large, events $A \in \mathfrak{A}_1$, $B \in \mathfrak{A}_2$ can be approximated using events measurable with respect to the random variables $(\xi_1, \ldots, \xi_n)$, $(\eta_1, \ldots, \eta_n)$. Returning to the definition of the information given by (1.6) we deduce that

$$I_{\xi\eta} = \lim_{n\to\infty} I(\xi_1, \ldots, \xi_n; \eta_1, \ldots, \eta_n),$$

where

$$I(\xi_1, \ldots, \xi_n; \eta_1, \ldots, \eta_n) = -\frac{1}{2}\sum_1^n \ln(1 - \rho_i^2),$$

thus proving Theorem 7.

It follows from this theorem that the random process $\xi(t)$ is regular with respect to information only if the operator B_τ has finite trace for $\tau > \tau_0$; it also follows that (3.1) holds true. We establish in proving Theorem 4 that $\lim_\tau \operatorname{Sp} B_\tau = 0$ only if $\operatorname{Sp} B_{\tau_0} < \infty$ for some τ_0. Thus, we have proved Theorem 6.

Comparing Theorem 4 and Theorem 6 we can convince ourselves that conditions for absolute regularity are equivalent to those for information regularity and that, for example,

$$\frac{1}{\pi^2} \leqslant \underline{\lim} \frac{\beta^2(\tau)}{I(\tau)} \leqslant \overline{\lim} \frac{\beta^2(\tau)}{I(\tau)} \leqslant \frac{1}{2\pi}$$

for $I(\tau) \to 0$.

Note. Using the note to Theorem 4, we conclude that the stationary Gaussian process $\xi(t)$ with discrete time is regular with respect to information if and only if the operator B_1 has finite trace.

IV.4 Conditions for Absolute Regularity and Processes with Discrete Time

Sections IV.4 and IV.5 contain a complete and sufficiently thorough description of absolutely regular stationary Gaussian processes (processes satisfying (1.3)). In accord with the results of Section IV.3 this description will involve regular (with respect to information) processes.

Theorem 8.* *A stationary Gaussian process $\xi(t)$ with discrete time $t = 0, \pm 1, \ldots$ is absolutely regular (and, consequently, information regular) if and only if this process has a spectral density $f(\lambda)$ representable as*

$$f(\lambda) = |P(e^{i\lambda})|^2 a(\lambda), \tag{4.1}$$

where $P(z)$ is the polynomial with roots on the circle $|z| = 1$; the coefficients a_j of the Fourier series $\sum a_j e^{i\lambda j}$ of the function $\ln a(\lambda)$ are such that

$$\sum |j||a_j|^2 < \infty. \tag{4.2}$$

The proof of Theorem 8 is based on Theorem 4, more precisely, on the note to Theorem 4. Naturally, the analytic formulation of Theorem 8 requires the transition from a space $H(-\infty, \infty)$ to an isometric space of functions $L(F)$. To avoid new notation we denote by $B_\tau^\pm$ operators in $L(F)$ similar to operators $B_\tau^\pm$.

Each absolutely regular process is regular and therefore (see Section II.2) it has a spectral density $f(\lambda)$ permitting a representation of the form

$$f(\lambda) = |g(e^{i\lambda})|^2, \tag{4.3}$$

where $g(z)$ is the outer function of Hardy's class $\mathcal{H}^2$ on the circle.

Let us introduce the function $c(\lambda) = \overline{g(e^{i\lambda})}/g(e^{i\lambda})$. We denote by $\sum_{j=-\infty}^{\infty} c_j e^{ij\lambda}$ its Fourier series.

Lemma 6. *A process $\xi(t)$ is absolutely regular if and only if the series*

$$\sum_{-\infty}^{0} |j||c_j|^2 \tag{4.4}$$

is convergent.

Proof. According to the note to Theorem 4 (see the end of Section IV.3) we need to prove that the convergence of (4.4) is equivalent to the fact that the operator B_1 is nuclear. It will be more convenient to deal with the unitary equivalent operator $B_1^+ = \mathcal{P}_0^+ \mathcal{P}_1^- \mathcal{P}_0^+$, where $\mathcal{P}_\tau^+$ denotes the projector on $L(F)$ into $L_{(\tau, \infty)}(F)$ and $\mathcal{P}_\tau^-$ the projector on $L(F)$ into $L_{(-\infty, -\tau)}(F)$. We

* I. A. Ibragimov and V. N. Solev, "A condition for regularity of a Gaussian stationary process," *Doklady Akademii Nauk SSSR* **185**, 371–375 (English translation).

need only to consider B_1^+ as the operator on $L_{(0,\infty)}(F) = L^+(F)$ into $L^+(F)$. It is known* that the operator B_1^+ is a nuclear operator (that is, it has finite trace) if and only if the series $\sum_j (B_1^+ e_j, e_j)_F$ is convergent for some orthonormal basis $\{e_j\}$ in $L^+(F)$. If this series is convergent for some orthonormal basis, it will be convergent for any basis of this kind, the sum of such a basis being equal to the trace of the operator B_1^+.

Let us take the functions $e_k(\lambda) = e^{i\lambda k} g^{-1}(e^{i\lambda})$, $k = 0, 1, \ldots$, as an orthonormal basis in $L^+(F)$. It is obvious that the system $\{e_k\}$ is orthonormal; we shall prove that this system is a basis.

In fact, we showed in Section II.2 that each element $\varphi \in L^+(F)$ can be written as φ_1/g where $\varphi_1 \in \mathscr{H}^2$. If $\sum c_k e_k$ is a linear combination of e_k, then†

$$\left\|\varphi - \sum c_k e_k\right\|_F = \left\|\varphi_1 - \sum c_k e^{i\lambda k}\right\|^{(2)}.$$

Since the linear combinations $\sum c_k e^{i\lambda k}$ are dense in $\mathscr{H}^2$, the last equality implies that the linear combinations $\sum c_k e_k$ are dense in $L^+(F)$, that is, $\{e_k\}$ is the basis in $L^+(F)$.

As in the proof of Lemma 2 in Chapter II we can prove the following equalities:

$$\mathscr{P}_\tau^+ = g^{-1}\Pi_\tau^+ g, \qquad \mathscr{P}_\tau^- = \bar{g}^{-1}\Pi_\tau^- \bar{g},$$

where Π_τ^+ and Π_τ^- are projectors in $\mathscr{L}^2(-\pi, \pi)$ on $e^{i\lambda\tau}\mathscr{H}^2$ and $e^{-i\lambda\tau}\mathscr{H}^{2-}$, respectively. Hence

$$\begin{aligned}
\operatorname{Sp} B_1^+ &= \sum_{k=0}^{\infty} (B_1^+ e_k, e_k)_F = \sum_{k=0}^{\infty} (\mathscr{P}_0^+ \mathscr{P}_1^- \mathscr{P}_0^+ e_k, e_k)_F \\
&= \sum_{k=0}^{\infty} (\mathscr{P}_1^- e_k, e_k)_F = \sum_{k=0}^{\infty} \left(\Pi_1^- \frac{\bar{g}}{g} e^{i\lambda k}, \frac{\bar{g}}{g} e^{i\lambda k}\right)^{(2)} \\
&= \sum_{k=0}^{\infty} \left(\left\|\Pi_1^- \frac{\bar{g}}{g} e^{i\lambda k}\right\|^{(2)}\right)^2 = \sum_{k=0}^{\infty} \int_{-\pi}^{\pi} \left|\sum_{j=-\infty}^{-1} c_{j-k} e^{i\lambda j}\right|^2 d\lambda \\
&= \sum_{j=-\infty}^{0} |c_j|^2 |j|,
\end{aligned} \tag{4.5}$$

which proves the lemma. □

It remains only to prove that the conditions of Theorem 8 are equivalent to those of Lemma 6.

Let us verify that the fulfillment of the conditions of Theorem 8 implies that the conditions of Lemma 6 are satisfied. It follows from (4.2) that $\ln a \in \mathscr{L}^2(-\pi, \pi)$; in fact, $\ln a \in \mathscr{L}^1(-\pi, \pi)$. Hence the function $a(\lambda)$ can be written as $a(\lambda) = |g_1(e^{i\lambda})|^2$, where g_1 is the outer function of class $\mathscr{H}^2$. Therefore, $f(\lambda) = |g(e^{i\lambda})|^2$, where the outer function $g(z) = P(z)g_1(z)$.

* See, for instance, [7], p. 55; we recall that each operator B_τ^+ is positive.

† We recall that $\|\varphi\|^{(p)}$ denotes the norm in the space $\mathscr{L}^p(-\pi, \pi)$.

If $e^{i\lambda_1}, \ldots, e^{i\lambda_\nu}$ are zeros of the polynomial $P(z)$, it is seen that

$$\frac{\overline{P(e^{i\lambda})}}{P(e^{i\lambda})} = e^{i\alpha} \prod_{s=1}^{\nu} \frac{e^{-i\lambda} - e^{-i\lambda_s}}{e^{i\lambda} - e^{i\lambda_s}} = e^{i\alpha} e^{-i\nu\lambda}(-1)^\nu,$$

where α is a real number. Hence it suffices to investigate the function $c_1(\lambda) = \overline{g_1(e^{i\lambda})}/g_1(e^{i\lambda}) = \sum c_{1j} e^{i\lambda j}$ and to prove that the series $\sum_{j<0} |c_{1j}|^2 |j|$ is convergent. We shall even prove that a series $\sum_j |j||c_{1j}|^2$ is convergent. To this end we need the following lemma.

Lemma 7. *Let a function $h(\lambda) \sim \sum h_j e^{i\lambda j}$ belong to $\mathscr{L}^2(-\pi, \pi)$. We denote by $\omega(\delta; h)$ the continuity index of the function h in the $\mathscr{L}^2$ metric, that is,*

$$\omega(\delta; h) = \sup_{|\theta| \leqslant \delta} \left(\int_{-\pi}^{\pi} |h(\lambda + \theta) - h(\lambda)|^2 \, d\lambda \right)^{1/2}.$$

Then the inequalities

$$\sum_1^\infty \omega^2\left(\frac{1}{n}; h\right) < \infty \tag{4.6}$$

and

$$\sum |j||a_j|^2 < \infty \tag{4.7}$$

are equivalent.

Proof. In fact

$$\begin{aligned}
\sum_n \omega^2\left(\frac{1}{n}; h\right) &= \sum_{n=1}^\infty \sup_{\delta \leqslant 1/n} 4 \sum_{j=-\infty}^{\infty} |h_j|^2 \sin^2 \frac{j\delta}{2} \\
&\leqslant \sum_{n=1}^\infty \frac{1}{n^2} \sum_{|j| \leqslant n} |h_j|^2 j^2 + 4 \sum_{n=1}^\infty \sum_{|j|>n} |h_j|^2 \\
&= \sum_{j=-\infty}^{\infty} |h_j|^2 j^2 \sum_{n>|j|} \frac{1}{n^2} + 4 \sum_{j=-\infty}^{\infty} |j||h_j|^2 \leqslant 5 \sum_j |j||h_j|^2.
\end{aligned}$$

Otherwise, for $|x| < 1$, $(\sin x)/x > 1 - (x^2/6)$ and hence

$$\begin{aligned}
\sum_n \omega^2\left(\frac{1}{n}; h\right) &\geqslant 4 \sum_{n=1}^\infty \sum_{|j| \leqslant n} |h_j|^2 \sin^2 \frac{j}{2n} \\
&\geqslant \frac{1}{2} \sum_{n=1}^\infty \frac{1}{n^2} \sum_{|j| \leqslant n} |h_j|^2 j^2 \\
&= \frac{1}{2} \sum_{j=-\infty}^{\infty} |h_j|^2 j^2 \sum_{n>|j|} \frac{1}{n^2} > \frac{1}{4} \sum |j||h_j|^2,
\end{aligned}$$

thus proving the lemma. □

Returning to the proof of Theorem 8 we note that $g_1 = \exp\{\frac{1}{2}(\ln a + i \widetilde{\ln a})\}$ and that, therefore, $\overline{g}_1/g_1 = \exp\{-i \widetilde{\ln a}\}$ where, in general, a function (harmonically) conjugate to h is denoted by $\tilde{h}$. Absolute values of the Fourier coefficients of the functions $\ln a$ and $\tilde{\ln}\, a$ coincide and hence $\omega(\delta; \ln a) = \omega(\delta; \tilde{\ln}\, a)$. Further,

$$\begin{aligned} &|\exp\{-i\tilde{\ln}\, a(\lambda+\delta)\} - \exp\{-i\tilde{\ln}\, a(\lambda)\}| \\ &= |\exp\{-i(\tilde{\ln}\, a(\lambda+\delta) - \tilde{\ln}\, a(\lambda))\} - 1| \\ &\leqslant |\tilde{\ln}\, a(\lambda+\delta) - \tilde{\ln}\, a(\lambda)|, \end{aligned}$$

and hence

$$\omega\left(\delta; \frac{\bar{g}_1}{g_1}\right) \leqslant \omega(\delta; \tilde{\ln}\, a) = \omega(\delta; \ln a).$$

From Lemma 7 we can see that convergence of the series $\sum |j||a_j|^2$ implies convengence of the series $\sum_j |j||c_{1j}|^2$. We have thus proved that the conditions given by (4.1) and (4.2) of Theorem 8 are sufficient for absolute regularity of the Gaussian process $\xi(t)$.

It is more difficult to prove that these conditions are necessary.

Let the process $\xi(t)$ be absolutely regular. Then it is regular as well and it has a spectral density $f(\lambda)$, $\ln f \in \mathscr{L}^1(-\pi, \pi)$. We introduce the coefficient of (complete) regularity

$$\rho(\tau; f) = \sup_{\varphi, \psi} \left| \int_{-\pi}^{\pi} \varphi(e^{i\lambda})\overline{\psi(e^{i\lambda})} f(\lambda)\, d\lambda \right| = \sup |(\varphi, \psi)_F|,$$

where the supremum is taken over all φ and ψ belonging to unit spheres of spaces $L_{(0,\infty)}(F)$ and $L_{(-\infty,-\tau)}(F)$, respectively.

As was noted in Section IV.1, any absolutely regular Gaussian process satisfies a strong mixing condition. Therefore, this process is completely regular, that is, $\lim_{\tau\to\infty} \rho(\tau) = 0$. In particular, there is arbitrary k such that $\rho(k) < 1$. By Theorem 4 in Chapter V the spectral density $f(\lambda)$ can be written as

$$f(\lambda) = |P(e^{i\lambda})|^2 a(\lambda),$$

where $P(z)$ is a polynomial with roots on $|z| = 1$, and $a = e^{u+\tilde{v}}$, $\|u\|^{(\infty)} < \infty$, $\|v\|^{(\infty)} < \pi/2$. By the same theorem the regularity coefficients $\rho(\tau; a)$ and $\rho(\tau; 1/a)$ constructed from the spectral densities a and $1/a$ are such that

$$\rho(1; a) < 1, \qquad \rho\left(1; \frac{1}{a}\right) = \rho < 1. \tag{4.8}$$

Obviously, the function $a(\lambda)$ can be written as $|g_1(e^{i\lambda})|^2$ where g_1 is an exterior function from $\mathscr{H}^2$. If we assume here as in proving the sufficiency of the conditions of Theorem 8 that

$$c_1(\lambda) = \frac{\overline{g_1(e^{i\lambda})}}{g_1(e^{i\lambda})} = \sum_j c_{1j} e^{i\lambda j}$$

we easily deduce from Lemma 6, using the equality $\overline{P(e^{i\lambda})}/P(e^{i\lambda}) = e^{i\alpha}e^{-i\nu\lambda}(-1)^\nu$, that the series $\sum_{-\infty}^{\infty} |j||c_{1j}|^2$ is convergent. If this is the case, then

$$\sum_{n=0}^{\infty} \inf_{A \in H^2} (\|c_1(\lambda) - e^{-in\lambda}A(\lambda)\|^2)^2$$

$$\leqslant \sum_{n=0}^{\infty} \sum_{-\infty}^{-(n+1)} |c_{1j}|^2 = \sum_{-\infty}^{0} |j||c_{1j}|^2 < \infty. \tag{4.9}$$

Let us choose a sequence of polynomials $A_n(z)$ such that

$$\sum_{n=0}^{\infty} (\|c_1(\lambda) - e^{-in\lambda}A_n(e^{i\lambda})\|^{(2)})^2 < \infty, \tag{4.10}$$

and let us assume that $e^{-in\lambda}g_1 A_n = Q_n(e^{-i\lambda}) + B_n$ where $Q_n(z)$ is a polynomial of degree not exceeding n, and $B_n \in \mathscr{H}^2$. By virtue of (4.8) we have

$$(\|c_1(\lambda) - e^{-in\lambda}A_n(\lambda)\|^{(2)})^2$$

$$= \int_{-\pi}^{\pi} |(\overline{g_1(e^{i\lambda})} - Q_n(e^{-i\lambda})) - B_n(e^{i\lambda})|^2 \frac{d\lambda}{a(\lambda)}$$

$$\geqslant \int_{-\pi}^{\pi} |\overline{g_1(e^{i\lambda})} - Q_n(e^{-i\lambda})|^2 \frac{d\lambda}{a(\lambda)} + \int_{-\pi}^{\pi} |B_n(e^{i\lambda})|^2 \frac{d\lambda}{a(\lambda)}$$

$$-2\left|\int_{-\pi}^{\pi} (\overline{g_1(e^{i\lambda})} - Q_n(e^{-i\lambda}))\overline{B_n(e^{i\lambda})} \frac{d\lambda}{a(\lambda)}\right|$$

$$\geqslant \int_{-\pi}^{\pi} |g_1(e^{i\lambda}) - Q_n(e^{-i\lambda})|^2 \frac{d\lambda}{a(\lambda)} + \int_{-\pi}^{\pi} |B_n(e^{i\lambda})|^2 \frac{d\lambda}{a(\lambda)}$$

$$-2\rho\left(\int_{-\pi}^{\pi} |\overline{g_1(e^{i\lambda})} - Q_n(e^{-i\lambda})|^2 \frac{d\lambda}{a(\lambda)} \int_{-\pi}^{\pi} |B_n(e^{i\lambda})|^2 \frac{d\lambda}{a(\lambda)}\right)^{1/2}$$

$$\geqslant (1-\rho)\int_{-\pi}^{\pi} |\overline{g_1(e^{i\lambda})} - Q_n(e^{-i\lambda})|^2\, d\lambda.$$

Hence, along with the series given by (4.10), the series

$$\sum_{n=0}^{\infty} \inf_{P_n} \int_{-\pi}^{\pi} |g_1(e^{i\lambda}) - P_n(e^{i\lambda})|^2 \frac{d\lambda}{a(\lambda)}$$

$$= \sum_{n=0}^{\infty} \inf_{P_n} \int_{-\pi}^{\pi} |g_1(e^{i\lambda}) - P_n(e^{i\lambda})|^2 \frac{d\lambda}{a(\lambda)} \tag{4.11}$$

is convergent, where inf is taken over each polynomial $P_n(e^{i\lambda})$ of degree not exceeding n.

Our problem has thus been reduced to the problem of investigating the properties of the best approximations of the functions $g_1(\lambda)$ by polynomials in the $\mathscr{L}^2$ metric with the weight $1/a$. For this purpose it will be natural to introduce polynomials $\varphi_\nu(z; 1/a) = \varphi_\nu(z)$, $\nu = 0, 1, \ldots$ orthogonal with the weight $1/a$. A number of properties of such polynomials will be formulated as lemmas, which follow for the convenience of the reader. The proofs of these lemmas can be found in many books dealing with orthogonal polynomials. From now on we shall refer, in particular, to Grenander and Szego [13], which is perhaps the most familiar to specialists in probability theory.

Let $\omega(\lambda)$ be a nonnegative summable function on $[-\pi, \pi]$. We assume in addition that $\ln \omega \in \mathscr{L}^1(-\pi, \pi)$; therefore $\omega(\lambda) = |\gamma(e^{i\lambda})|^2$ where $\gamma(z)$ is an exterior function of class $\mathscr{H}^2$. By orthogonal polynomials $\varphi_0(z; \omega)$, $\varphi_1(z; \omega), \ldots, \varphi_\nu(z; \omega), \ldots$ with weight ω we call polynomials such that:

(a) φ_ν is a polynomial of degree ν with a positive coefficient in a higher term;

(b) $\dfrac{1}{2\pi}\displaystyle\int_{-\pi} \varphi_\nu(z; w)\overline{\varphi_\mu(z; w)} w(\lambda)\, d\lambda = \delta_{\nu\mu}, \qquad z = e^{i\lambda}.$

Next let $\varphi_\nu^*(z; \omega) = z^n \bar{\varphi}_\nu(z^{-1})$. The polynomials φ_ν, φ_ν^* are related as follows (see [13], p. 58):

$$\begin{aligned} k_n z \varphi_n(z) &= k_{n+1}\varphi_{n+1}(z) - l_{n+1}\varphi_{n+1}^*(z), \\ k_n \varphi_{n+1}(z) &= k_{n+1} z \varphi_n(z) + l_{n+1}\varphi_n^*(z), \end{aligned} \tag{4.12}$$

where k_n denotes the coefficient of z^n and l_n denotes the constant term of the polynomial φ_n.

Let us denote by c_ν the νth Fourier coefficient of a function ω and let

$$D_n(\omega) = \det\|c_{\nu-\mu}\|, \qquad \nu, \mu = 0, 1, \ldots, n.$$

The determinants $D_n(\omega)$ are called *Toeplitz determinants corresponding to the weight* (to the function) ω. We have the equalities ([13], p. 54)

$$k_n(\omega) = \left(\frac{D_{n-1}(\omega)}{D_n(\omega)}\right)^{1/2}, \qquad n = 1, \ldots\ . \tag{4.13}$$

Let us define the geometric mean $G(\omega)$ of the function ω by the equality

$$G(\omega) = \exp\left\{\frac{1}{2\pi}\int_{-\pi}^{\pi} \ln \omega(\lambda)\, d\lambda\right\}.$$

Lemma 8. *Suppose a function $\omega(\lambda)$ is definite and summable on $[-\pi, \pi]$, where $\omega \geqslant 0$ and $\ln \omega \in \mathscr{L}^{(1)}$. As before, let $\omega(\lambda) = |\gamma(e^{i\lambda})|^2$ where γ is an exterior function from $\mathscr{H}^2$. Finally, let $\varphi_\nu(z) = \varphi_\nu(z; \omega)$ be orthogonal polynomials associated with the weight ω. Then*

$$s(z; \omega) = s(z) = \sum_{\nu=0}^{\infty} \overline{\varphi_\nu(0)}\varphi_\nu(z) = \frac{1}{\overline{\gamma(0)}}\frac{1}{\gamma(z)}, \qquad |z| < 1, \tag{4.14}$$

$$\lim_{n\to\infty} k_n(\omega) = [G(\omega)]^{-1/2} = \frac{1}{\gamma(0)}, \tag{4.15}$$

uniformly on any circle $|z| \leqslant r < 1$, with

$$\lim_{n\to\infty} \varphi_n^*(z) = \gamma(z). \tag{4.16}$$

Lemma 9. *By the hypotheses of Lemma 8*

$$k_n^2(\omega) = \frac{D_{n-1}(\omega)}{D_n(\omega)} = \sum_0^n |\varphi_\nu(0;\omega)|^2,$$
$$G(\omega) = \left(\sum_0^\infty |\varphi_\nu(0,\omega)|^2\right)^{-2}. \tag{4.17}$$

Proof. For the proof of the first equality in (4.17) see [13], p. 56; the second equality in (4.17) follows from the first equality and (4.15). □

Lemma 10. *If* $\omega(\lambda) = 1/|\Gamma(e^{i\lambda})|^2$ *where* $\Gamma(z)$ *is a polynomial of degree* p, *all zeros of which lie in a circle* $|z| < 1$, *then for* $\nu \geqslant p$

$$\varphi_\nu(z) = z^{\nu-p}\Gamma(z).$$

Proof. If $\nu \geqslant p$ and $\mu < \nu$, then

$$\int_{-\pi}^{\pi} e^{-i\mu\lambda}\varphi_\nu(e^{i\lambda})\omega(\lambda)\,d\lambda = \frac{1}{i}\int_{|z|=1} \frac{z^{\nu-\mu-1}}{z^p\overline{\Gamma(z)}} = 0, \tag{4.18}$$

since the integrand is analytic in the circle $|z| \leqslant 1$. In fact, if $\Gamma(z) = \sum_0^p \gamma_j z^j$, then on $|z| = 1$ we have

$$z^p\overline{\Gamma(z)} = \sum_0^p \bar{\gamma}_j z^{p-j},$$

and therefore the function $(z^p\overline{\Gamma(z)})^{-1}$ permits an analytic extension in the circle $|z| < 1$. (4.18) implies that the polynomials $\varphi_\nu(e^{i\lambda})$ are orthogonal to all functions $e^{i\mu\lambda}$, $\mu < \nu$. The equality

$$\frac{1}{2\pi}\int_{-\pi}^{\pi} |\varphi_\nu(e^{i\lambda})|^2\omega(\lambda)\,d\lambda = 1$$

is obvious. Thus we have proved the lemma.

Lemma 11. *By the hypotheses of Lemma 10, for all* $\nu \geqslant p$ *we have*

$$\frac{D_\nu(\omega)}{[G(\omega)]^{\nu+1}} = \frac{D_p(\omega)}{[G(\omega)]^{p+1}} = \exp\left\{\frac{1}{\pi}\iint_{|\sigma|\leqslant 1} \left|\frac{\gamma'(z)}{\gamma(z)}\right|^2 d\sigma\right\}, \tag{4.19}$$

where $\gamma(z)$ *is an outer function of class* $\mathscr{H}^2$ *given by the equality* $\omega = |\gamma|^2$. *The integration is carried out to the right over the circle* $|z| \leqslant 1$.

Proof. For the proof of Lemma 10 see [13], p. 102. □

Now we have all necessary information about orthogonal polynomials and hence we can return to proving our basic Theorem 8.

Proof of Theorem 8. Let $\varphi_\nu(z) = \varphi_\nu(z; 1/a)$ be orthogonal polynomials corresponding to the weight $1/a$. Note that $1/|a(\lambda)|^2 = |\gamma(e^{i\lambda})|^2$, where the exterior function $\gamma(e^{i\lambda}) = 1/g_1(e^{i\lambda})$. From this it follows, due to (4.14), that the Fourier coefficients of the function $g_1 = 1/\gamma$ are $\overline{\varphi_\nu(0)}\,\overline{\gamma(0)}$ in the orthogonal system $\{\varphi_\nu(e^{i\lambda})\}$. Hence

$$\inf_{P_n} \int_{-\pi}^{\pi} |g_1(e^{i\lambda}) - P_n(e^{i\lambda})|^2 \frac{d\lambda}{a(\lambda)} = |\gamma(0)|^2 \sum_{n+1}^{\infty} |\varphi_\nu(0)|^2.$$

Convergence of the series given by (4.11) enables us to assert that

$$\sum_{n=0}^{\infty} \sum_{\nu=n+1}^{\infty} |\varphi_\nu(0)|^2 = \sum_{\nu} \nu |\varphi_\nu(0)|^2 < \infty. \tag{4.20}$$

as well.

Next we have to show that convergence of (4.20) implies convergence of (4.2). To this end we shall consider in addition to the polynomials $\varphi_n(z)$ orthogonal polynomials $\varphi_{\nu n}(z) = \varphi_n(z; |\varphi_n|^{-2})$ associated with the weight $|\varphi_n(e^{i\lambda})|^{-2}$. It is known (see [13], p. 57) that zeros of orthogonal polynomials $\varphi_n(z; \omega)$ corresponding to any weight ω lie in the domain $|z| < 1$. Therefore, zeros of polynomials $\varphi_n^*(z) = z^n(\overline{\varphi}_n(z^{-1})$ lie outside of the circle $|z| \leqslant 1$, so that the $\varphi_n^*(z)$ are outer functions of $\mathscr{H}^2$. Furthermore, we have the equality $|\varphi_n|^2 = |\varphi_n^*|^2$ on $|z| = 1$; therefore $(\varphi_n^*)^{-1}$ is an outer function corresponding to the weight $|\varphi_n|^{-2}$.

Next, by Lemma 10, we obtain

$$\varphi_{\nu n}(z) = z^{n-\nu}\varphi_n(z), \qquad \nu \geqslant n.$$

In particular, $\varphi_{nn}(z) = \varphi_n(z)$. From this equality and (4.12) it follows that

$$\varphi_{\nu n}(z) = \varphi_\nu(z), \qquad \nu \leqslant n. \tag{4.21}$$

The orthogonal polynomials $\varphi_\nu(z)$ are naturally linearly independent. Therefore the functions $e^{ik\lambda}$ are linear combinations of polynomials $\varphi_\nu(z)$, $0 \leqslant \nu \leqslant k$, for all k. From this and (4.21) it follows that the first $n + 1$ Fourier coefficients $c_0, \ldots, c_n$ (and therefore $c_{-n}, \ldots, c_{-1}, c_0$) of the functions $1/a(\lambda)$ and $1/|\varphi_n(e^{i\lambda})|^{-2}$ are equivalent. Therefore,

$$D_s(|\varphi_n|^{-2}) = D_s\left(\frac{1}{a}\right), \qquad s \leqslant n.$$

Using Lemmas 8 and 9 we can find with the aid of the above equality that for all $s \geqslant n$

$$\begin{aligned}
\ln \frac{D_s(|\varphi_n|^{-2})}{G(|\varphi_n|^{-2})^{s+1}} &= \ln \frac{D_n(|\varphi_n|^{-2})}{G(|\varphi_n|^{-2})^{n+1}} \\
&= \sum_{1}^{n} \ln \frac{D_k(|\varphi_n|^{-2})}{D_{k-1}(|\varphi_n|^{-2})} + \ln D_0(|\varphi_n|^{-2}) - (n+1)\ln G(|\varphi_n|^{-2}) \\
&= \sum_{k=0}^{n} \ln \frac{\sum_{0}^{n} |\varphi_\nu(0)|^2}{\sum_{0}^{k} |\varphi_\nu(0)|^2} = -\sum_{k=0}^{n-1} \ln\left(1 - \frac{\sum_{k+1}^{n} |\varphi_\nu(0)|^2}{\sum_{0}^{n} |\varphi_\nu(0)|^2}\right).
\end{aligned} \tag{4.22}$$

Next, assuming $\sum_{k+1}^{n} |\varphi_\nu(0)|^2 / \sum_0^n |\varphi_\nu(0)|^2 = \alpha$, we have that

$$-\ln\left(1 - \frac{\sum\limits_{k+1}^{n} |\varphi_\nu(0)|^2}{\sum\limits_{0}^{n} |\varphi_\nu(0)|^2}\right) = \sum_{s=1}^{\infty} \frac{\alpha^s}{s}$$

$$\leqslant \alpha + \frac{\alpha^2}{1-\alpha}$$

$$\leqslant \frac{2\alpha}{1-\alpha} \leqslant \alpha \cdot 2 \frac{\sum\limits_{0}^{\infty} |\varphi_\nu(0)|^2}{|\varphi_0(0)|^2} = C_1 \alpha,$$

C_i denoting constants here and further on. From this and (4.22) the inequality

$$\ln \frac{D_s(|\varphi_n|^{-2})}{G(|\varphi_n|^{-2})^{s+1}} \leqslant \frac{C_1}{\sum\limits_{0}^{n} |\varphi_\nu(0)|^2} \sum_{k=0}^{n-1} \sum_{\nu=k+1}^{n} |\varphi_\nu(0)|^2$$

$$\leqslant \frac{C_1}{|\varphi_0(0)|^2} \sum_{\nu=1}^{\infty} \nu |\varphi_\nu(0)|^2 = C_2 \sum_{\nu=1}^{\infty} \nu |\varphi_\nu(0)|^2$$

follows.

By Lemma 11 we have that

$$\iint\limits_{|z| \leqslant 1} \left| \frac{(\varphi_n^*(z))'}{\varphi_n^*(z)} \right|^2 d\sigma \leqslant C_3 \sum_{\nu=1}^{\infty} \nu |\varphi_\nu(0)|^2. \tag{4.23}$$

By (4.16) of Lemma 8 we infer that $\lim_n \varphi_n^*(z) = 1/g_1(z)$ uniformly on any circle $|z| \leqslant r < 1$. The derivatives $(\varphi_n^*(z))'$ as well as the analytic functions $\varphi_n^*(z)$ converge uniformly to $(1/g_1(z))'$ on this circle. Hence taking the limit in (4.23), as $n \to \infty$, we obtain by Fatou's lemma

$$\iint\limits_{|z| \leqslant 1} \left| \frac{g_1'(z)}{g_1(z)} \right|^2 d\sigma \leqslant C_3 \sum_{\nu=1}^{\infty} \nu |\varphi_\nu(0)|^2. \tag{4.24}$$

Let us express the integral in the left-hand side of (4.24) in terms of Fourier coefficients of a summable function $\ln g_1(e^{i\lambda})$.

We note first that the function $\ln g_1(e^{i\lambda})$ is the boundary value on the circle $|z| = 1$ of the function $\ln g_1(z)$ analytic in the domain $|z| < 1$ (the exterior function $g_1(z)$ has no zero on the circle $|z| < 1$). Hence a Fourier expansion of the function $\ln g_1(e^{i\lambda})$ contains only nonnegative powers of $e^{i\lambda}$. Similarly, a Fourier series of the function $\ln \overline{g_1(e^{i\lambda})}$ contains only nonpositive powers of $e^{i\lambda}$. Furthermore,

$$\ln \overline{g_1(e^{i\lambda})} + \ln g_1(e^{i\lambda}) = \ln a(\lambda).$$

Therefore, the Fourier coefficients of the function $\ln g_1(e^{i\lambda})$ are equivalent to Fourier coefficients of a function $\ln a(\lambda)$, i.e.,

$$\ln g_1(e^{i\lambda}) \sim \frac{1}{2} a_0 + \sum_{s=1}^{\infty} a_s e^{is\lambda}.$$

Computing the integral in (4.24) we note first that for $|z| \leqslant r < 1$

$$\frac{g_1'(z)}{g_1(z)} = (\ln g_1(z))' = \sum_{s=1}^{\infty} s a_s z^{s-1}.$$

Hence

$$\begin{aligned}\iint_{|z| \leqslant r} \left|\frac{g_1'(z)}{g_1(z)}\right|^2 d\sigma &= \int_0^r \rho\, d\rho \int_{-\pi}^{\pi} \left|\sum_{s=1}^{\infty} s a_s \rho^{s-1} e^{i\lambda(s-1)}\right|^2 d\lambda \\ &= 2\pi \sum_{s=1}^{\infty} s^2 |a_s|^2 \int_0^r \rho^{2s-1}\, d\rho \\ &= \pi \sum_{s=1}^{\infty} s|a_s|^2 r^{2s} \to \pi \sum_{s=1}^{\infty} s|a_s|^2,\end{aligned}$$

from which and (4.24) follows

$$\sum_{s=-\infty}^{\infty} |s||a_s|^2 < \infty,$$

thus completing the proof of Theorem 8. □

IV.5 Condition for Absolute Regularity and Processes with Continuous Time

In the case of processes with continuous time we can prove only a result weaker than Theorem 8 and similar to Lemma 6.

Let $\xi(t)$ be a stationary process with continuous time. An absolutely regular process is regular. Therefore we can assume without loss of generality that the process $\xi(t)$ has a spectral density $f(\lambda)$ and that $f(\lambda) = |g(\lambda)|^2$, where $g \in \mathscr{H}^2$ in the upper half-plane. Let us denote by $c_\varepsilon(u)$ the Fourier transform of the function

$$\frac{\overline{g}(\lambda)}{g(\lambda)} \frac{i}{i + \varepsilon\lambda}.$$

Theorem 9. *A stationary Gaussian process $\xi(t)$ is absolutely regular if and only if this process is regular for any $T \geqslant 0$:*

$$I(T) = \overline{\lim_{\varepsilon \to 0}} \int_{-\infty}^{-T} |u||c_\varepsilon(u)|^2\, du < \infty. \tag{5.1}$$

We prove Theorem 9 essentially in a similar way as Lemma 6. As before we shall rely on Theorem 4. According to this theorem it suffices to establish that the inequality $I(T) < \infty$ is equivalent to the fact that the operator B_T

is nuclear. As in Section IV.4, it is more convenient to consider here operators B_T^+ unitarily equivalent to operators B_T. If, as in Section IV.4, we retain the notation for a standard isometric mapping on $H(-\infty, \infty)$ into $L(F)$, then

$$B_T^+ = \mathscr{P}_T^+ \mathscr{P}_0^- \mathscr{P}_T^+,$$

where $\mathscr{P}_T^+$ is the projector on $L(F)$ into $L_{(T,\infty)}(F)$ and $\mathscr{P}_0^-$ is the projector on $L(F)$ into $L_{(-\infty,0)}(F)$. It suffices to regard the operator B_T^+ as an operator on $L_{(T,\infty)}(F)$ into $L_{(T,\infty)}(F)$.

The equalities that follow can be proved in a similar way as Lemma 2 in Chapter II:

$$\mathscr{P}_T^+ = g^{-1} \Pi_T^+ g, \qquad \mathscr{P}_0^- = \bar{g}^{-1} \Pi_0^- \bar{g}, \tag{5.2}$$

where Π_T^+ is the projector on $\mathscr{L}^2(-\infty, \infty)$ into $e^{i\lambda T}\, \mathscr{H}^2$ and Π_0^- is the projector on $\mathscr{L}^2(-\infty, \infty)$ into $\mathscr{H}^{2-}$.

Let us prove that the conditions of Theorem 9 are necessary. To this end we introduce the functions

$$e_x(\lambda; \varepsilon) = e_x(\lambda) = e_x = i \frac{e^{ix\lambda}}{g(\lambda)(i + \varepsilon\lambda)}, \qquad x \geqslant 0.$$

It is seen that $e_x \in L_{(x,\infty)}(F)$. (We recall that $L_{(X,\infty)}(F) = (e^{ix\lambda}/\sigma)\mathscr{H}^2$ according to the results obtained in Section II.2).

Lemma 12. *The integrals* $S_T = \int_T^\infty \langle B_T e_x, e_x \rangle_F \, dx < \infty$ *if and only if*

$$\int_{-\infty}^{-T} |u| |c_\varepsilon(u)|^2 \, du < \infty.$$

Furthermore,

$$S_T = \int_{-\infty}^{-T} (|u| - T) |c_\varepsilon(u)|^2 \, du. \tag{5.3}$$

Proof. By (5.2) we have for $x, y \geqslant T$ that

$$\begin{aligned} \langle B_T^+ e_x, e_y \rangle_F &= \langle \mathscr{P}_0^- e_x, e_y \rangle_F \\ &= \int_{-\infty}^{\infty} \pi_0^- \left(\frac{\overline{g(\lambda)}}{g(\lambda)} \frac{ie^{i\lambda x}}{i + \varepsilon\lambda} \right) \overline{\frac{\overline{g(\lambda)}}{g(\lambda)} \frac{ie^{iy\lambda}}{i + \varepsilon\lambda}} \, d\lambda. \end{aligned}$$

Next,

$$\frac{\overline{g(\lambda)}}{g(\lambda)} \frac{ie^{ix\lambda}}{i + \varepsilon\lambda} = \int_{-\infty}^{\infty} e^{iu\lambda} c_\varepsilon(u - x) \, du,$$

$$\frac{\overline{g(\lambda)}}{g(\lambda)} \frac{ie^{iy\lambda}}{i + \varepsilon\lambda} = \int_{-\infty}^{\infty} e^{iu\lambda} c_\varepsilon(u - y) \, dy,$$

$$\Pi_0^- \frac{\bar{g}}{g} \frac{ie^{ix\lambda}}{1 + \varepsilon\lambda} = \int_{-\infty}^{0} e^{iu\lambda} c_\varepsilon(u - x) \, du.$$

Therefore, for $x, y \geqslant T$

$$\langle B_T^+ e_x, e_y \rangle_F = \int_{-\infty}^{0} c_\varepsilon(u - x) \overline{c_\varepsilon(u - y)}. \tag{5.4}$$

In particular, if $x \geqslant T$, then

$$\langle B_T^+ e_x, e_x \rangle_F = \int_{-\infty}^{-x} |c_\varepsilon(u)|^2 \, du, \tag{5.5}$$

and therefore

$$\int_T^\infty \langle B_T^+ e_x, e_x \rangle_F \, dx = \int_T^\infty dx \int_{-\infty}^{-x} |c_\varepsilon(u)|^2 \, du. \tag{5.6}$$

If the integrals in (5.6) are finite, the function $C(x) = \int_{-\infty}^{-x} |c_\varepsilon(u)|^2 \, du$ is integrable. However this function is monotone and hence $C(x) = o(x^{-1})$ as $x \to \infty$.* Hence, integrating the right-hand side of (5.6) by parts, we find

$$\begin{aligned} \int_T^\infty dx \int_{-\infty}^{-x} |c_\varepsilon(u)|^2 \, du &= -TC(T) + \int_T^\infty u |c_\varepsilon(-u)|^2 \, du \\ &= \int_{-\infty}^{-T} (|u| - T) |c_\varepsilon(u)|^2 \, du. \end{aligned}$$

Conversely, if the integral on the right-hand side of (5.3) is convergent, we have

$$C(x) \leqslant \frac{1}{|x|} \int_{-\infty}^{-x} |u| |c_\varepsilon(u)|^2 \, du = o(x^{-1}),$$

and we can take an integral in (5.6) by parts. The lemma is proved. □

Thus, let the process $\xi(t)$ be absolutely regular. By Theorem 4, there is a number τ, $0 < \tau < \infty$, such that the operator B_τ^+ is a completely continuous operator with finite trace. Let $\varphi_j(\lambda) = \varphi_j$ be (normalized) eigenvectors of the operator B_τ^+, and let μ_j be the corresponding eigenvalues. For all $\varepsilon > 0$, $x \geqslant \tau$,

$$\langle B_\tau^+ e_x, e_x \rangle_F = \sum_j |\beta_j(x)|^2 \langle B_\tau^+ \varphi_j, \varphi_j \rangle_F = \sum \mu_j |\beta_j(x)|^2, \tag{5.7}$$

where $\beta_j(x) = \langle e_x, \varphi_j \rangle_F$. Since

$$\varphi_j \in L_{(\tau, \infty)}(F) = \frac{e^{i\lambda\tau}}{g} \mathscr{H}^2,$$

the functions φ_j can be expressed as $\varphi_j = e^{i\lambda\tau} \psi_j / g$, $\psi_j \in \mathscr{H}^2$. Therefore

$$\beta_j(x) = \langle e_x, \varphi_j \rangle_F = i \int_{-\infty}^\infty \frac{e^{i\lambda(x-\tau)} \overline{\psi_j(\lambda)}}{i + \varepsilon\lambda} \, d\lambda,$$

$$\int_{-\infty}^\infty |\beta_j(x)|^2 \, dx = \int_{-\infty}^\infty \frac{|\psi_j(\lambda)|^2}{1 + \varepsilon^2 \lambda^2} \, d\lambda = \left\| \frac{i\varphi_j}{1 + \varepsilon\lambda} \right\|_F^2,$$

from which and (5.7) the inequality

$$\int_\tau^\infty \langle B_\tau^+ e_x, e_x \rangle_F \, dx = \sum_j \mu_j \left\| \frac{i\varphi_j}{i + \varepsilon\lambda} \right\|_F^2 \leqslant \sum_j \mu_j \|\varphi_j\|_F^2 = \sum_j \mu_j = \operatorname{Sp} B_\tau^+$$

follows.

* Let $C(x) \geqslant 0$, $C(x)\downarrow$, and $\int_T^\infty C(x) \, dx < \infty$. For large x we have $C(x) \leqslant (2/x) \int_{x/2}^x C(y) \, dy = o(x^{-1})$.

By Lemma 12 we have

$$\overline{\lim_{\varepsilon\to 0}} \int_{-\infty}^{-\tau} (|u| - \tau)|c_\varepsilon(u)|^2 \, du = \overline{\lim_{\varepsilon\to 0}} \int_{\tau}^{\infty} \langle B_\tau^+ e_x, e_x \rangle_F \, dx \leqslant \operatorname{Sp} B_\tau^+.$$

Let $T = 2\tau$. The last inequality implies that

$$\overline{\lim_{\varepsilon\to 0}} \int_{-\infty}^{-T} |u| \, |c_\varepsilon(u)|^2 \, du \leqslant \overline{\lim_{\varepsilon\to 0}} \, 2 \int_{-\infty}^{-\tau} (|u| - \tau)|c_\varepsilon(u)|^2 \, du \leqslant \operatorname{Sp} B_\tau^+ < \infty. \quad (5.8)$$

We have proved that the conditions of Theorem 9 are necessary.

Further we shall prove that they are sufficient. We need to construct an orthonormal basis $\{\alpha_j\}$ in $L_{(T,\infty)}(F)$ and to prove that a series $\sum_j \langle B_T^+ \alpha_j, a_j \rangle_F$ is convergent. The sum of this series will be equal to the trace of the operator B_T^+.* As such a basis we choose the functions

$$\alpha_j(\lambda) = e^{i\lambda T} \frac{a_j(\lambda)}{g(\lambda)}, \qquad a_j(\lambda) = \frac{1}{\sqrt{\pi}} \frac{1}{i + \lambda} \left(\frac{i - \lambda}{i + \lambda} \right)^j, \qquad j = 0, 1, \ldots .$$

Lemma 13. *The functions* $\alpha_j(\lambda)$, $j = 0, 1, \ldots$, *form a complete orthonormal system in* $L_{(\tau,\infty)}(F)$.

Proof. It suffices to consider the case $T = 0$. Since $L_{(0,\infty)}(F) = L^+(F) = (1/g)\mathscr{H}^2$ (see Section II.2), we need to prove that the functions $a_j(\lambda)$ form a complete orthonormal system in $\mathscr{H}^2$.

For the sake of definiteness, let $k \geqslant l$. Then†

$$(a_k, a_l)^{(2)} = \frac{1}{\pi} \int_{-\infty}^{\infty} \frac{1}{1 + \lambda^2} \left(\frac{i - \lambda}{i + \lambda} \right)^{k-l} d\lambda.$$

The integrand in this integral is analytic in the upper half-plane, with the exception of the pole at the point i. The residue of the integrand at this point is equal to zero if $k - l > 0$, and is equal to $1/2i$ if $k - l = 0$. Therefore,

$$(a_k, a_l)^{(2)} = \delta_{kl}.$$

We have established that the system a_j, $j = 0, 1, \ldots$, as well as the system α_j, $j = 0, 1, \ldots$, are orthonormal.

It remains only to verify that the system a_j, $j = 0, 1, \ldots$, is complete in $\mathscr{H}^2$. Let $\varphi(\lambda) - \varphi \subset \mathscr{H}^2$. We need to prove that since $(\varphi, a_j)^{(2)} = 0$ for all $j = 0, 1, \ldots$, then $\varphi(\lambda) \equiv 0$. The function $\varphi \in \mathscr{H}^2$ is a boundary value of the function $\varphi(z)$, $z = \lambda + i\mu$, which is analytic in the upper half-plane $\mu > 0$ and is such that

$$\sup_\mu \int_{-\infty}^{\infty} |\varphi(\lambda + i\mu)|^2 \, d\lambda < \infty.$$

Hence the integral

$$\int_{-\infty}^{\infty} \varphi(\lambda)\overline{a_j(\lambda)} \, d\lambda = (\varphi, a_j)^{(2)}$$

* See [7], p. 55; we note that B_T^+ is a positive operator.

† We recall that $(\varphi, \psi)^{(2)}$ denotes the scalar product in the Hilbert space $\mathscr{L}^2(-\infty, \infty) \supseteq \mathscr{H}^2$.

can be computed with the aid of the residue theory. The integrand has in the upper half-plane a unique pole of order $j+1$ at the point $z = i$. Therefore, for all $j = 0, 1, \ldots$ we have the equalities

$$\int_{-\infty}^{\infty} \varphi(\lambda)\overline{a_j(\lambda)}\, d\lambda = \frac{2\sqrt{\pi}}{j!}(-1)^{j+1} \frac{d^j}{dz^j} \varphi(z)(z+i)^{j+1}\Big|_{z=i} = 0.$$

Considering these equalities subsequently for $j = 0, 1, \ldots,$ we obtain that all derivatives $\varphi^{(j)}(i)$, $j = 0, 1, \ldots,$ of the analytic function $\varphi(z)$ are equal to zero. Hence $\varphi(z) \equiv 0$, $\tilde{\varphi}(\lambda) \equiv 0$, which proves the lemma. □

Proof of Theorem 9. Let us compute the sum $\sum_j \langle B_\tau^+ \alpha_j, \alpha_j \rangle_F$. We denote by $A_j(x)$ the Fourier transform of the function $a_j(\lambda)$:

$$a_j(\lambda) = \frac{1}{\sqrt{2\pi}} \int_0^\infty e^{i\lambda x} A_j(x)\, dx.$$

Further, let

$$\alpha_{j\varepsilon}(\lambda) = i\frac{\alpha_j(\lambda)}{i + \varepsilon\lambda} = \frac{i \cdot a_j(\lambda)}{1 + \varepsilon\lambda} \frac{e^{i\lambda T}}{g(\lambda)}.$$

Then, by virtue of (5.6),

$$\langle B_\tau^+ \alpha_{j\varepsilon}, \alpha_{j\varepsilon} \rangle_F = \int_0^\infty \int_0^\infty \langle B_T^+ e_{x+\tau}, e_{x+\tau} \rangle A_j(x)\overline{A_j(y)}\, dx\, dy$$
$$= \int_{-\infty}^{0} du \left| \int_0^\infty c_\varepsilon(u - x - \tau) A_j(x)\, dx \right|^2.$$

Using Lemma 13, we can easily show that the functions $A_j(x)$ form a complete orthonormal system in $\mathscr{L}^2(0, \infty)$. In fact, the Fourier transform of the functions from $\mathscr{H}^2$ vanish on the half-line $(-\infty, 0)$. Conversely, if we define a function in $\mathscr{L}^2(0, \infty)$ more precisely, using zero on the half-line $(-\infty, 0)$, this function can be regarded as the Fourier transform of a function in $\mathscr{H}^2$ (this assertion is, in fact, the Paley–Wiener theorem). Therefore, the operator $U: \mathscr{H}^2 \to \mathscr{L}^2(0, \infty)$ associating each function in $\mathscr{H}^2$ with the Fourier transform of this function embodies the isometric correspondence between $\mathscr{H}^2$ and $\mathscr{L}^2(0, \infty)$. In particular, since $\{a_j\}$ is a complete orthonormal system in $\mathscr{H}^2$, the functions A_j form a complete orthonormal system in $\mathscr{L}^2(0, \infty)$.*

Let us regard the function $c_\varepsilon(u - x - T)$ for fixed u, T as an element of the space $\mathscr{L}^2(0, \infty)$. We write $c_\varepsilon(u - x - T)$ as

$$c_\varepsilon(u - x - T) = \sum_{j=0}^{\infty} \gamma_j(u) A_j(x),$$

where

$$\gamma_j(u) = \int_0^\infty c_\varepsilon(u - x - T) A_j(x)\, dx$$

* It is easy to compute that $A_j(x) = e^{-x} L_j(2x)$ where $L_j(x) = \sum_{\nu=0}^{j} c_j^\nu((-x)^\nu/\nu)$ is the Laguerre polynomial of order j (see [26]).

and for all u

$$\sum_{j=0}^{\infty} |\gamma_j(u)|^2 = \int_0^{\infty} |c_\varepsilon(u - x - T)|^2 \, dx = \int_{-\infty}^{u-T} |c_\varepsilon(v)|^2 \, dv.$$

Therefore, for all $\varepsilon > 0$

$$\begin{aligned}\sum_{j=0}^{\infty} \langle B_T \alpha_{j\varepsilon}, \alpha_{j\varepsilon} \rangle_F &= \sum_{j=0}^{\infty} \int_{-\infty}^{0} du \left| \int_0^{\infty} c_\varepsilon(u - x - \tau) A_j(x) \, dx \right|^2 \\ &= \sum_{j=0}^{\infty} \int_{-\infty}^{0} |\gamma_j(u)|^2 \, du \\ &= \int_{-\infty}^{0} \int_{-\infty}^{u-T} |c_\varepsilon(v)|^2 \, dv \\ &= \int_{-\infty}^{-T} dx \int_{-\infty}^{x} |c_\varepsilon(u)|^2 \, du = \int_{-\infty}^{-T} (|u| - T)|c_\varepsilon(u)|^2 \, du.\end{aligned}$$

Letting $\varepsilon \to 0$, we find that for all n

$$\begin{aligned}\sum_{j=0}^{n} \langle B_T \alpha_j, \alpha_j \rangle_F &= \lim_{\varepsilon \to 0} \sum_{j=0}^{n} \langle B_T \alpha_{j\varepsilon}, \alpha_{j\varepsilon} \rangle_F \\ &= \underline{\lim}_{\varepsilon \to 0} \sum_{j=0}^{n} \langle B_T \alpha_{j\varepsilon}, \alpha_{j\varepsilon} \rangle_F \\ &\leqslant \underline{\lim}_{\varepsilon \to 0} \sum_{j=0}^{\infty} \langle B_T \alpha_{j\varepsilon}, \alpha_{j\varepsilon} \rangle_F \\ &= \underline{\lim}_{\varepsilon \to 0} \int_{-\infty}^{-T} (|u| - T)|c_\varepsilon(u)|^2 \, du.\end{aligned}$$

Hence also

$$\begin{aligned}\operatorname{Sp} B_T &= \sum_{j=0}^{\infty} \langle B_T \alpha_j, \alpha_j \rangle_F \\ &\leqslant \underline{\lim}_{\varepsilon \to 0} \int_{-\infty}^{-T} (|u| - T)|c_\varepsilon(u)|^2 \, du \\ &\leqslant \overline{\lim}_{\varepsilon \to 0} \int_{-\infty}^{-T} |u||c_\varepsilon(u)|^2 \, du < \infty, \end{aligned} \tag{5.9}$$

thus proving the theorem. □

Note. If we compare (5.8) and (5.9) it can be readily seen that actually the limit $\lim_{\varepsilon \to 0} \int_{-\infty}^{-T} |u||c_\varepsilon(u)|^2 \, du$ (perhaps infinite) exists and that

$$\operatorname{Sp} B_T = \lim_{\varepsilon \to 0} \int_{-\infty}^{-T} (|u| - T)|c_\varepsilon(u)|^2 \, du.$$

CHAPTER V

Complete Regularity and Processes with Discrete Time

V.1 Definitions and Preliminary Constructions with Examples

We consider in this chapter a wide-sense stationary process $\xi(t)$ with discrete time $t = 0, \pm 1, \ldots$. Here we deal only with the concepts formulated in terms of the second-order statistics; hence it does not really matter whether the process $\xi(t)$ is Gaussian or not.

We recall (see Chapter IV) that the process $\xi(t)$ is said to be completely regular if the regularity coefficient

$$\rho(\tau) = \sup_{\eta_1 \in H(\tau, \infty),\, \eta_2 \in H(-\infty, 0)} |\mathsf{M}\eta_1\bar{\eta}_2| = \sup|(\eta_1, \eta_2)| \xrightarrow[\tau \to \infty]{} 0$$

(sup is taken over η_1 and η_2 that satisfy the normalization condition $\|\eta_1\| = \|\eta_2\| = 1$).

We investigate also spectral characteristics of a completely regular process $\xi(t)$; for this purpose we need to express the regularity coefficient in terms of the spectral representation.

Let us recall that each completely regular process is (linearly) regular. Hence:

(a) this process has a spectral density $f(\lambda)$ representable as

$$f(\lambda) = |g(e^{i\lambda})|^2, \tag{1.1}$$

where g is a function of Hardy class $\mathscr{H}^2$ on the unit circle;

(b) $\int_{-\pi}^{\pi} |\ln f(\lambda)\, d\lambda| < \infty.$ (1.2)

In fact, (1.1) and (1.2) are equivalent and

$$g(z) = \exp\left\{\frac{1}{4\pi}\int_{-\pi}^{\pi}\frac{e^{i\lambda}+z}{e^{i\lambda}-z}\ln f(\lambda)\,d\lambda\right\}, \qquad |z|<1,$$

from which and Theorem 1 in Chapter II it follows immediately that

$$\rho(\tau) = \sup_{\varphi,\,\psi}\left|\int_{-\pi}^{\pi} e^{i\lambda\tau}\varphi(e^{i\lambda})\psi(e^{i\lambda})f(\lambda)\,d\lambda\right| = \sup_{\varphi,\,\psi}|\langle e^{i\lambda\tau}\varphi, \bar{\psi}\rangle_F|, \tag{1.3}$$

where sup is taken over all functions $\varphi(e^{i\lambda})$, $\psi(e^{i\lambda})$ belonging to the subspace $L^+(F) = (1/g)\mathscr{H}^2$ and satisfying the normalization condition

$$\begin{aligned} \|\varphi\|_F &= \left(\int_{-\pi}^{\pi}|\varphi(e^{i\lambda})|^2 f(\lambda)\,d\lambda\right)^{1/2} = 1,\\ \|\psi\|_F &= \left(\int_{-\pi}^{\pi}|\psi(e^{i\lambda})|^2 f(\lambda)\,d\lambda\right)^{1/2} = 1. \end{aligned} \tag{1.4}$$

It is useful to note that the value $\rho(\tau)$ does not change if sup is taken in (1.3) over any set (dense in $L^+(F)$) of functions satisfying (1.4); for example, over polynomials or functions from $\mathscr{H}^2$.

If $\varphi, \psi \in \mathscr{H}^2$, then $\theta = \varphi\psi \in \mathscr{H}^1$ and $\|\theta\|_F^{(1)} \leqslant \|\varphi\|_F \cdot \|\psi\|_F$. Conversely, if the function $\theta \in \mathscr{H}^1$, it can be expressed (see Section II.1) as the product $\theta = \varphi\psi$ of two functions $\varphi, \psi \in \mathscr{H}^2$; for almost all λ $|\varphi| = |\psi| = |\theta|^{1/2}$, and therefore $\|\theta\|_F^{(1)} = \|\varphi\|_F = \|\psi\|_F$. Hence we have in addition to (1.3) the following expression for $\rho(\tau)$:

$$\rho(\tau) = \sup_{\theta}\left|\int_{-\pi}^{\pi} e^{i\lambda\tau}\theta(\lambda)f(\lambda)\,d\lambda\right|, \tag{1.5}$$

where sup is taken over all $\theta \in \mathscr{H}^1$, $\|\theta\|_F^{(1)} \leqslant 1$.

To obtain another expression necessary for our discussion, we need to invoke Beurling's theorem, according to which the aggregate of functions $\{\varphi g\}$ where $\varphi(z)$ runs through all the polynomials is dense in $\mathscr{H}^2$. Starting from (1.3) we can calculate that

$$\begin{aligned} \rho(\tau) &= \sup_{\varphi,\,\psi}\left|\int_{-\pi}^{\pi}\varphi(e^{i\lambda})\psi(e^{i\lambda})e^{i\lambda\tau}f(\lambda)\,d\lambda\right|\\ &= \sup_{\varphi,\,\psi}\left|\int_{-\pi}^{\pi}(\varphi g)(\psi g)e^{i\lambda\tau}\frac{\bar{g}}{g}\,d\lambda\right|\\ &= \sup_{\varphi_1,\,\psi_1}\left|\int_{-\pi}^{\pi}\varphi_1(e^{i\lambda})\psi_1(e^{i\lambda})e^{i\lambda\tau}\frac{\overline{g(e^{i\lambda})}}{g(e^{i\lambda})}\,d\lambda\right|, \end{aligned} \tag{1.6}$$

sup being taken over all φ_1, ψ_1 from the unit sphere of the space $\mathscr{H}^2$ in the last integral.

Finally, just as we obtained (1.5) from (1.3) we can deduce from (1.6) the equality

$$\rho(\tau) = \sup_{\theta} \left| \int_{-\pi}^{\pi} \theta(e^{i\lambda}) e^{i\tau\lambda} \frac{g(e^{i\lambda})}{\overline{g(e^{i\lambda})}} \, d\lambda \right|, \tag{1.7}$$

where θ runs through the unit sphere of the space $\mathscr{H}^1$.

From now on in this chapter we shall deal only with absolutely continuous spectral functions given by their (spectral) densities. To avoid unnecessary notation we shall use $L(f)$, $L^+(g)$, $\|\cdot\|_h$, $\langle\cdot,\cdot\rangle_f$, etc., throughout the present chapter, instead of $L(F)$, $L^+(G)$, $\|\cdot\|_H$, $\langle\cdot,\cdot\rangle_F$, etc., where $f = F'$, $g = G'$, $h = H'$, etc.

Returning to the problem of describing completely regular processes we note that we can formulate this as follows: namely, to describe the class of nonnegative functions $f(\lambda)$ summable on $[-\pi, \pi]$ for which the values of $\rho(\tau)$ defined by (1.3) and (1.5) vanish as $\tau \to \infty$. This analytic formulation of our problem will be the basis of our further investigation.

Before making a more detailed analysis we list a few assertions that follow immediately from (1.3) or (1.5) but that, nevertheless, enable us to have a general idea about the structure of spectral densities of completely regular processes.

Assuming $\theta(\lambda) \equiv 1$ in (1.5) we find that the nth Fourier coefficient of $f(\lambda)$ satisfies the inequality

$$|B(n)| \leqslant \rho(n) \int_{-\pi}^{\pi} f(\lambda) \, d\lambda.$$

Therefore, the complete regularity conditions impose some restrictions on the smoothness of $f(\lambda)$; in particular, $f(\lambda)$ is continuous if $\sum \rho(n) < \infty$. We shall improve this rather rough approach to obtain stronger results.

It follows from (1.2) that $f(\lambda)$ can have no zeros of exceedingly high order. We shall show in Section V.5 that even more rigorous restrictions must, in fact, be imposed on zeros of $f(\lambda)$ (roughly speaking, $f(\lambda)$ is a product of the function without zeros and the square of a polynomial).

Conversely, if the spectral density $f(\lambda)$ is positive and sufficiently smooth, the corresponding random process will be completely regular. The following theorem holds true.

Theorem 1. *If a spectral density $f(\lambda)$ is continuous* and strictly positive, $f(\lambda) \geqslant m > 0$, the stationary process corresponding to $f(\lambda)$ is completely regular, and*

$$\rho(\tau) \leqslant \frac{1}{m} E_{\tau-1}(f). \tag{1.8}$$

Here and below we denote by $E_n(h)$ the best approximation of a function $h(\lambda)$ by trigonometric polynomials of degree less than n on the interval $[-\pi, \pi]$ in the uniform metric.

* We recall that the points π and $-\pi$ are identified as the same.

In fact, for any trigonometric polynomial $\theta(\lambda)$ of degree less than $\tau - 1$ and for any function $\theta \in \mathscr{H}^1$ we have

$$\int_{-\pi}^{\pi} e^{i\lambda\tau}\theta(\lambda)Q(\lambda)\,d\lambda = 0.$$

Hence, if $P(\lambda)$ is a polynomial of best approximation for f of degree $\leqslant \tau - 1$, by virtue of (1.5) we obtain

$$\rho(\tau) = \sup_{\theta}\left|\int_{-\pi}^{\pi} e^{i\lambda\tau}\theta(\lambda)[f(\lambda) - P(\lambda)]\,d\lambda\right|$$
$$\leqslant E_{\tau-1}(f) \sup_{\|\theta\|_f^{(1)}=1}\left|\int_{-\pi}^{\pi} |\theta(\lambda)|\,d\lambda\right| \leqslant \frac{1}{m} E_{\tau-1}(f).$$

The restriction $f \geqslant m$ can be weakened if we note that the following result holds.

Theorem 2. *If $w(\lambda)$ is a spectral density of a completely regular process, and if $P(z)$ is a polynomial of degree n, then*

$$f(\lambda) = |P(e^{i\lambda})|^2 w(\lambda) \tag{1.9}$$

is a spectral density of a completely regular process. In this case we have

$$\rho(\tau; f) \leqslant \rho(\tau - n; w). \tag{1.10}$$

Indeed, $e^{i\lambda n}|P|^2 \in \mathscr{H}^1$. In addition, $\|\theta P\|_w^{(1)} = \|\theta\|_f^1$. Hence

$$\rho(\tau; f) = \sup_{\theta}\left|\int_{-\pi}^{\pi} e^{i(\tau-n)\lambda}\theta(\lambda)[e^{in\lambda}|P(\lambda)|^2]w(\lambda)\,d\lambda\right|$$
$$\leqslant \sup_{\|\theta\|_w^{(1)}\leqslant 1}\left|\int_{-\pi}^{\pi} e^{i(\tau-n)\lambda}\theta(\lambda)w(\lambda)\,d\lambda\right| = \rho(\tau - n; w).$$

We have proved the theorem. We shall often use further decompositions of the form given by (1.9).

V.2 The First Method of Study: Helson–Sarason's Theorem

We present in this section the description of a class of spectral densities of completely regular processes given by Helson and Sarason,* whose method essentially involves the entry into the complex plane. In this connection it is convenient to consider that the spectral density $f(\lambda)$ is given on the circle C: $|z| = 1$, $z = \mathrm{r}e^{i\lambda}$.

Theorem 3. *A stationary process $\xi(t)$ with discrete time is completely regular if and only if this process has spectral density $f(\lambda)$ representable as*

$$f(\lambda) = w(\lambda)|P(e^{i\lambda})|^2. \tag{2.1}$$

* Helson and Sarason, "Past and future," *Math. Scand.* **21** (1967), 5–16.

Here $P(z)$ is a polynomial with roots on $|z| = 1$, and the function $w(\lambda)$ is representable for any $\varepsilon > 0$ as

$$w = \exp\{r_\varepsilon + u_\varepsilon + \tilde{v}_\varepsilon\}, \tag{2.2}$$

where r_ε is continuous on C and $\|u_\varepsilon\|^{(\infty)} + \|v_\varepsilon\|^{(\infty)} \leqslant \varepsilon$.

Proof. Let us first consider sufficiency. It follows from Theorem 2 that $P \equiv 1$. Using the Weierstrass theorem we choose a trigonometric polynomial Q_ε of degree τ_0 such that

$$e^{r_\varepsilon} = Q_\varepsilon(1 + \theta), \qquad \max_\lambda |\theta(\lambda)| < \varepsilon.$$

Then ($P \equiv 1$!)

$$f(\lambda) = e^{r_\varepsilon + u_\varepsilon + \tilde{v}_\varepsilon} = Q_\varepsilon e^{\tilde{v}_\varepsilon - iv_\varepsilon}(1 + \theta_\varepsilon), \tag{2.3}$$

where $\|\theta_\varepsilon\|^{(\infty)} \leqslant 7_\varepsilon$ if $\varepsilon < 1$.

Set $f_\varepsilon = |Q_\varepsilon|\, |e^{\tilde{v}_\varepsilon - iv_\varepsilon}|$. By (2.3) we have

$$\frac{1}{2}\|\varphi\|_{f_\varepsilon} \leqslant \|\varphi\|_f \leqslant 2\|\varphi\|_{f_\varepsilon}$$

for ε sufficiently small. Further, the function $e^{i\lambda\tau_0}Q_\varepsilon(\lambda)e^{\tilde{v}_\varepsilon(\lambda) - iv_\varepsilon(\lambda)}$ is summable and constitutes the boundary value (on the circle $|z| = 1$) of a function analytic on the circle $|z| < 1$. Therefore, for any two polynomials $\varphi(z)$ and $\psi(z)$ and all $\tau > \tau_0$ we have

$$\int_{-\pi}^{\pi} \varphi(e^{i\lambda})\psi(e^{i\lambda})e^{i\lambda\tau}Q_\varepsilon(\lambda)e^{\tilde{v}_\varepsilon - iv_\varepsilon}\, d\lambda = 0.$$

Hence, for all $\tau > \tau_0$, by virtue of (2.3) we obtain

$$\begin{aligned}\rho(\tau) &= \sup_{\varphi,\psi}\left|\int_{-\pi}^{\pi} \varphi(e^{i\lambda})\psi(e^{i\lambda})e^{i\lambda\tau}f(\lambda)\, d\lambda\right| \\ &\leqslant \sup_{\varphi,\psi}\int_{-\pi}^{\pi} |\varphi(e^{i\lambda})||\psi(e^{i\lambda})||f_\varepsilon(\lambda)||\theta_\varepsilon(\lambda)|\, d\lambda \\ &\leqslant 7\varepsilon\|\varphi\|_{f_\varepsilon}\|\psi\|_{f_\varepsilon} \leqslant 28\varepsilon,\end{aligned} \tag{2.4}$$

thus proving the sufficiency of the theorem. □

As to necessity, starting from the equality given by (1.7) we prove the following lemma.

Lemma 1. *The regularity coefficient can be characterized as*

$$\rho(\tau) = \inf_A \left\|\frac{\bar{g}}{g}e^{i\lambda\tau} - A\right\|^{(\infty)}, \tag{2.5}$$

where inf is taken over all functions $A \in \mathscr{H}^\infty$, $A(0) = 0$.

Proof. Lemma 1 is a particular case of the general duality principle for analogous extremal problems based on the well-known Hahn–Banach theorem.

In fact, the integral in the right-hand side of (1.7) determines a linear functional $l(\theta)$ in $\mathscr{L}^1$. This functional, considered only in $\mathscr{H}^1 \subset \mathscr{L}^1(-\pi, \pi)$, has the norm equal to $\rho(\tau)$ (according to (1.7)). Any extension $l_1(\theta)$ of this functional $l(\theta)$ from the space $\mathscr{H}^1$ to all $\mathscr{L}^1$ is

$$l_1(\theta) = l(\theta) - l^*(\theta),$$

where $l^*(\theta)$ vanishes in $\mathscr{H}^1$. All extensions have norm not less than $\rho(\tau)$; by the Hahn–Banach theorem these extensions include l_1 with the norm equal to $\rho(\tau)$ exactly. Therefore

$$\rho(\tau) = \inf_{l^*} \|l - l^*\|,$$

where $\|\cdot\|$ denotes the norm of a functional in $\mathscr{L}^1$. Each functional L in $\mathscr{L}^1(-\pi, \pi)$ can be expressed as

$$L(\theta) = \int_{-\pi}^{\pi} \theta(e^{i\lambda}) A(e^{i\lambda})\, d\lambda, \qquad A \in \mathscr{L}^\infty,\ \|L\| = \|A\|^{(\infty)},$$

as is well known. The condition $l^*(\theta) = 0$ for all $\theta \in \mathscr{H}^1$ is equivalent to the fact that

$$\int_{-\pi}^{\pi} A(e^{i\lambda}) e^{in\lambda}\, d\lambda = l^*(e^{in\lambda}) = 0, \qquad n \geqslant 0,$$

for the function $A \in \mathscr{L}^\infty(-\pi, \pi)$ determining the functional l^*.

The above equalities are, in turn, equivalent to the fact that $A \in \mathscr{H}^\infty$ and $A(0) = 0$. From this and (1.7) it follows that

$$\rho(\tau) = \inf_{l^*} \|l - l^*\| = \inf_{A \in \mathscr{H}^\infty,\ A(0) = 0} \left\| \frac{\overline{g}}{g} e^{i\tau\lambda} - A \right\|^\infty,$$

thus proving the lemma.

From Lemma 1 almost immediately we have □

Lemma 2. *A necessary and sufficient condition for the process $\xi(t)$ to be completely regular is that for any $\varepsilon > 0$ there be a function $A \in \mathscr{H}^\infty$ such that*

$$-\varepsilon < \ln|A| < \varepsilon, \qquad -\varepsilon < \arg(Ag^2 e^{-i\tau\lambda}) < \varepsilon \ (\mathrm{mod}\ 2\pi). \tag{2.6}$$

Proof. In fact, $\overline{g}/g = \exp\{-i\arg(g^2)\}$. Hence (2.5) can be written as

$$\rho(\tau) = \inf_A \|1 - e^{\ln|A|} \exp\{-i\arg(Ag^2 e^{-i\tau\lambda})\}\|^\infty,$$

from which (2.6) follows due to the fact that $\rho(\tau) \to 0$. □

We shall need another lemma on analytic extension; this lemma as well as the duality principle form the basis for the proof of the Helson–Sarason theorem.

Lemma 3. *Let the function $S(z)$ be analytic on a circle $|z| < 1$ with the exception of the point $z = 0$, at which this function can have a pole of order τ. If $z^\tau S(z) \in \mathscr{H}^{1/2}$, and if the function S is real and nonnegative on $|z| = 1$, this function is analytically extendable across $|z| = 1$ in $|z| > 1$ and the function extended is a polynomial of z and $1/z$.*

We shall prove the lemma at the end of the section. Here we complete proving Theorem 3 assuming Lemma 3 to be proved.

Proof of Theorem 3. Relying on (2.6) we choose a function $s(e^{i\lambda}) \in \mathscr{L}^\infty$ such that

$$|s| \leqslant \varepsilon, \ \arg(Ag^2 e^{-i\tau\lambda}) + s \equiv 0 (\operatorname{mod} 2\pi).$$

Let us consider the function

$$S(z) = A(z)g^2(z)z^{-\tau}e^{(is - \tilde{s})(z)}.$$

It is seen that the function $z^\tau S(z)$ is analytic on $|z| < 1$; on $|z| = 1$ we have

$$S(e^{i\lambda}) = |A(\lambda)|e^{-\tilde{s}(e^{i\lambda})}f(\lambda) \geqslant 0. \tag{2.7}$$

It is a well-known fact that restrictions on the growth of the function $s(\lambda)$ involve corresponding restrictions on the growth of the conjugate function $\tilde{s}$. In particular, it follows from the inequality $|s| \leqslant \varepsilon$ (see [28], p. 404) that $\exp\{k|\tilde{s}|\} \in \mathscr{L}^1(-\pi, \pi)$ for all $k < \pi/2\varepsilon$. Therefore, for $\varepsilon < 2/\pi$ we have

$$\int_{-\pi}^{\pi} |S(e^{i\lambda})|^{1/2}\, d\lambda < \sqrt{2}\left(\int_{-\pi}^{\pi} f(\lambda)\, d\lambda \int_{-\pi}^{\pi} e^{|\tilde{s}(e^{i\lambda})|}\, d\lambda\right)^{1/2} < \infty.$$

Therefore, all the conditions of Lemma 3 are satisfied by the function $S(z)$; hence this function is a polynomial of z and $1/z$.

Next, the function $S(z)$ is a nonnegative trigonometric polynomial on the circle $|z| = 1$. By the Fejer–Riesz theorem, a nonnegative trigonometric polynomial is the square of a polynomial of $e^{i\lambda}$ (see [13], p. 33). Therefore,

$$S(e^{i\lambda}) = |P_1(e^{i\lambda})|^2,$$

where $P_1(z)$ is a polynomial. Let us write P_1 as $P_1 = P \cdot Q$ where Q and P are polynomials with roots located respectively outside of $|z| = 1$ and on $|z| = 1$. It follows from (2.7) that

$$\begin{aligned} f(\lambda) &= S(e^{i\lambda})|A|^{-1}e^{\tilde{s}} \\ &= |P(e^{i\lambda})|^2 \exp\{\ln|Q|^2 - \ln|A| + \tilde{s}\} \\ &= |P|^2 \exp\{r_\varepsilon + u_\varepsilon + \tilde{v}_\varepsilon\}, \end{aligned} \tag{2.8}$$

where: (1) $r_\varepsilon = \ln|Q|^2$ is a continuous function; (2) $u_\varepsilon = -\ln|A|$ and $\|u_\varepsilon\|^{(\infty)} \leqslant \varepsilon$; (3) $v_\varepsilon = s$ and, by definition of s, $\|v_\varepsilon\|^{(\infty)} \leqslant \varepsilon$.

To complete proving the theorem, it remains only to verify that P is independent of ε. Let

$$f(\lambda) = |P'|^2 \exp\{r_{\varepsilon'} + u_{\varepsilon'} + v_{\varepsilon'}\}. \tag{2.9}$$

We shall show that the polynomials P and P' coincide within a constant multiplier. As was noted above, $e^{|\tilde{v}_\varepsilon|} \in \mathscr{L}^1$, $e^{|\tilde{v}_{\varepsilon'}|} \in \mathscr{L}^1$. Hence all functions

$$\left(\frac{f}{|P|^2}\right)^{\pm 1}, \quad \left(\frac{f}{|P'|^2}\right)^{\pm 1}$$

are summable. By the Schwarz inequality we infer that

$$\begin{aligned} \int_{-\pi}^{\pi} \left|\frac{P}{P'}\right| d\lambda &\leqslant \left(\int_{-\pi}^{\pi} \frac{|P|^2}{f} d\lambda \int_{-\pi}^{\pi} \frac{f}{|P'|^2} d\lambda\right)^{1/2} < \infty, \\ \int_{-\pi}^{\pi} \left|\frac{P'}{P}\right| d\lambda &\leqslant \left(\int_{-\pi}^{\pi} \frac{|P'|^2}{f} d\lambda \int_{-\pi}^{\pi} \frac{f}{|P|^2} d\lambda\right)^{1/2} < \infty. \end{aligned} \tag{2.10}$$

The inequalities given by (2.10) imply that the polynomials P and P' are multiples of each other, the roots of both polynomials lying on $|z| = 1$, and therefore that $P/P' =$ constant.

We have proved Theorem 3; we need only to make sure that Lemma 3 holds true. □

Proof of Lemma 3. The function $z^\tau S \in \mathscr{H}^{1/2}$ can be written as a product $z^\tau S = b\theta_1\theta_2$ where b is a Blaschke product, θ_1 is an inner function analytic on $|z| < 1$, and θ_2 is an exterior function on $|z| < 1$. We have $|b| = |\theta_1| = 1$ on $|z| = 1$; the function $\theta_2^{1/2} \in \mathscr{H}^1$. Hence, assuming $S_1 = z^{-\tau}b\theta_1\theta_2^{1/2}$ and $S_2 = \theta_2^{1/2}$, we represent S as the product S_1S_2 where $z^\tau S_1, S_2 \in \mathscr{H}^1$ and also $|S_1| = |S_2|$ for $|z| = 1$. From this and the fact that S is real on the circle $|z| = 1$ it follows that (on $|z| = 1$) $S_1 = \bar{S}_2$. Therefore, the functions $S_1 + S_2$ and $i(S_1 - S_2)$ are real and summable on the circle $|z| = 1$, and, by the well-known symmetry principle, analytically extendable across $|z| = 1$ (values of extended functions for $|z| > 1$ are determined by the equalities $(S_1 + S_2)(z) = \overline{(S_1 + S_2)(\bar{z}^{-1})}$, $i(S_1 - S_2)(z) = \overline{i(S_1 - S_2)(\bar{z}^{-1})}$.* Then the functions S_1 and S_2 (therefore, $S = S_1S_2$ also) are analytically extendable across $|z| = 1$. In this case the extension technique implies that the extended function S is analytic everywhere with the exception of points $z = 0$ and $z = \infty$, where the function S may have poles of order less than τ. By Liouville's theorem, $S(z)$ is necessarily a polynomial of z and $1/z$. We have proved Lemma 3, thus completing proving Theorem 3. □

* To prove that the extended function is analytic in a neighborhood of the circle $|z| = 1$ we need to use the equality

$$\lim_{r \uparrow 1} \int_\alpha^\beta \gamma(re^{i\lambda})\, d\lambda = \int_\alpha^\beta \gamma(e^{i\lambda})\, d\lambda,$$

which holds for all functions $\gamma \in \mathscr{H}^1$.

Theorem 3 proved above leads to some simple corollaries enabling us to have a fairly good picture of the structure of spectral densities of completely regular processes. We shall deduce such corollaries in Section V.5.

The method used in this case allows us to describe completely the class of random processes for which $\rho(1) < 1$ (and not necessarily $\rho(\tau) \to 0$).

As was noted before, the inequality $\rho(1) < 1$ is a stronger regularity condition than one of linear regularity and has the geometric meaning that the minimum angle between subspaces L^+ and $e^{i\lambda}L^-$ is positive. We shall obtain further a finer result* and list the conditions under which $\rho(\tau) < 1$ for the first time for $\tau = k$.

Theorem 4. *For a stationary process* $\xi(t)$, $t = 0, \pm 1, \ldots$, *the relations*

$$\rho(k-1) = 1, \qquad \rho(k) < 1$$

can be satisfied if and only if this process has a spectral density $f(\lambda)$ *representable as*

$$f(\lambda) = |P(e^{i\lambda})|^2 e^{u+\tilde{v}(\lambda)},$$

where $P(z)$ *is a polynomial of degree* $k-1$ *with roots on* $|z| = 1$, *and where* $u(\lambda)$ *and* $v(\lambda)$ *are real bounded functions,* $\|v\|^{(\infty)} < \pi/2$.

Proof. Let us prove first that the conditions mentioned above are sufficient. By virtue of (1.10)

$$\rho(k; f) \leqslant \rho(1; e^{u+\tilde{v}}),$$

we can restrict ourselves to the case where $P \equiv 1$. It is seen that $\ln f = u + \tilde{v} \in \mathscr{L}^1$, so that $f(\lambda) = |g(e^{i\lambda})|^2$ where $g \in \mathscr{H}^2$. Let $A(z)$ be an exterior function of class $\mathscr{H}^\infty$; $|A(e^{i\lambda})| = e^{-u}$ for this function. Further, let us construct the function $\psi = e^{\tilde{v}-iv}$. This exterior function is of class $\mathscr{H}^1$ since, as was noted above, $e^{\tilde{v}} \in \mathscr{L}^1$ because $|\tilde{v}| < \pi/2$. We have the equality $|\psi| = |Ag^2|$ on $|z| = 1$. Exterior functions whose values coincide on $|z| = 1$ differ only by a constant multiplier, and hence we may assume that $\psi = Ag^2$.

Let us take a positive number γ such that $\gamma|A| \leqslant 1$. The values of the function $\zeta = \gamma e^{-u}e^{-iv}$ lie within the domain

$$\mathfrak{a} = \left\{0 < \gamma \inf_{\lambda} e^{-u} \leqslant |\zeta| \leqslant 1, |\arg \zeta| < \sup_{\lambda} |v| < \pi/2\right\}$$

for all λ. It is easy to show that $\rho_1 = \inf_{\zeta \in \mathfrak{A}} |1 - \zeta| < 1$. By virtue of (2.5) we have

$$\rho(1) \leqslant \left\|\frac{\bar{g}}{g} - \gamma A\right\|^{(\infty)} = \|1 - \gamma e^{\ln|A|}\exp\{-i\arg(Ag^2)\}\|^{(\infty)}$$

$$= \|1 - \gamma e^{-u}e^{-iv}\|^{(\infty)} \leqslant \rho_1 < 1.$$

* This result for $k = 1$ was proved in H. Helson and G. Szegö, "A problem in prediction theory," *Ann. Math. Pure Appl.* **51** (1960), 107–138. The same result for $k > 1$ was obtained by Helson and Sarason, "Past and future," *Math. Scand.* **21** (1967), 5–16.

We have thus proved that the conditions of the theorem are sufficient. Let us prove that they are necessary.

We recall that the inequality $\rho(k) < 1$ implies that the process $\xi(t)$ is regular. Hence the spectral density f is factorizable: $f = |g|^2$, $g \in \mathscr{H}^2$, and $\rho(k)$ can be found from (2.5). Using formula (2.5), as in Lemma 2 we deduce that since $\rho(k) < 1$ there is necessarily a function $A \in \mathscr{H}^\infty$ with properties

$$|A| \geqslant \varepsilon, \qquad |\arg(Ag^2 e^{-i(k-1)\lambda})| \leqslant \frac{\pi}{2} - \varepsilon, \qquad \varepsilon > 0$$

(the second inequality holds in mod 2π).

Let us define the function v, $|v| \leqslant (\pi/2) - \varepsilon$, so that

$$v + \arg(Ag^2 e^{-i(k-1)\lambda}) \equiv 0 \qquad (\text{mod } 2\pi).$$

The function

$$S(z) = Ag^2 z^{-(k-1)} e^{-\tilde{v} + iv}$$

is analytic on $|z| < 1$ with the exception of the pole of order $k - 1$ at the point $z = 0$. Similarly to what was done in the proof of Theorem 3, we deduce here that $Sz^{k-1} \in \mathscr{H}^{1/2}$ and next that $S = |P|^2$, where P is a polynomial of degree less than $k - 1$. Finally, assuming $u = -\ln|A|$, we find that

$$f = |P|^2 e^{u + \tilde{v}}$$

It remains to show that the degree N of polynomial P not exceeding $k - 1$ is equal, in fact, to $k - 1$. According to what has been proved $\rho(1; e^{u+\tilde{v}}) < 1$. Hence, if $N < k - 1$ then

$$\rho(k - 1; f) \leqslant \rho(k - 1 - N; e^{u+\tilde{v}}) \leqslant \rho(1; e^{u+\tilde{v}}) < 1,$$

although $\rho(k - 1; f) = 1$, thus proving the theorem. □

V.3 The Second Method of Study: Local Conditions

In Sections V.3 and V.4 we present a new approach to investigating completely regular sequences. Unlike the methods described in Section V.2, this approach is purely real and imposes local conditions on $f(\lambda)$. There is, however, a certain divergence between the necessary conditions and the sufficient conditions given here in the local form.

Theorem 5. *A necessary condition for the positive function $f(\lambda)$, summable on $[-\pi, \pi]$, to be the spectral density of a completely regular stationary sequence is that this function $f(\lambda)$ be represented as*

$$f(\lambda) = |P(e^{i\lambda})|^2 w(\lambda), \tag{3.1}$$

where $P(z)$ is a polynomial with roots on $|z| = 1$ and where the primitive $W(\lambda)$ of the function $w(\lambda)$ satisfies the condition

$$\lim_{\delta \to 0} \omega_W(\delta) = 0, \tag{3.2}$$

where

$$\omega_W(\delta) = \sup_{\lambda} \sup_{|x| \leqslant \delta} \frac{|W(\lambda + x) + W(\lambda - x) - 2W(\lambda)|}{|W(\lambda + x) - W(\lambda - x)|}.$$

A simple converse of the theorem may be stated as follows.

Theorem 6. *Let the spectral density $f(\lambda)$ of a sequence $\xi(t)$ be of the form given by* (3.1), *where $P(z)$ is a polynomial of degree k and $w(\lambda)$ has the following properties*:

1. $\sum_n \omega_W^2(2^{-n}) < \infty$;

2. $0 < m \leqslant w(\lambda) \leqslant M < \infty$.

Then the sequence $\xi(t)$ is completely regular, and

$$\rho(\tau) \leqslant 40 \left(\frac{M}{m}\right)^{5/2} \left(\sum_1^{\infty} \omega_W^2 \left(\frac{1}{2^{n-1}(\tau - k) - 1}\right)\right)^{1/2}. \tag{3.3}$$

The proof of Theorem 5 is rather cumbersome, based on the analysis of the functions

$$\gamma(N; \mu) = \int_{-\pi}^{\pi} \frac{\sin^2 N \dfrac{\lambda - \mu}{2}}{\sin^2 \dfrac{\lambda - \mu}{2}} f(\lambda)\, d\lambda,$$

given below in the present section.*

We shall prove Theorem 5, using the analysis mentioned above, as well as Theorem 6 in Section V.4.

To investigate $\gamma(N; \mu)$ we note first that

$$\gamma(N; \mu) = \mathsf{M} \left| \sum_0^{N-1} e^{i\mu t} \xi(t) \right|^2. \tag{3.4}$$

In fact, if the spectral representation of the sequence $\xi(t)$ is

$$\xi(t) = \int_{-\pi}^{\pi} e^{i\lambda t} \Phi(d\lambda)$$

(where $\Phi(d\lambda)$ is an orthogonal random measure, $\mathsf{M}|\Phi(d\lambda)|^2 = f(\lambda)\, d\lambda$), then

$$\mathsf{M} \left| \sum_0^{N-1} e^{i\mu t} \xi(t) \right|^2 = \mathsf{M} \left| \int_{-\pi}^{\pi} \frac{e^{iN(\lambda - \mu)} - 1}{e^{i(\lambda - \mu)} - 1} \Phi(d\lambda) \right|^2$$

$$= \int_{-\pi}^{\pi} \frac{\sin^2 N \dfrac{\lambda - \mu}{2}}{\sin^2 \dfrac{\lambda - \mu}{2}} f(\lambda)\, d\lambda.$$

* Our study of the function $\gamma(N; \mu)$ follows essentially V. P. Leonov, "On the variance of the time-dependent means of a stationary stochastic process," *Teoria Verojatn. i Primen.* **6** (1961), 87–93 (English translation).

Lemma 4. *As $N \to \infty$ the value*

$$\gamma(N; 0) = \gamma(N) = \mathsf{M}\left|\sum_{1}^{N} \xi(t)\right|^2 = \int_{-\pi}^{\pi} \frac{\sin^2 \dfrac{N\lambda}{2}}{\sin^2 \dfrac{\lambda}{2}} f(\lambda)\, d\lambda$$

either tends to ∞ or is bounded, this value being bounded if and only if

$$\int_{-\pi}^{\pi} \frac{f(\lambda)}{\sin^2 \dfrac{\lambda}{2}}\, d\lambda < \infty. \tag{3.5}$$

Proof. Let U denote the unitary operator on the space $H = H(-\infty, \infty)$ corresponding to a sequence $\xi(t)$: $U(t) = \xi(t + 1)$. We regard the sums $S_N = \sum_0^{N-1} \xi(t)$ as elements of the space H. The assertion $\underline{\lim}_N \mathsf{M}|S_N|^2 < \infty$ implies that for some subsequence $N_j \, \|S_{N_j}\| = \mathsf{M}^{1/2}|S_{N_j}|^2 < C < \infty$. Since a bounded sphere in the Hilbert space H is weakly compact we can derive from $\{N_j\}$ a subsequence $\{n_k\}$ such that S_{n_k} is weakly convergent to $-\eta \in H$, that is, for all $\zeta \in H$

$$\lim_{n_k} (S_{n_k}, \zeta) = \lim_{n_k} \mathsf{M}S_{n_k}\bar{\zeta} = -(\eta, \zeta) = -\mathsf{M}\eta\bar{\zeta}.$$

Then

$$\lim_{n_k} (US_{n_k}, \zeta) = -(U\eta, \zeta),$$

and therefore, for all $\xi \in H$,

$$\begin{aligned}(U\eta - \eta, \zeta) &= \lim_{n_k} (S_{n_k} - US_{n_k}, \zeta)\\ &= (\xi(0), \zeta) - \lim_{n_k} (\xi(n_k), \zeta)\\ &= \mathsf{M}\xi(0)\bar{\zeta} - \lim_{n_k} \mathsf{M}\xi(n_k)\bar{\zeta}.\end{aligned}$$

Because of the isometry between H and $L(f)$ the random variable ζ is associated with a function $\varphi \in L(f)$ such that

$$\mathsf{M}\xi(n_k)\bar{\xi} = \int_{-\pi}^{\pi} e^{i\lambda n_k}\overline{\varphi(\lambda)} f(\lambda)\, d\lambda. \tag{3.6}$$

It is obvious that $\bar{\varphi} f \in \mathscr{L}^1(-\pi, \pi)$, and by the Riemann–Lebesgue theorem the integral in the right-hand side of (3.6) vanishes as $n_k \to \infty$. Hence if $\underline{\lim}\, \gamma(N) < \infty$, for all $\zeta \in H$

$$\mathsf{M}(U\eta - \eta)\bar{\zeta} = \mathsf{M}\xi(0)\bar{\zeta},$$

and therefore

$$\xi(0) = U\eta - \eta = \eta(1) - \eta(0),$$

where $\eta(0) = \eta$, $\eta(k) = U^k\eta$. In this case $\xi(t) = \eta(t + 1) - \eta(t)$ and for all N

$$S_N = \eta(N) - \eta(0),$$
$$\gamma(N) = \mathsf{M}|S_N|^2 = \mathsf{M}|\eta(N) - \eta(0)|^2 \leqslant 4\mathsf{M}|\eta|^2 < \infty.$$

Thus we have proved that either $\gamma(N) \to \infty$ or $\sup_N \gamma(N) < \infty$, the last inequality holding only if $\xi(t)$ can be expressed as the difference

$$\xi(t) = U^{t+1}\eta - U^t\eta = \eta(t+1) - \eta(t), \qquad \eta \in H. \tag{3.7}$$

Obviously, it follows from (3.7) that $\sup_N \gamma(N) \leqslant 4\mathsf{M}|\eta^2| < \infty$.

It remains to verify that (3.7) is equivalent to (3.5). Let (3.7) be satisfied. We denote by $f_\eta(\lambda)$ the spectral density of a sequence $\eta(t)$. By virtue of (3.7) we have

$$f_\eta(\lambda) = \frac{f(\lambda)}{|e^{i\lambda} - 1|^2} = \frac{f(\lambda)}{4\sin^2\dfrac{\lambda}{2}};$$

therefore

$$\int_{-\pi}^{\pi} \frac{f(\lambda)}{\sin^2\dfrac{\lambda}{2}}\, d\lambda = 4\int_{-\pi}^{\pi} f_\eta(\lambda)\, d\lambda < \infty.$$

Conversely, if the integral given by (3.5) is finite, by the definition of $\gamma(N)$

$$\gamma(N) \leqslant \int_{-\pi}^{\pi} \frac{f(\lambda)}{\sin^2\dfrac{\lambda}{2}}\, d\lambda < \infty,$$

thus proving the lemma. □

Lemma 5. *The function $\gamma(N, \mu)$ either tends to ∞ or is bounded for each μ as $N \to \infty$. The limit function is bounded if and only if*

$$\int_{-\pi}^{\pi} \frac{f(\lambda)}{\sin^2\dfrac{\lambda - \mu}{2}}\, d\lambda < \infty.$$

Proof. Writing the function $\gamma(N; \mu)$ as

$$\gamma(N; \mu) = \int_{-\pi}^{\pi} \frac{\sin^2\dfrac{N\lambda}{2}}{\sin^2\dfrac{\lambda}{2}} f(\lambda + \mu)\, d\lambda,$$

we convince ourselves that $\gamma(N; \mu)$ is equivalent to $\mathsf{M}|\sum_1^n \xi'(t)|^2$, where the wide-sense stationary sequence $\{\xi'(t)\}$ has spectral density $f(\lambda + \mu)$. □

Lemma 4 implies the following.

Lemma 6. *As $N \to \infty$ either $\inf_\mu \gamma(N, \mu) \to \infty$, or there exists a point $\theta \in [-\pi, \pi]$ such that*

$$\int_{-\pi}^{\pi} \frac{f(\lambda)}{\sin^2\dfrac{\lambda - \theta}{2}}\, d\lambda < \infty.$$

Proof. The proof of Lemma 6 differs slightly from that of Lemma 4. Let $\underline{\lim}_N \inf_\mu \gamma(N, \mu) < \infty$. We can distinguish a sequence N_k and a convergent sequence $\{\theta_k\}$ of points from $[-\pi, \pi]$ with a limit θ so that

$$\lim_k \gamma(N_k; \theta_k) < \infty.$$

Assuming $S_k = \sum_0^{N_k - 1} e^{it\theta_k}\xi(t)$ and arguing as in the proof of Lemma 4, we arrive at the existence of a limit element $\eta \in H$ such that for all $\zeta \in H$

$$\lim_k \mathsf{M} S_k \bar{\zeta} = M\eta\bar{\zeta},$$

$$\lim_k \mathsf{M} e^{-i\theta_k} U S_k \bar{\zeta} = e^{-i\theta} \mathsf{M} U S_k \bar{\zeta}.$$

As above, we deduce from these equalities the representation

$$\xi(t) = \eta(t) - e^{-i\theta}\eta(t + 1),$$

from which the assertion of the lemma follows. □

Lemma 7. *The spectral density $f(\lambda)$ of the sequence $\{\xi(t)\}$ satisfying the complete regularity condition can be represented as*

$$f(\lambda) = w(\lambda)|P(e^{i\lambda})|^2, \tag{3.8}$$

where $P(z)$ is a polynomial with roots on $|z| = 1$, and where $w(\lambda)$ has the property that

$$\lim_N \inf_\mu \int_{-\pi}^{\pi} \frac{\sin^2 N \dfrac{\lambda - \mu}{2}}{\sin^2 \dfrac{\lambda - \mu}{2}} w(\lambda)\, d\lambda = \infty.$$

Proof. Let

$$\sigma^2(n) = \min \mathsf{M}|\xi(0) - \eta_n|^2, \tag{3.9}$$

where min is taken over all $\eta_n \in H(|t| > n)$. Denote by η_n^* a random variable from $H(|t| > n)$ on which a minimum is attained in (3.9). It is seen that

$$\mathsf{M}|\xi(0)|^2 + \mathsf{M}|\eta_n^*|^2 - 2\sqrt{2}\rho(n)\mathsf{M}^{1/2}|\xi(0)|^2\mathsf{M}^{1/2}|\eta_n^*|^2 \leqslant \sigma^2(n) \leqslant \mathsf{M}|\xi(0)|^2. \tag{3.10}$$

Hence for large n we have

$$\sigma^2(n) \geqslant \frac{1}{2}\mathsf{M}|\xi(0)|^2 > 0, \tag{3.11}$$

that is, the process $\xi(t)$ is noninterpolatable, which fact implies* existence of a polynomial $Q(z)$ such that

$$\int_{-\pi}^{\pi} \frac{|Q(e^{i\lambda})|^2}{f(\lambda)}\, d\lambda < \infty. \tag{3.12}$$

* See, for instance, [24], p. 142.

Let us denote by $Q_0(z)$ a polynomial $Q(z)$ of the smallest degree satisfying (3.12) and having coefficient one for the term of highest degree. (3.12) implies also that among the polynomials $Q(z)$ with real roots for which $\int_{-\pi}^{\pi} (f(\lambda)/|Q(e^{i\lambda})|^2)\, d\lambda < \infty$, there exists a polynomial $P(z)$ of maximum (finite) degree. In fact, from the inequality

$$\left(\int_{-\pi}^{\pi} \left|\frac{Q_0(e^{i\lambda})}{P(e^{i\lambda})}\right|^2 d\lambda\right)^2 \leqslant \int_{-\pi}^{\pi} \frac{|Q_0(e^{i\lambda})|^2}{f(e^{i\lambda})}\, d\lambda \int_{-\pi}^{\pi} \frac{f(e^{i\lambda})}{|P(e^{i\lambda})|^2}\, d\lambda < \infty$$

it follows that the polynomial Q_0 is divisible by P.

We next set $f(\lambda) = |P(e^{i\lambda})|^2 w(\lambda)$ and assume that

$$\liminf_{N \quad \mu} \int_{-\pi}^{\pi} \frac{\sin^2 N \dfrac{\lambda - \mu}{2}}{\sin^2 \dfrac{\lambda - \mu}{2}}\, w(\lambda)\, d\lambda < \infty.$$

By Lemma 6 there is a point $\theta \in [-\pi, \pi]$ for which

$$\int_{-\pi}^{\pi} \frac{w(\lambda)}{|1 - e^{i(\lambda - \theta)}|^2}\, d\lambda = \int_{-\pi}^{\pi} \frac{f(\lambda)\, d\lambda}{|P(e^{i\lambda})(1 - e^{i(\lambda - \theta)})|^2} < \infty.$$

The last inequality obviously contradicts the assumption that the polynomial $P(z)$ is maximal. The lemma is proved.

Lemma 8. *If $f(\lambda)$ is the spectral density of the sequence $\{\xi(t)\}$ satisfying the complete regularity condition, then at the points μ where $\lim_\mu \gamma(N;\mu) = \infty$, the function $\gamma(N;\mu)$ is representable as $\gamma(N;\mu) = Nh(N;\mu)$, where $h(N;\mu)$ is a function N varying slowly (in the sense of Karamata), that is, for all $k > 0$*

$$\lim_{N \to \infty} \frac{h(kN;\mu)}{h(N;\mu)} = 1.$$

Proof. For the sake of brevity let

$$\gamma(N;\mu) = \gamma(N), \qquad h(N;\mu) = h(N).$$

Further, let

$$z_j = \sum_{s=1}^{N} \xi[(j-1)N + (j-1)r + s] \exp\{-i\mu[(j-1)N + (j-1)r + s]\}, \qquad j = 1, \ldots, k;$$

$$y_j = \sum_{s=1}^{r} \xi[jN + (j-1)r + s] \exp\{-i\mu[jN + (j-1)r + s]\}, \qquad j = 1, \ldots, k-1; \tag{3.13}$$

$$y_k = -\sum_{s=1}^{(k-1)r} \xi(Nk + s) \exp\{-i\mu(Nk + s)\},$$

where $r = r(N) \to \infty$ as $N \to \infty$, but slowly so that

$$\frac{\gamma(r)}{\gamma(N)} \to 0, \qquad \frac{\gamma((k-1)r)}{\gamma(N)} \to 0.$$

Then

$$\gamma(N) = \mathsf{M}|z_1 + y_1 + \ldots + z_k + y_k|^2$$
$$= \sum_{j=1}^{k} \mathsf{M}|z_j|^2 + \sum_{i,j} (\mathsf{M}z_i\bar{y}_j + \mathsf{M}\bar{z}_i y_j) + \sum_{i \neq j} \mathsf{M}z_i\bar{z}_j + \sum_{i,j} \mathsf{M}y_i\bar{y}_j. \quad (3.14)$$

The first sum to the right in (3.14) is equal to $k\gamma(N)$ because it is stationary; the second sum and the fourth sum in (3.14) do not exceed $k^2(\gamma(N)\gamma(r))^{1/2} = o(\gamma(N))$; the third sum in (3.14) does not exceed $k^2\gamma(N)\rho(r) = o(\gamma(N))$ because of the complete regularity condition. Thus, $\gamma(kN) = k\gamma(N)(1 + o(1))$, thus proving the lemma. □

We shall show further that the function $h(N)$ can be extended from integers to all positive numbers while retaining the property of slowly varying. Such extension is not, in general, possible for arbitrary (slowly varying) functions on integers. We note first the properties of $h(N)$ making this extension possible. In deducing these properties we assume $\mu = 0$. We consider the case $\mu \neq 0$ in a similar way; we need only to replace the variables $\xi(t)$ by $\xi(t)e^{-i\mu t}$.

1. If k is fixed and $N \to \infty$, then $\lim(h(N + k)/h(N)) = 1$. In fact, since $\gamma(N) \to \infty$ as $N \to \infty$, then

$$\begin{aligned} \gamma(N + k) &= \gamma(N) + \gamma(k) + 2\mathsf{M}(\xi(1) + \ldots + \xi(N))(\bar{\xi}(N + 1) + \ldots \\ &\quad + \bar{\xi}(N + k)), \\ |(\xi(1) + \ldots &+ \xi(N))(\bar{\xi}(N + 1) + \ldots + \bar{\xi}(N + k))| \leqslant (\gamma(N)\gamma(k))^{1/2}, \end{aligned} \quad (3.15)$$

and

$$\begin{aligned} \frac{\gamma(N + k)}{\gamma(N)} &= \frac{N + k}{N}\frac{h(N + k)}{h(N)} = 1 + o(1), \\ \frac{h(N + k)}{h(N)} &= 1 + o(1) \end{aligned} \quad (3.16)$$

2. For all $\varepsilon > 0$, $\lim_{N\to\infty} N^{\varepsilon}h(N) = \infty$ and $\lim_{N\to\infty} N^{-\varepsilon}h(N) = 0$. In fact, making use of the relation $h(2N) \sim h(N)$ and the property mentioned above, we have

$$\ln h(N) = \sum_j \ln \frac{h\left(\left[\frac{N}{2^j}\right]\right)}{h\left(\left[\frac{N}{2^{j+1}}\right]\right)} = o(\ln N).$$

3. If $r \leqslant p \leqslant 2r$, for sufficiently large r we have

$$\sup_p \frac{h(p)}{h(r)} \leqslant 4.$$

We fix m chosen large so that $\rho(m) < 1/16$. Assume that $p > 3r/2$ (the case $p \leqslant 3r/2$ is treated in a similar way). From the representation

$$\sum_1^{p+m} \xi(t) = \sum_1^{r} \xi(t) + \sum_{r+1}^{r+m} \xi(t) + \sum_{r+m+1}^{p+m} \xi(t)$$

we find that

$$(p + m)h(p + m) = rh(r) + (p - r)h(p - r) + \theta,$$

where

$$\begin{aligned} |\theta| \leqslant 2[\rho(m)(r(p - r)h(r)h(p - r))^{1/2} &+ (rmh(r)h(m))^{1/2} \\ &+ ((p - r)mh(m)h(p - r))^{1/2}] + mh(m). \end{aligned}$$

Obviously,

$$2(r(p - r)h(r)h(p - r))^{1/2} \leqslant rh(r) + (p - r)h(p - r).$$

Hence for large r we have $h(p + m) = \theta_1 h(r) + \theta_2 h(p - r) + R_p$, where $R_p = O(p^{-1/4})$, $\theta_1 > 15/32$ and $\theta_2 > 0$; therefore, for large r (see the property given above),

$$\theta_1 \frac{h(r)}{h(r + m)} < \frac{3}{2}, \qquad \theta_1 \frac{h(r)}{h(p)} < \frac{4}{3}, \qquad \frac{h(r)}{h(p)} < 4.$$

4. For all sufficiently small c and sufficiently large N we have $h(cN)/h(N) < c^{-1/2}$.

Naturally, we take into account only those N for which cN is an integer. Using properties (1) and (2) we have

$$\begin{aligned} \ln \frac{h(cN)}{h(N)} \sum_{k=0}^{[-\ln c/\ln 2]} \left(\ln h\left(\left[\frac{N}{2^{k+1}}\right]\right) - \ln h\left(\left[\frac{N}{2^k}\right]\right)\right) \\ + \ln h(cN) - \ln h\left(\left[\frac{N}{2^{[-\ln c/\ln 2]}}\right]\right) < \frac{1}{2} \ln|c| = -\frac{1}{2} \ln c. \end{aligned} \tag{3.17}$$

Let us note that the inequality we have proved above holds uniformly over all sufficiently small c, $c < c_0$. We extend next the functions $\gamma(N)$ and $h(N)$ to all positive x, assuming

$$\begin{aligned} \gamma(x; \mu) &= \int_{-\pi}^{\pi} \frac{\sin^2 \dfrac{x\lambda}{2}}{\sin^2 \dfrac{\lambda}{2}} f(\lambda + \mu)\, d\lambda, \\ h(x; \mu) &= \frac{1}{x} \gamma(x, \mu). \end{aligned} \tag{3.18}$$

Lemma 9. *For all μ for which $\gamma(N) \to \infty$, the function $h(x)$ is a function varying slowly, that is, for all $y > 0$*

$$\lim_{x\to\infty} \frac{h(xy)}{h(x)} = 1. \tag{3.19}$$

Proof. We shall prove (3.19) first for rational y. It is seen that for $x \to \infty$

$$\gamma(x) = \gamma([x])(1 + o(1)). \tag{3.20}$$

Hence for $y = k$, where k is an integer, by property (1) of the function $h(N)$ we have

$$\frac{\gamma(kx)}{\gamma(x)} = \frac{[kx]}{[x]} \frac{h([kx])}{h([t])} (1 + o(1)) = k(1 + o(1)). \tag{3.21}$$

If $y = p/q$, where p, q are integers, we find, taking into account (3.21), that

$$\lim_{x\to\infty} \frac{\gamma\left(\frac{p}{q} x\right)}{\gamma(x)} = \lim \frac{\gamma\left(p \frac{x}{q}\right)}{\gamma\left(\frac{x}{q}\right)} \frac{\gamma\left(\frac{x}{q}\right)}{\gamma\left(q \frac{x}{q}\right)} = \frac{p}{q}. \tag{3.22}$$

Further, let y be an arbitrary real number, and let

$$\psi_1(y) = \underline{\lim}_{x\to\infty} \frac{\gamma(x, y)}{\gamma(x)}, \qquad \psi_2(y) = \overline{\lim}_{x\to\infty} \frac{\gamma(x, y)}{\gamma(x)}$$

For rational y, $\psi_1(y) = \psi_2(y) = y$. Hence it suffices to show that the functions $\psi_1(y)$, $\psi_2(y)$ are continuous. Since

$$\frac{|\gamma((y + \varepsilon)x) - \gamma(yx)|}{\gamma(x)}$$

$$\leqslant \frac{1}{\gamma(x)} \int_{-\pi}^{\pi} \frac{\sin^2 \frac{\varepsilon x \lambda}{2}}{\sin^2 \frac{\lambda}{2}} f(\lambda + \mu)\, d\lambda + \frac{1}{2\gamma(x)} \left| \int_{-\pi}^{\pi} \frac{\sin \varepsilon x \lambda \sin yx\lambda}{\sin^2 \frac{\lambda}{2}} f(\lambda + \mu)\, d\lambda \right|$$

$$\leqslant \frac{\gamma(\varepsilon x)}{\gamma(x)} + \frac{1}{2} \left(\frac{\gamma(yx)}{\gamma(x)}\right)^{1/2} \left(\frac{\gamma(\varepsilon x)}{\gamma(x)}\right)^{1/2}, \tag{3.23}$$

it suffices to prove that $\psi_1(y)$ and $\psi_2(y)$ are continuous at zero. Applying property (3) of the function $h(N)$ we have

$$\frac{\gamma(\varepsilon x)}{\gamma(x)} = \frac{[\varepsilon x]}{[x]} \frac{h\left(\frac{[\varepsilon x]}{[x]} [x]\right)}{h([x])} (1 + o(1)) \leqslant \varepsilon^{1/2}(1 + o(1)). \tag{3.24}$$

It follows from (3.23) and (3.24) that

$$|\psi_1(y+\varepsilon)-\psi_1(y)| = O(\varepsilon^{1/4}), \qquad |\psi_2(y+\varepsilon)-\psi_2(y)| = O(\varepsilon^{1/4}),$$

thus proving the lemma.

Lemma 10. *If* $\gamma_0(N) = \inf_\mu \gamma(N;\mu) \xrightarrow[N\to\infty]{} \infty$, *the relation*

$$\lim_{x\to\infty} \frac{h(xy)}{h(x)} = 1$$

holds uniformly over all μ *and* y *such that* $0 < y_0 < y < y_1 < \infty$.

Proof. Let $h(xy)/h(x) = 1 + R(\mu; y; x)$. First let $x = N$ and $y = k$ be integers. It follows from the proof of Lemma 8 that

$$R(\mu; k; N) = O\left[k\rho(r) + k\left(\frac{\gamma(r;\mu)}{\gamma_0(N)}\right)^{1/2} + k\left(\frac{\gamma((k-1)r;\mu)}{\gamma_0(N)}\right)^{1/2}\right].$$

For all μ, $\gamma(N;\mu) < N^2 \int_{-\pi}^{\pi} f(\lambda)\,d\lambda$. Hence, letting $r = \ln\gamma_0(N)$ we have that

$$R(\mu; k; N) = O\left[k\rho(r) + \left(\frac{\ln^2\gamma_0(N)}{\gamma_0(N)}\right)^{1/2}\right].$$

Further, we derive from the proof of (3.16) that

$$\frac{\gamma(N+k;\mu)}{\gamma(N;\mu)} = 1 + O\left(\frac{k^2}{\gamma_0(N)}\right),$$

from which it follows that uniformly in μ

$$\ln h(N;\mu) = \sum\left(\ln h\left(\left[\frac{N}{2^{k-1}}\right];\mu\right) - \ln h\left(\left[\frac{N}{2^k}\right];\mu\right)\right) + O(1) = o(\ln N);$$

therefore, $\lim_N h(N;\mu)N^{-\varepsilon} = 0$ uniformly in μ. From what has been said above we easily obtain that properties given by (3) and (4) of the function $h(N;\mu)$, and the equalities given by (3.20) and (3.24) hold uniformly in μ.

Next, let $y \in [y_0, y_1]$. Suppose an integer q is given. Uniformly in μ and p, such that $y_0 < p/q < 1_1$,

$$\lim_x \frac{h(px/q)}{h(x)} = 1.$$

Let y' be a number of the form p/q nearest to y. It follows from (3.23) and (3.24) that

$$\left|\frac{\gamma(yx)}{\gamma(x)} - y\right| \leqslant \left|\frac{\gamma(y'x)}{\gamma(x)} - y'\right| + |y'-y| + \left|\frac{\gamma(yx)}{\gamma(x)} - \frac{\gamma(y'x)}{\gamma(x)}\right| \leqslant Cq^{-1/4} + o(1),$$

where the constant C and $o(1)$ are independent of μ, q. The lemma is proved. □

V.4 Local Conditions (continued)

We now proceed to prove Theorem 5. This theorem can be regarded as an assertion of the Tauberian type in which, we draw conclusions about the behavior of

$$\frac{1}{x}\int_0^x f(\lambda+\mu)\,d\lambda \qquad \text{as } x \to 0$$

on the basis of the behavior of the function

$$\gamma(N;\mu) = \int_{-\pi}^{\pi} \frac{\sin^2 N\dfrac{\lambda-\mu}{2}}{\sin^2\dfrac{\lambda}{2}} f(\lambda)\,d\lambda \qquad \text{as } N \to \infty.$$

For the proof of this theorem we shall use Karamata's method for proving theorems of the Tauberian type.

Before proving the theorem we show that for all μ the function $(\gamma(N;\mu)$ can be expressed as

$$\gamma(N;\mu) = N\cdot h(N;\mu),$$

where $h(N;\mu)$ is a slowly varying function, which fact follows from Lemma 7 and Lemma 11.

Lemma 11. *If $f(\lambda)$ is the spectral density of a sequence $\{\xi(t)\}$ satisfying the complete regularity condition, and if each root of the polynomial $P(z)$ is equal to* 1, *and if further the function $w(\lambda) = f(\lambda)/|P(e^{i\lambda})|^2$ is integrable on $[-\pi,\pi]$, then $w(\lambda)$ is the spectral density of a sequence $\{\eta(t)\}$ satisfying the complete regularity condition, and $\rho(\tau; w) \leqslant \rho(\tau, f)$.*

Proof. Let the functions $\varphi, \psi \in \mathscr{H}^2$. If the polynomial $P(z)$ has roots only on the circle $|z| = 1$, then $\varphi/P \in \mathscr{H}^\delta$ and $\psi/\bar{P} \in \mathscr{H}^\delta$ for sufficiently small δ. Hence (1.3) yields

$$\begin{aligned}\rho(\tau; w) &= \sup_{\varphi,\psi}\left|\int_{-\tau}^{\tau}\varphi(\lambda)\psi(\lambda)e^{i\lambda\tau}w(\lambda)\,d\lambda\right| \\ &= \sup_{\varphi,\psi}\left|\int_{-\tau}^{\tau}\frac{\varphi(\lambda)}{P(\lambda)}\frac{\psi(\lambda)}{\bar{P}(\lambda)}e^{i\lambda\tau}w(\lambda)|P(e^{i\lambda})|^2\,d\lambda\right| \\ &\leqslant \sup_{\varphi,\psi}\left|\int_{-\tau}^{\tau}\varphi(\lambda)\psi(\lambda)e^{i\lambda\tau}f(\lambda)\,d\lambda\right| = \rho(\tau; f),\end{aligned}$$

the supremum being taken over all $\varphi, \psi \in L^+(w)$ with $\|\varphi\|_w = \|\psi\|_w = 1$ in the first integral, and over all $\varphi, \psi \in L^+(f)$ with $\|\varphi\|_f = \|\psi\|_f = 1$ in the last integral. The lemma is proved. □

Here and through the proof of Lemma 14 we shall assume that $f(\lambda)$ denotes the spectral density of a stationary sequence $\{\xi(t)\}$ satisfying the complete regularity condition with regularity coefficient $\rho(\tau)$. Furthermore, we shall assume $f(\lambda)$ to be such that $\lim_N \inf_\mu \gamma(N;\mu) = \infty$.

Lemma 12. *Let $a(\lambda)$ be an even function with a bounded third derivative vanishing outside of the interval $[-1, 1]$. Then, as $x \to \infty$ uniformly in μ*

$$\frac{1}{x}\int_{-\pi}^{\pi} \frac{\sin^2 \frac{x\lambda}{2}}{\sin^2 \frac{\lambda}{2}} f(\lambda+\mu)a(x\lambda)\,d\lambda = \frac{1}{\pi}\int_0^1 \frac{\sin^2 \frac{\lambda}{2}}{\left(\frac{\lambda}{2}\right)^2} a(\lambda)\,d\lambda \cdot h(x)(1+o(1)). \tag{4.1}$$

Proof. Let $|a'''(\lambda)| \leqslant C < \infty$. Next, let

$$\begin{aligned} a(\lambda) &= \sqrt{\frac{2}{\pi}}\int_0^\infty \cos \lambda z A(z)\,dz, \\ A(z) &= \sqrt{\frac{2}{\pi}}\int_0^\infty \cos \lambda z a(\lambda)\,d\lambda. \end{aligned} \tag{4.2}$$

We have $|A(z)| \leqslant C/(|z|^3 + 1)$ because $a(\lambda)$ is smooth. Given $\varepsilon > 0$ we choose s and δ such that

$$\int_s^\infty |A(z)|\,dz + \int_{|1-z|\leqslant\delta} |A(z)|\,dz + \int_0^\delta |A(z)|\,dz < \varepsilon. \tag{4.3}$$

Let $B = \{z\colon \delta \leqslant z \leqslant s, |1-z| > \delta\}$. Then

$$\begin{aligned} &\frac{1}{x}\int_{-\pi}^{\pi} \frac{\sin^2 \frac{x\lambda}{2}}{\sin^2 \frac{\lambda}{2}} f(\lambda+\mu)a(x\lambda)\,d\lambda \\ &= \frac{1}{x}\sqrt{\frac{2}{\pi}}\int_{-\pi}^{\pi} \frac{\sin^2 \frac{x\lambda}{2}}{\sin^2 \frac{\lambda}{2}} f(\lambda+\mu)\,d\lambda \int_0^\infty \cos x\lambda z \cdot A(z)\,dz \\ &= \sqrt{\frac{2}{\pi}}\int_B A(z)\,dz\,\frac{1}{x}\int_{-\pi}^{\pi} \frac{\sin^2 \frac{x\lambda}{2}}{\sin^2 \frac{\lambda}{2}} \cos x\lambda z \cdot f(\lambda+\mu)\,d\lambda + R(x), \end{aligned} \tag{4.4}$$

where $|R(x)| \leqslant \varepsilon h(x)$. Using the identity

$$\cos x\lambda z \sin^2 \frac{x\lambda}{2} = \frac{1}{2}\left[\sin^2 \frac{x\lambda}{2}(z+1) + \sin^2 \frac{x\lambda}{2}(z-1) - 2\sin^2 \frac{x\lambda}{2} z\right],$$

we rewrite the right-hand side of (4.4) as

$$\sqrt{\frac{1}{2\pi}}\left[\int_{1+\delta}^{s} z[h(x(z+1)) + h(x(z-1)) - 2h(xz)]A(z)\,dz\right.$$

$$+ \int_{1+\delta}^{s} [h(x(z+1)) - h(x(z-1))]A(z)\,dz$$

$$+ \int_{\delta}^{1-\delta} z[h(x(z+1)) - h(x(1-z)) - 2h(xz)]A(z)\,dz$$

$$\left. + \int_{\delta}^{1-\delta} [h(x(1+z)) + h(x(1-z))]A(z)\,dz\right] + R(x). \tag{4.5}$$

Noting that $\lim(h(xz)/h(z)) = 1$ uniformly in $\mu \in [-\pi, \pi]$, $\delta \leqslant z \leqslant s+1$ (Lemma 10), we conclude that the first integral and the second integral in (4.5) are $o(h(x))$ uniformly in μ. The sum of the remaining terms, with the exception of $R(x)$, becomes

$$\sqrt{\frac{2}{\pi}} \int_{\delta}^{1-\delta} (1-z)A(z)\,dz \cdot h(x)(1+o(1))$$

uniformly in μ, as $x \to \infty$.

δ and ε being arbitrary, we see that, uniformly in μ,

$$\frac{1}{x}\int_{-\pi}^{\pi} \frac{\sin^2 \frac{x\lambda}{2}}{\sin^2 \frac{\lambda}{2}} f(\lambda+\mu)a(x\lambda)\,d\lambda$$

$$= \sqrt{\frac{2}{\pi}} \int_0^1 (1-z)A(z)\,dz \cdot h(x)(1+o(1))$$

$$= \frac{2}{\pi}\int_0^1 a(\lambda)\,d\lambda \int_0^1 (1-z)\cos\lambda z\,dz \cdot h(x)(1+o(1) \tag{4.6}$$

$$= \frac{1}{\pi}\int_0^1 \frac{\sin^2 \frac{\lambda}{2}}{\left(\frac{\lambda}{2}\right)^2} a(\lambda)\,d\lambda \cdot h(x)(1+o(1)). \qquad \square$$

Lemma 13. *Let $a(\lambda)$ be an odd thrice-differentiable function with a bounded third derivative vanishing outside of the interval $[-1, 1]$. Then as $x \to \infty$,* *uniformly in* μ,

$$\frac{1}{x}\int_{-\pi}^{\pi}\frac{\sin^2\frac{x\lambda}{2}}{\sin^2\frac{\lambda}{2}}f(\lambda+\mu)a(x\lambda)\,d\lambda = o(h(x)). \tag{4.7}$$

Proof. We assume first that $x \to \infty$, with x having integral values only, and write N instead of x to indicate this. As above, we introduce here the Fourier transform of the function $a(\lambda)$:

$$\begin{aligned} A(z) &= \sqrt{\frac{2}{\pi}}\int_0^\infty \sin\lambda z a(\lambda)\,d\lambda, \\ a(\lambda) &= \sqrt{\frac{2}{\pi}}\int_0^\infty \sin\lambda z A(z)\,dz. \end{aligned} \tag{4.8}$$

Given $\varepsilon > 0$ we define the set

$$B_1 = \{z \geqslant s\} \bigcup \{z\colon |z-n| \leqslant \delta, n = 0, 1, \ldots\}$$

so that $\int_{B_1}|A(z)|\,dz < \varepsilon$. Let us denote by B the complement of the set B_1 on the half-line $[0, \infty]$. Then

$$\frac{1}{N}\int_{-\pi}^{\pi}\frac{\sin^2\frac{N\lambda}{2}}{\sin^2\frac{\lambda}{2}}f(\lambda+\mu)a(N\lambda)\,d\lambda$$

$$= \frac{1}{N}\int_B A(z)\,dz\int_{-\pi}^{\pi}\frac{\sin^2\frac{N\lambda}{2}}{\sin^2\frac{\lambda}{2}}\sin N\lambda z f(\lambda+\mu)\,d\lambda + R(N), \qquad |R(N)| \leqslant \varepsilon h(N). \tag{4.9}$$

For integral N,

$$\frac{\sin^2\frac{N\lambda}{2}}{\sin^2\frac{\lambda}{2}} = \left|\sum_0^{N-1} e^{ij\lambda}\right|^2.$$

Expanding $\sin N\lambda z$ in powers of $e^{i\lambda}$ we find that

$$\begin{aligned} \sin N\lambda z &= \frac{\sin Nz\pi}{2\pi i}\sum_{j=1}^{\infty}(-1)^j(e^{ij\lambda}-e^{-ij\lambda})\left(\frac{1}{Nz-j}-\frac{1}{Nz+j}\right) \\ &= \sum_{j=1}^{\infty} a_j(z)(e^{ij\lambda}-e^{-ij\lambda}). \end{aligned} \tag{4.10}$$

We can now rewrite the first term in the right-hand side of (4.9) as

$$\int_B A(z)\,dz\,\frac{1}{N}\int_{-\pi}^{\pi}[\Phi_1(\lambda; z) + \Phi_2(\lambda; z)]f(\lambda + \mu)\,d\lambda$$

$$-\frac{1}{N}\int_B A(z)\,dz\int_{-\pi}^{\pi}[\Phi_1(-\lambda; z) + \Phi_2(-\lambda; z)]f(\lambda + \mu)\,d\lambda$$

$$+\frac{1}{N}\int_B A(z)\,dz\int_{-\pi}^{\pi}\Phi_0(\lambda; z)f(\lambda + \mu)\,d\lambda,$$

where

$$\Phi_1(\lambda; z) = \sum_{k=0}^{N-1} e^{ik\lambda}\sum_{j=1}^{N-1} a_j(z)\sum_{m=1}^{j} e^{im\lambda},$$

$$\Phi_2(\lambda; z) = \sum_{j=1}^{N-1} a_j(z)\sum_{k=N-j}^{N-1} e^{ik\lambda}\sum_{m=0}^{N-j-1} e^{-imk}, \tag{4.11}$$

$$\Phi_0(\lambda; z) = 2i\,\frac{\sin^2\dfrac{N\lambda}{2}}{\sin^2\dfrac{\lambda}{2}}\sum_{j=N}^{\infty} a_j(z)\sin j\lambda.$$

Next we shall estimate the integrals

$$\frac{1}{N}\int_{-\pi}^{\pi}\Phi_j(\lambda; z)f(\lambda + \mu)\,d\lambda.$$

1. Estimation of $(1/N)\int_{-\pi}^{\pi}\Phi_0(\lambda; z)f(\lambda + \mu)\,d\lambda$. We note that

$$|\Sigma(-1)^k\sin k\lambda| \leqslant C$$

for all n, where C is a constant. Taking the Abel sum, we obtain

$$\sum_{j=N}^{\infty} a_j(z)\sin j\lambda = \frac{\sin Nz\pi}{2\pi}\sum_{j=N}^{\infty}\left[\left(\frac{1}{Nz-j} - \frac{1}{Nz-j-1}\right)\right.$$

$$\left.-\left(\frac{1}{Nz+j} - \frac{1}{Nz+j+1}\right)\right]\sum_{k=N}^{j}(-1)^k\sin k\lambda. \tag{4.12}$$

It follows readily from (4.12) that for $z < 1 - \delta$

$$|\Phi_0(\lambda; z)| \leqslant \frac{C}{\delta N}. \tag{4.13}$$

Let $z > 1 + \delta$. We denote by n a large fixed number. Using the Abel sum again we find that for sufficiently large N

$$\left|\sum_{|j-Nz|>n} a_j(z)\sin j\lambda\right| \leqslant C\left(\frac{1}{N\delta} + \frac{1}{n}\right). \tag{4.14}$$

It follows from the complete regularity condition that

$$\left| \frac{1}{N} \int_{-\pi}^{\pi} \frac{\sin^2 \frac{N\lambda}{2}}{\sin^2 \frac{\lambda}{2}} \sum_{|j-Nz| \leqslant n} a_j(z)(e^{ij\lambda} - e^{-ij\lambda}) f(\lambda + \mu)\, d\lambda \right|$$

$$= \left| \sum_{|j-Nz| \leqslant n} \frac{a_j(z)}{N} \int_{-\pi}^{\pi} \left[\left(\sum_{0}^{N-1} e^{ik(\lambda-\mu)} \right)^2 e^{i(j-N+1)(\lambda-\mu)} \right.\right.$$

$$\left.\left. + \left(\sum_{0}^{N-1} e^{-ik(\lambda-\mu)} \right)^2 e^{-i(j-N+1)(\lambda-\mu)} \right] f(\lambda)\, d\lambda \right|$$

$$\leqslant \frac{4n}{N} \rho(N\delta - n) \int_{-\pi}^{\pi} \frac{\sin^2 \frac{N\lambda}{2}}{\sin^2 \frac{\lambda}{2}} f(\lambda + \mu)\, d\lambda = o(h(N)). \tag{4.15}$$

(4.13)–(4.15) yield, uniformly in μ,

$$\frac{1}{N} \left| \int_B A(z)\, dz \int_{-\pi}^{\pi} \Phi_0(\lambda; z) f(\lambda + \mu)\, d\lambda \right| = o(h(N)). \tag{4.16}$$

2. Estimation of $(1/N) \int_{-\pi}^{\pi} \Phi_1(\lambda; z) f(\lambda + \mu)\, d\lambda$. Let us denote by n a large fixed number and let us divide the exterior sum in the formula for $\Phi_1(\lambda; z)$ (see (4.11)) into two sums: $\sum_{k=0}^{n}$ and $\sum_{k=n+1}^{N-1}$. The integral of the first sum does not exceed

$$\frac{n}{N} \int_{-\pi}^{\pi} \left| \sum_{j=1}^{N-1} a_j(z) \sum_{m=0}^{j-1} e^{im\lambda} \right| f(\lambda + \mu)\, d\lambda$$

$$= \frac{n}{2N} \int_{-\pi}^{\pi} \frac{1}{\left| \sin \frac{\lambda}{2} \right|} \left| \sum_{1}^{N} a_j(z)(e^{ij\lambda} - 1) \right| f(\lambda + \mu)\, d\lambda$$

$$\leqslant \frac{Cn}{N} \int_{-\pi}^{\pi} \frac{1}{\left| \sin \frac{\lambda}{2} \right|} \left| \sum_{j=1}^{N-1} a_j(z) \sin j\lambda \right| f(\lambda + \mu)\, d\lambda$$

$$+ \frac{Cn}{N} \int_{-\pi}^{\pi} \frac{1}{\left| \sin \frac{\lambda}{2} \right|} \left| \sum_{j=1}^{N} a_j(z) \sin^2 \frac{j\lambda}{2} \right| f(\lambda + \mu)\, d\lambda. \tag{4.17}$$

Using (4.10) and (4.13)–(4.15), we easily find that the right-hand side of (4.17), uniformly in μ, does not exceed

$$\frac{Cn}{N} \int_{-\pi}^{\pi} \left| \frac{\sin N\lambda z}{\sin \frac{\lambda}{2}} \right| f(\lambda + \mu)\, d\lambda + o(h(N)) \leqslant C \frac{n\sqrt{z}}{\sqrt{N}} h(Nz) + o(h(N)). \tag{4.18}$$

For the second term, taking

$$\int_{-\pi}^{\pi} \Phi_1(\lambda; z) f(\lambda + \mu)\, d\lambda,$$

we have, taking the Abel sum,

$$\frac{1}{N}\int_{-\pi}^{\pi}\left(\sum_{k=n}^{N-1} e^{ik\lambda} \sum_{j=1}^{N-1} a_j(z) \sum_{1}^{j} e^{im\lambda}\right) f(\lambda + \mu)\, d\lambda$$

$$= -\frac{1}{N}\int_{-\pi}^{\pi} e^{in\lambda}\left(\sum_{k=0}^{N-n-1} e^{ik\lambda} \sum_{j=1}^{N-2}\left(\sum_{s=1}^{j} a_s(z)\right) e^{i(j+1)\lambda}\right) f(\lambda + \mu)\, d\lambda$$

$$+ \frac{1}{N}\int_{-\pi}^{\pi} e^{in\lambda}\left(\sum_{k=0}^{N-n-1} e^{ik\lambda} \sum_{m=1}^{N-1} e^{im\lambda} \sum_{j=1}^{N-1} a_j(z)\right) f(\lambda + \mu)\, d\lambda. \tag{4.19}$$

We note that, for $z \leqslant 1 - \delta, j \leqslant Nz$,

$$\sum_{s=1}^{j} a_s(z) = \frac{1}{2\pi i}\sum_{s=1}^{j} \frac{(-1)^s \sin Nz\pi}{Nz - s} + O\left(\frac{1}{N\delta}\right) = O\left(\frac{1}{N\delta} + \frac{|\sin Nz\pi|}{|Nz - j|}\right); \tag{4.20}$$

for $z \leqslant 1 - \delta, j > Nz$,

$$\sum_{s=1}^{j} a_s(z) = \frac{1}{2\pi i}\sum_{-\infty}^{\infty} \frac{(-1)^s \sin Nz\pi}{Nz - s} + O\left(\frac{1}{N\delta} + \frac{1}{|Nz - j|}\right); \tag{4.21}$$

finally, for $z > 1 + \delta$,

$$\sum_{s=1}^{j} a_s(z) = O\left(\frac{1}{N\delta}\right). \tag{4.22}$$

Let $z \leqslant 1 - \delta$; then, using (4.20) and (4.21) and the complete regularity condition, we obtain for the magnitude of the right-hand side of (4.19) the upper bound

$$\frac{C}{N}\left(\frac{1}{\delta} + \ln N\right)\int_{-\pi}^{\pi}\left|\frac{\sin\dfrac{N-n}{2}\lambda}{\sin\dfrac{\lambda}{2}}\right| f(\lambda + \mu)\, d\lambda$$

$$+ \frac{C\theta(z)}{N}\rho(n + Nz)\left(\int_{-\pi}^{\pi}\frac{\sin^2\dfrac{N-n}{2}\lambda}{\sin^2\dfrac{\lambda}{2}} f(\lambda + \mu)\, d\lambda\right)^{1/2}$$

$$\times\left(\int_{-\pi}^{\pi}\frac{\sin^2 N\dfrac{1-z}{2}\lambda}{\sin^2\dfrac{\lambda}{2}} f(\lambda + \mu)\, d\lambda\right)^{1/2}$$

$$= o(h(N)), \tag{4.23}$$

where

$$\theta(z) = \frac{1}{2\pi i} \sum_{-\infty}^{\infty} \frac{(-1)^s \sin Nz\pi}{Nz - s}.$$

Similarly, for $z > 1 + \delta$ we obtain for (4.19) the upper bound

$$\frac{C}{N\delta} \int_{-\pi}^{\pi} \left| \frac{\sin \dfrac{N-n}{2} \lambda}{\sin \dfrac{\lambda}{2}} \right| f(\lambda + \mu)\, d\lambda = o(h(N)). \tag{4.24}$$

It is seen that (4.23) and (4.24) are satisfied uniformly in μ and $z \in B$. From (4.18) and (4.24) we have that uniformly in μ and $z \in B$

$$\frac{1}{N} \left| \int_{-\pi}^{\pi} \Phi_1(\lambda; z) f(\lambda + \mu)\, d\lambda \right| = o(h(N)).$$

3. Estimation of $\int_{-\pi}^{\pi} \Phi_2(\lambda; z) f(\lambda + \mu)\, d\lambda$. Fixing a large number n we obtain

$$\begin{aligned}
&\frac{1}{N} \left| \int_{-\pi}^{\pi} \Phi_2(\lambda; z) f(\lambda + \mu)\, d\lambda \right| \\
&= \frac{1}{N} \left| \int_{-\pi}^{\pi} \left(\sum_{j=1}^{N-1} a_j(z) \sum_{k=0}^{N-1} e^{ik\lambda} \sum_{m=1}^{N-1} e^{im\lambda} \right) f(\lambda + \mu)\, d\lambda \right| \\
&\leqslant \frac{1}{N} \left| \int_{-\pi}^{\pi} \left(\sum_{j=1}^{n} a_j(z) \sum_{k=0}^{j-1} e^{ik\lambda} \sum_{m=1}^{N-j} e^{im\lambda} \right) f(\lambda + \mu)\, d\lambda \right| \\
&\quad + \frac{1}{N} \left| \int_{-\pi}^{\pi} \left(\sum_{j=n+1}^{N-1} a_j(z) \sum_{m=1}^{N-j} e^{im\lambda} \right) \sum_{k=0}^{n} e^{ik\lambda} f(\lambda + \mu)\, d\lambda \right| \\
&\quad + \frac{1}{N} \left| \int_{-\pi}^{\pi} e^{in\lambda} \left(\sum_{j=n+1}^{N-1} a_j(z) \sum_{k=1}^{j-n} e^{ik\lambda} \sum_{m=1}^{N-j} e^{im\lambda} \right) f(\lambda + \mu)\, d\lambda \right|.
\end{aligned} \tag{4.25}$$

It is obvious that the first term in the right-hand side of (4.25) does not exceed

$$\frac{n}{N} \sum_{j=1}^{n} |a_j(z)| \int_{-\pi}^{\pi} \left| \frac{\sin \dfrac{N-j}{2} \lambda}{\sin \dfrac{\lambda}{2}} \right| f(\lambda + \mu)\, d\lambda = O\left(\frac{h(N)}{\sqrt{N}} \right) = o(h(N)). \tag{4.26}$$

Taking the Abel sum of the second and third terms in the right-hand side of (4.25) and arguing in the same way as in deducing the estimates for (4.19), we find that the terms indicated are $o(h(N))$ uniformly in μ and $z \in B$. Thus,

$$\frac{1}{N}\left|\int_{-\pi}^{\pi} \Phi_2(z; \lambda) f(\lambda + \mu)\, d\lambda\right| = o(h(N)). \tag{4.27}$$

We have thus proved Lemma 13 for integral x. Let x now be an arbitrary real number. As was seen above (see Section V.3), $h(x) = h([x])(1 + o(1))$ uniformly in μ. Hence, uniformly in μ,

$$\left|\frac{1}{x}\int_{-\pi}^{\pi} \frac{\sin^2 \frac{x\lambda}{2}}{\sin^2 \frac{\lambda}{2}} f(\lambda + \mu) a([x]\lambda)\, d\lambda - \frac{1}{[x]}\int_{-\pi}^{\pi} \frac{\sin^2 \frac{[x]\lambda}{2}}{\sin^2 \frac{\lambda}{2}} f(\lambda + \mu) a([x]\lambda)\, d\lambda\right|$$

$$\leqslant Cx^{-1/4} = o(h(x)). \tag{4.28}$$

Finally,

$$\left|\frac{1}{x}\int_{-\pi}^{\pi} \frac{\sin^2 \frac{x\lambda}{2}}{\sin^2 \frac{\lambda}{2}} f(\lambda + \mu)(a([x]\lambda) - a(x\lambda))\, d\lambda\right|$$

$$\leqslant \left|\frac{\max_{\lambda} |a'(\lambda)|}{x} \int_{-\pi}^{\pi} \frac{\sin^2 \frac{x\lambda}{2}}{\sin^2 \frac{\lambda}{2}} |\lambda| f(\lambda + \mu)\, d\lambda\right|$$

$$\leqslant C\left(\frac{h(x)}{x}\right)^{1/2} = o(h(x)), \tag{4.29}$$

thus completing the proof of Lemma 13. □

Lemma 14. *Let $a(\lambda)$ be a function of bounded variation vanishing outside of the interval $[-1, 1]$. Then as $x \to \infty$, uniformly in μ,*

$$\frac{1}{x}\int_{-\pi}^{\pi} \frac{\sin^2 \frac{x\lambda}{2}}{\sin^2 \frac{\lambda}{2}} f(\lambda + \mu) a(x\lambda)\, d\lambda = \frac{h(x)}{2\pi}\int_{-1}^{1} \frac{\sin^2 \frac{\lambda}{2}}{\left(\frac{\lambda}{2}\right)^2} a(\lambda)\, d\lambda + o(h(x)). \tag{4.30}$$

Proof. First let the function $a(\lambda)$ have a bounded third derivative. Applying Lemma 12 to the even function $(a(\lambda) + a(-\lambda))/2$, and applying Lemma 13 to the odd function $(a(\lambda) - a(-\lambda))/2$, we obtain

$$\frac{1}{x}\int_{-\pi}^{\pi} \frac{\sin^2 \frac{x\lambda}{2}}{\sin^2 \frac{\lambda}{2}} f(\lambda + \mu) a(x\lambda)\, d\lambda$$

$$= \frac{1}{x}\int_{-\pi}^{\pi} \frac{\sin^2 \frac{x\lambda}{2}}{\sin^2 \frac{\lambda}{2}} f(\lambda + \mu) \frac{a(x\lambda) + a(-x\lambda)}{2}\, d\lambda$$

$$+ \frac{1}{x}\int_{-\pi}^{\pi} \frac{\sin^2 \frac{x\lambda}{2}}{\sin^2 \frac{\lambda}{2}} f(\lambda + \mu) \frac{a(x\lambda) - a(-x\lambda)}{2}\, d\lambda$$

$$= \frac{h(x)}{\pi}\int_{0}^{1} \frac{\sin^2 \frac{\lambda}{2}}{\left(\frac{\lambda}{2}\right)^2} \frac{a(\lambda) + a(-\lambda)}{2}\, d\lambda + o(h(x))$$

$$= \frac{h(x)}{2\pi}\int_{-1}^{1} \frac{\sin^2 \frac{\lambda}{2}}{\left(\frac{\lambda}{2}\right)^2} a(\lambda)\, d\lambda + o(h(x)). \tag{4.31}$$

Next, let $a(\lambda)$ have bounded variation. Given any $\varepsilon > 0$ one can always choose functions $\underline{a}(\lambda)$ and $\overline{a}(\lambda)$ satisfying the conditions of the lemma, having a bounded third derivative and such that

$$\underline{a}(\lambda) \leqslant a(\lambda) \leqslant \overline{a}(\lambda),$$

$$\int_{-\pi}^{\pi} (\overline{a}(\lambda) - \underline{a}(\lambda)) \frac{\sin^2 \frac{\lambda}{2}}{\left(\frac{\lambda}{2}\right)^2}\, d\lambda < \varepsilon. \tag{4.32}$$

In fact, this is obvious for the function $a(\lambda)$ with a finite number of jumps. If the number of jumps in $a(\lambda)$ is infinite, we can select functions $\underline{b}(\lambda)$ and $\overline{b}(\lambda)$ satisfying inequalities of the type given by (4.32) and having only a finite number of jumps.

If the functions $\underline{a}(\lambda)$ and $\overline{a}(\lambda)$ are selected according to (4.32), then

$$\frac{1}{x}\int_{-\pi}^{\pi}\frac{\sin^2\frac{x\lambda}{2}}{\sin^2\frac{\lambda}{2}}f(\lambda+\mu)\underline{a}(x\lambda)\,d\lambda$$

$$\leqslant\frac{1}{x}\int_{-\pi}^{\pi}\frac{\sin^2\frac{x\lambda}{2}}{\sin^2\frac{\lambda}{2}}f(\lambda+\mu)a(x\lambda)\,d\lambda$$

$$\leqslant\frac{1}{x}\int_{-\pi}^{\pi}\frac{\sin^2\frac{x\lambda}{2}}{\sin^2\frac{\lambda}{2}}f(\lambda+\mu)\overline{a}(x\lambda)\,d\lambda. \tag{4.33}$$

We have (4.30) for the functions $\underline{a}(\lambda)$ and $\overline{a}(\lambda)$. Hence (4.33) can be rewritten as

$$\frac{h(x)}{2\pi}\int_{-\pi}^{\pi}\frac{\sin^2\frac{\lambda}{2}}{\left(\frac{\lambda}{2}\right)^2}\underline{a}(\lambda)\,d\lambda(1+o(1))$$

$$\leqslant\frac{1}{x}\int_{-\pi}^{\pi}\frac{\sin^2\frac{\lambda}{2}}{\left(\frac{\lambda}{2}\right)^2}f(\lambda+\mu)a(x\lambda)\,d\lambda$$

$$\leqslant\frac{h(x)}{2\pi}\int_{-\pi}^{\pi}\frac{\sin^2\frac{\lambda}{2}}{\left(\frac{\lambda}{2}\right)^2}\overline{a}(\lambda)\,d\lambda(1+o(1)). \tag{4.34}$$

The extremal terms in (4.34) differ from one another and from

$$\frac{h(x)}{2\pi}\int_{-\pi}^{\pi}\frac{\sin^2\frac{\lambda}{2}}{\left(\frac{\lambda}{2}\right)^2}a(\lambda)\,d\lambda$$

by not more than $\varepsilon h(x)(1+o(1))$, and since ε is arbitrary we can assert that

$$\frac{1}{x}\int_{-\pi}^{\pi}\frac{\sin^2\frac{x\lambda}{2}}{\sin^2\frac{\lambda}{2}}f(\lambda+\mu)a(x\lambda)\,d\lambda=\frac{h(x)}{2\pi}\int_{-1}^{1}\frac{\sin^2\frac{\lambda}{2}}{\left(\frac{\lambda}{2}\right)^2}a(\lambda)\,d\lambda(1+o(1)).$$

Lemma 14 is proved. (Strictly speaking, one must investigate in more detail the function with discontinuities at points ± 1; we shall leave this to the reader.) □

Let us now complete the proof of Theorem 5.

Proof of Theorem 5. Let $f(\lambda)$ be the spectral density of a Gaussian sequence $\xi(t)$ satisfying the complete regularity condition. Using Lemma 7 we can find a maximal polynomial $P(z)$ for which the function $w(\lambda) = f(\lambda)/|P(e^{i\lambda})|^2$ is integrable and

$$\inf_{\mu} \int_{-\pi}^{\pi} \frac{\sin^2 \dfrac{N\lambda}{2}}{\sin^2 \dfrac{\lambda}{2}} w(\lambda + \mu)\, d\lambda \to \infty.$$

By Lemma 11, $w(\lambda)$ is the spectral density of a sequence satisfying the complete regularity condition. Applying Lemma 14 to $w(\lambda)$, setting $a(\lambda) = (\lambda/2)^2/\sin^2(\lambda/2)$ if $|\lambda| \leqslant 1$, and $a(\lambda) = 0$ if $|\lambda| > 1$, we obtain that uniformly in μ

$$\frac{1}{x}\int_{-1/x}^{1/x} w(\lambda + \mu)\, d\lambda = \frac{1}{\pi} h(x)(1 + o(1)). \tag{4.35}$$

Let us apply Lemma 14 again, now choosing however $a(\lambda) = (\lambda/2)^2/\sin^2(\lambda/2)$ for $0 < \lambda \leqslant 1$ and $a(\lambda) = 0$ for $\lambda > 1$; in this case $a(\lambda) = -a(-\lambda)$ if $\lambda < 0$. It follows that uniformly in μ

$$\frac{1}{x}\int_{0}^{1/x} w(\lambda + \mu)\, d\lambda - \frac{1}{x}\int_{-1/x}^{0} w(\lambda + \mu)\, d\lambda = o(h(x)) = o\left(\frac{1}{x}\int_{-1/x}^{1/x} w(\lambda + \mu)\, d\lambda\right). \tag{4.36}$$

If we denote by $W(\lambda)$ the primitive of $w(\lambda)$ and replace $1/x$ by δ, we can write (4.36) in an equivalent form: uniformly in μ as $\delta \to 0$

$$W(\mu + \delta) + W(\mu - \delta) - 2W(\mu) = o(W(\mu + \delta) - W(\mu - \delta)),$$

or

$$W(\mu + \delta) + W(\mu - \delta) - 2W(\mu) = o(W(\mu + \delta) - W(\mu)). \tag{4.37}$$

With the usual notation $\Delta_\delta W(\mu) = W(\mu + \delta) - W(\mu)$, $\Delta_\delta^2 W(\mu) = \Delta_\delta \Delta_\delta W(\mu)$, it can readily be seen that (4.37) is equivalent to the following relation: uniformly in μ,

$$\Delta_\delta^2 W(\mu) = o(\Delta_\delta W(\mu)). \tag{4.38}$$

Finally, using the notation $\omega_W(\delta)$ introduced in the statement of Theorem 5, we see that (4.37) implies $\omega_W(\delta) \to 0$ as $\delta \to 0$. Theorem 5 is proved.

Some corollaries to this theorem will be given in Section V.5.

Now we proceed to the proof of Theorem 6. It follows from Theorem 2 that it suffices to restrict oneself to estimating $\rho(\tau, w)$ in proving Theorem 6. Hence we shall assume the polynomial P in (3.1) to be equal to 1.

Let $W_1(\lambda) = W(\lambda) + a\lambda$ where a is such that $W_1(\pi) = W_1(-\pi)$. It is seen that

$$\sup_{\lambda, |x| \leqslant \delta} \frac{|W_1(\lambda + x) + W_1(\lambda - x) - 2W_1(\lambda)|}{|W(\lambda + x) - W(\lambda - x)|}$$
$$= \sup_{\lambda, |x| \leqslant \delta} \frac{|W(\lambda + x) + W(\lambda - x) - 2W(\lambda)|}{|W(\lambda + x) - W(\lambda - x)|} = \omega_W(\delta).$$

Lemma 15. *For any trigonometric polynomials $\psi, \varphi \in \mathscr{H}^2$ we have the equality*

$$\int_{-\pi}^{\pi} \varphi(\lambda)\psi(\lambda)e^{i\lambda\tau}f(\lambda)\,d\lambda = \int_{-\pi}^{\pi} [\varphi'(\lambda)\psi(\lambda) + \varphi(\lambda)\psi'(\lambda) + i\tau\varphi(\lambda)\psi(\lambda)]e^{i\tau\lambda}W_1(\lambda)\,d\lambda. \tag{4.39}$$

Proof. For all $\tau > 0$

$$\int_{-\pi}^{\pi} \varphi(\lambda)\psi(\lambda)e^{i\lambda\tau}\,d\lambda = 0.$$

Hence

$$\int_{-\pi}^{\pi} \varphi(\lambda)\psi(\lambda)e^{i\lambda\tau}f(\lambda)\,d\lambda = \int_{-\pi}^{\pi} \varphi(\lambda)\psi(\lambda)e^{i\lambda\tau}\,dW_1(\lambda).$$

Integrating by parts in the last integral we arrive at (4.39). □

From now on, to prove Theorem 6 we proceed as follows. Conditions (1) and (2) of the theorem imply that W_1 is a sufficiently smooth function and, therefore, can be approximated adequately by trigonometric polynomials. By approximating W_1 by these polynomials in (4.39) we obtain an estimate for $\rho(\tau)$ (compare with the proof of Theorem 1).

Let $\varphi(\lambda) = \sum_0^n a_j e^{i\lambda j}$ be a trigonometric polynomial. By the hypothesis of Theorem 6 we have the following inequalities:

$$\|\varphi'\|_f \leqslant C_1 n \|\varphi\|_f \qquad \text{(Zygmund's inequality);} \tag{4.40}$$

$$\left\| \sum_0^m a_j e^{i\lambda j} \right\|_f \leqslant C_2 \|\varphi\|_f, \qquad m \leqslant n \text{ (M. Riesz's inequality);} \tag{4.41}$$

$$\left\| \sum_0^r a_j e^{i\lambda j} \right\|_f^2 + \sum_p \left\| \sum_{2^{p-1}r+1}^{2^p r} a_j e^{i\lambda j} \right\|_f^2 \leqslant C_3 \|\varphi\|_f^2, \qquad r > 0$$

(Littlewood–Paley inequality) (4.42)

(here $a_j = 0$ if $j > n$). The constants C_1, C_2, C_3 depend only on f and do not depend on φ.

Due to the assumption that $0 < m = \min f(\lambda) \leqslant f(\lambda) \leqslant \max f(\lambda) = M < \infty$, these inequalities are obvious. For example

$$\|\varphi'\|_f \leqslant \sqrt{M}\|\varphi'\|^{(2)} \leqslant \sqrt{M}n\|\varphi\|^{(2)} \leqslant \sqrt{\frac{M}{m}}\, n\|\varphi\|_f.$$

(4.41) and (4.42) can be proved in a similar way; we may assume $C_2 = \sqrt{(M/m)}$, $C_3 = M/m$.

Lemma 16. *Let*

$$\varphi(\lambda) = \sum_{L}^{N} a_j e^{i\lambda j}, \qquad 0 \leqslant L \leqslant N,$$

$$\psi(\lambda) = \sum_{L_1}^{N_1} b_j e^{i\lambda j}, \qquad 0 \leqslant L_1 \leqslant N_1, 1 < L_1 + L.$$

Then

$$\left|\int_{-\pi}^{\pi} \varphi(\lambda)\psi(\lambda)f(\lambda)\,d\lambda\right| \leqslant C_4 \left(\frac{M}{m}\right)^{3/2} \frac{N + N_1}{L + L_1 - 1}\,\omega_W\left(\frac{1}{L + L_1 - 1}\right). \tag{4.43}$$

Proof. By virtue of (4.39) it suffices to show that

$$\left|\int_{-\pi}^{\pi} \varphi'(\lambda)\psi(\lambda)W_1(\lambda)\right| \leqslant C_4 \left(\frac{M}{m}\right)^{3/2} \frac{N\omega_W\left(\dfrac{1}{L + L_1 - 1}\right)}{L + L_1 - 1}, \tag{4.44}$$

$$\left|\int_{-\pi}^{\pi} \varphi(\lambda)\psi'(\lambda)W_1(\lambda)\,d\lambda\right| \leqslant C_4 \left(\frac{M}{m}\right)^{3/2} \frac{N_1}{L + L_1 - 1}\,\omega_W\left(\frac{1}{L + L_1 - 1}\right).$$

It follows from the conditions of Theorem 6 and from the definition of the function W_1 that

$$|W_1(\lambda + \delta) + W_1(\lambda - \delta) - 2W_1(\lambda)| \leqslant 2M\delta\omega_W(\delta),$$

that is, $W_1(\lambda)$ is a smooth function,* and by a theorem due to N. I. Akhieser (see [27], p. 274) the value $E_s(W_1)$ of best approximation of the function W_1 by trigonometric polynomials of degree less or equal to s satisfies the inequality

$$E_s(W_1) \leqslant C_4 \frac{M}{s}\,\omega_W\left(\frac{1}{s}\right). \tag{4.45}$$

Let Q be a polynomial of best approximation for W_1 of degree less than $L + L_1 - 1$. Then

$$\begin{aligned}
&\left|\int_{-\pi}^{\pi} \varphi'(\lambda)\psi(\lambda)W_1(\lambda)\,d\lambda\right| \\
&= \left|\int_{-\pi}^{\pi} \varphi'(\lambda)\psi(\lambda)[W_1(\lambda) - Q(\lambda)]\,d\lambda\right| \\
&\leqslant \frac{1}{m} E_{L+L_1-1}(W_1)\|\varphi'\|_f\|\psi\|_f \\
&\leqslant \frac{C_1C_4}{m} M \frac{N}{L + L_1 - 1}\,\omega_W\left(\frac{1}{L + L_1 - 1}\right) \\
&\leqslant C_4 \left(\frac{M}{m}\right)^{3/2} \frac{N\omega_W\left(\dfrac{1}{L + L_1 - 1}\right)}{L + L_1 - 1}.
\end{aligned}$$

The second inequality in (4.44) can be proved in a similar way. □

* A function $W(\lambda)$ is said to be *smooth* if the *smoothness index* of this function satisfies the condition $\sup_{\lambda,\,|x| \leqslant \delta} |W(\lambda + x) + W(\lambda - x) - 2W(\lambda)| = o(\delta)$, $\delta \to 0$.

Proof of Theorem 6. Let $\varphi,\ \psi \in \mathscr{H}^2$ be trigonometric polynomials with $\|\varphi\|_f = \|\psi\|_f = 1$ of the form

$$\varphi(\lambda) = \sum_{\tau+1}^{N} a_j e^{i\lambda j}, \qquad \psi(\lambda) = \sum_{0}^{N_1} b_j e^{i\lambda j}.$$

We can assume without loss of generality that $N = 2^p\tau$, $N_1 = 2^p\tau$, $p \geqslant 0$. For $j = 1, \ldots, p$ let

$$\varphi_j(\lambda) = \sum_{2^{j-1}\tau+1}^{2^j\tau} a_s e^{i\lambda s},$$

$$\Phi_j(\lambda) = \sum_{\tau+1}^{2^j\tau} a_s e^{i\lambda s},$$

$$\psi_j(\lambda) = \sum_{2^{j-1}\tau+1}^{2^j\tau} a_s e^{i\lambda s}, \qquad \psi_0(\lambda) = \sum_{0}^{\tau} a_s e^{i\lambda s},$$

$$\Psi_j(\lambda) = \sum_{0}^{2^j\tau} a_s e^{i\lambda s}.$$

Using (4.43), (4.42), and (4.41) we find

$$\begin{aligned}
&\left|\int_{-\pi}^{\pi} \varphi(\lambda)\psi(\lambda)f(\lambda)\,d\lambda\right| \\
&\leqslant \sum_{j=1}^{p}\left(\left|\int_{-\pi}^{\pi} \varphi_j(\lambda)\Psi_j(\lambda)f(\lambda)\,d\lambda\right| + \left|\int_{-\pi}^{\pi} \psi_j(\lambda)\Phi_j(\lambda)f(\lambda)\,d\lambda\right|\right) \\
&\leqslant 8C_4\left(\frac{M}{m}\right)^{3/2}\left[\max_j \|\Psi_j\|_f \sum_{j=1}^{p} \omega_W\left(\frac{1}{2^{j-1}\tau - 1}\right)\|\varphi_j\|_f\right. \\
&\quad \left. + \max_j \|\Phi_j\|_f \sum_{j=1}^{p} \omega_W\left(\frac{1}{2^{j-1}\tau - 1}\right)\|\psi_j\|_f\right] \\
&\leqslant 8C_4\left(\frac{M}{m}\right)^{2}\left(\sum_{j=1}^{\infty} \omega_W^2\left(\frac{1}{2^{j-1}\tau - 1}\right)\right)^{1/2}\left(\left(\sum_j \|\varphi_j\|_f^2\right)^{1/2} + \left(\sum_j \|\psi_j\|_f^2\right)^{1/2}\right) \\
&\leqslant 6C_4\left(\frac{M}{m}\right)^{5/2}\left(\sum_{j=1}^{\infty} \omega_W^2\left(\frac{1}{2^{j-1}\tau - 1}\right)\right)^{1/2}. \qquad (4.46)
\end{aligned}$$

Noting that $C_4 \leqslant 5/2$ ([27], p. 305), we arrive at (3.3). We have thus proved Theorem 6. □

V.5 Corollaries to the Basic Theorems with Examples

We give here some corollaries to the theorems from Sections V.2–4 enabling us to see the restrictions that the complete regularity condition imposes on spectral density. We note first the following.

The spectral density of a completely regular process with discrete time has no discontinuities of jumps of the first kind.

To prove this assertion we take the representation given by (3.1). It suffices to prove that $w(\lambda)$ has no jumps. In fact, if $w(\lambda)$ had a jump at a point λ_0, $w(\lambda)$ would be bounded in a neighborhood of this point λ_0. Therefore, the function $W(\lambda)$ would be smooth in this neighborhood because of (3.1), i.e., we would have the equality

$$W(\lambda + \delta) + W(\lambda - \delta) - 2W(\lambda) = o(\delta).$$

It is known ([3], p. 182) that the derivative of a smooth function having the Darboux property has no discontinuities of the first kind.

As was noted above in (2.1) and (3.1), the first polynomial factor contains all zeros of the spectral density $f(\lambda)$ while the second factor of $w(\lambda)$ is "almost" positive; hence all zeros of $f(\lambda)$ occur, roughly speaking, in pairs. To make this more precise we note first that the functions w in the expansions given by (2.1) and (3.1) are summable to power p, $-\infty < p < \infty$.

Let us consider the function w in (2.1). It is seen that for all $\varepsilon > 0$ we have $\|\exp\{\pm r \pm u_\varepsilon\}\|^{\infty} \leqslant \|\exp\{|r_\varepsilon| + |u_\varepsilon|\}\|^{(\infty)} < \infty$. We noted above (see Section V.2) that the inequality $\|v_\varepsilon\|^{(\infty)} < \varepsilon$ implies that $\exp\{k|\tilde{v}_\varepsilon|\} \in \mathscr{L}^1$ for all $k < \pi/2\varepsilon$. Therefore, $e^{\pm v_\varepsilon} \in \mathscr{L}^p$ if only $\varepsilon < \pi/2p$. Further, if we take the function w in (3.1) then $w^{-1} \in \mathscr{L}^1$ (as follows from the proof of Theorem 5). Comparing (2.1) and (3.1) we can see that the functions w are the same in these expansions.

We use the terms *order*, *upper order*, *lower order* for any zero λ_0 of the function $f(\lambda)$ to denote (respectively)

$$k(\lambda_0) = \lim_{\lambda \to \lambda_0} \frac{\ln f(\lambda)}{\ln|\lambda - \lambda_0|}, \qquad \overline{k}(\lambda_0) = \overline{\lim_{\lambda \to \lambda_0}} \frac{\ln f(\lambda)}{\ln|\lambda - \lambda_0|}, \qquad \underline{k}(\lambda_0) = \underline{\lim_{\lambda \to \lambda_0}} \frac{\ln f(\lambda)}{\ln|\lambda - \lambda_0|}.$$

The assertion we have proved, that the function w is summable, implies that the lower order of any zero of w is zero; hence the real order of zeros of $f(\lambda)$ must be even.

It is seen finally that the spectral density $f(\lambda)$ of a completely regular process $\xi(t)$ is summable to any positive power.

Let us construct now examples of completely regular processes or, more precisely, examples of spectral densities $f(\lambda)$ of these processes. Naturally, many examples can be constructed with the aid of Theorem 1. We shall obtain nontrivial examples if we violate the conditions of Theorem 1, that is, if $f(\lambda)$ vanishes in a nonpolynomial manner or has discontinuities (of the second kind, of course); similar examples can be constructed with the aid of Theorem 3 and Theorem 6.

We establish first the result that follows; we denote by $\widetilde{\ln} f$ the conjugate of $\ln f \in \mathscr{L}^1(-\pi, \pi)$.

Theorem 7. *If for at least one determination of* $\ln f$, $\widetilde{\ln} f$ *is continuous, then a stationary process* $\xi(t)$ *with spectral density* $f(\lambda)$ *is completely regular,* *and*

$$\rho(\tau) = O\left(\min \omega\left(\frac{1}{\tau}; \ln f\right), \omega\left(\frac{1}{\tau}; \widetilde{\ln} f\right)\right) \tag{5.1}$$

where $\omega(\delta; h)$ denotes the index of continuity of the function $h(\lambda)$.

Proof. In the case where the function $\ln f(\lambda)$ is continuous, we need only to invoke Theorem 1 since by Jackson's theorem on best approximation we have

$$E_{\tau-1}(\ln f) = O\left(\omega\left(\frac{1}{\tau}; \ln f\right)\right).$$

The complete regularity of the process $\xi(t)$ with continuous function $\widetilde{\ln} f$ follows from Theorem 3. In fact, for any $\varepsilon > 0$ we can find, using Weierstrass's theorem, a trigonometric polynomial Q_ε such that

$$\widetilde{\ln} f = Q_\varepsilon + v_\varepsilon, \qquad \|v_\varepsilon\|^{(\infty)} \leqslant \varepsilon.$$

It can be seen that the function $r_\varepsilon = \tilde{Q}_\varepsilon$ is also a trigonometric polynomial, and therefore $f(\lambda)$ permits the representation

$$f(\lambda) = e^{\ln f(\lambda)} = e^{r_\varepsilon + \tilde{v}_\varepsilon}, \qquad \|v_\varepsilon\|^{(\infty)} \leqslant \varepsilon$$

for all $\varepsilon > 0$. We invoke now Theorem 3.

To obtain the estimate given by (5.1) we shall give a direct proof, which is rather simple, however. Let $Q(\lambda)$ be a trigonometric polynomial of best approximation of degree larger or equal to $\tau - 1$ for the continuous function $e^{i \widetilde{\ln} f}$. Then for all $\theta \in \mathscr{H}^1$

$$\int_{-\pi}^{\pi} e^{i\lambda\tau}\theta(\lambda)Q(\lambda)\, d\lambda = 0.$$

By virtue of (1.7) we infer that

$$\begin{aligned}
\rho(\tau) &= \sup_{\theta \in H^1,\, \|\theta\|^1 = 1} \left|\int_{-\pi}^{\pi} e^{i\lambda\tau}\theta(\lambda)\frac{\overline{g(e^{i\lambda})}}{g(e^{i\lambda})}\, d\lambda\right| \\
&= \sup_{\theta} \left|\int_{-\pi}^{\pi} e^{i\lambda\tau}\theta(\lambda)[e^{i\widetilde{\ln} f} - Q]\, d\lambda\right| \\
&\leqslant E_{\tau-1}(e^{i\widetilde{\ln} f}) = O\left(\omega\left(\frac{1}{\tau}; e^{i\widetilde{\ln} f}\right)\right) = O\left(\omega\left(\frac{1}{\tau}; \widetilde{\ln} f\right)\right),
\end{aligned}$$

thus proving the theorem. □

We give here a few examples based on Theorem 7

Example 1. Let $\xi(t)$ be a stationary process with spectral density

$$f(\lambda) = \exp\left\{\sum_1^{\infty} \frac{\cos k\lambda}{k(\ln k + 1)}\right\}.$$

In this case the function

$$\ln f(\lambda) = \sum_{1}^{\infty} \frac{\sin k\lambda}{k(\ln k + 1)}$$

is continuous; by Theorem 7 the process $\xi(t)$ is completely regular. At the same time, using properties of a Fourier series with monotone coefficients,* we can show that as $\lambda \to 0$

$$\ln f(\lambda) \sim \int_{1}^{1/\lambda} \frac{dt}{t(\ln t + 1)} \sim \ln\ln\frac{1}{\lambda}.$$

Therefore, as $\lambda \to 0$ the function $f(\lambda)$ increases as the function $\ln(1/\lambda)$ does. Similarly, the process $\xi(t)$ with spectral density $1/f(\lambda)$ is completely regular as well; its spectral density at the point zero vanishes as $1/\ln(1/\lambda)$ does. It can be computed that the index of continuity $\omega(\delta)$ of the function $\widetilde{\ln f}$ does not exceed $O(1/\ln(1/\delta))$; in our example, therefore,

$$\rho(\tau) = O\left(\frac{1}{\ln \tau}\right).$$

Example 2. This example illustrates a completely regular process with upper and lower bounded spectral density having discontinuities (naturally, of the second kind). By Theorem 7 it suffices to construct a bounded discontinuous function the conjugate of which is continuous.

Let us consider the rectangular domain with cross-cuts, "Littlewood's hippopotamus,"† shown in Figure 2. Suppose the number of cross-cuts (teeth) is infinite, and the length of each cross-cut is 3/4. Let $G(z)$ be a function conformally mapping the unit circle onto the domain mentioned. If $G(e^{i\lambda}) = u(\lambda) + iv(\lambda)$, it is seen that the function $v(\lambda)$ is discontinuous, $0 \leqslant v \leqslant 1$, and that the function $u(\lambda) = \tilde{v}(\lambda)$ is continuous. We can take a function $f(\lambda) = e^{v(\lambda)}$ as the required spectral density. It would be of some interest to determine a good estimate for $\rho(\tau)$.

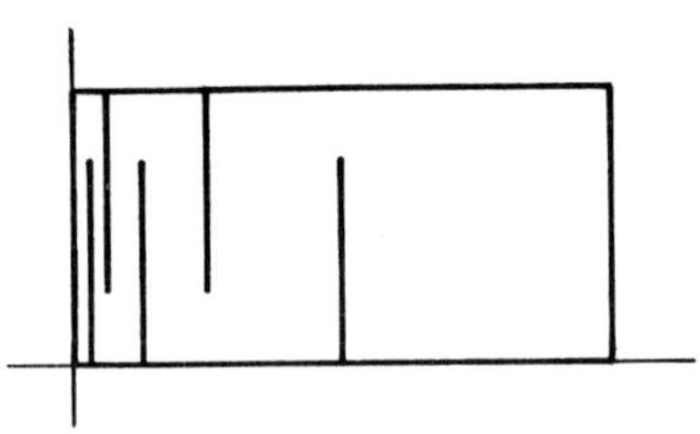

Figure 2

* See [28], p. 303.

† See J. E. Littlewood, *A Mathematician's Miscellany*, London, 1953. Domains of this type are well known in the theory of prime ends; see [11].

V.6 Strong Mixing

We have investigated above the problem of convergence of $\rho(\tau)$ to zero without discussing the question of the rate of the convergence. Our objective in Section V.6 is to find out when $\rho(\tau)$ decreases, as $\tau \to \infty$, sufficiently rapidly, let us say, not slower than $\tau^{-\varepsilon}$, $\varepsilon > 0$ (we note that this case is most interesting if we consider the applicability of the central limit theorem to a process satisfying the strong mixing condition).* It turns out that we can obtain in this case complete and final results.

Theorem 8. *A necessary and sufficient condition for*

$$\rho(\tau) = O(\tau^{-r-\beta}), \qquad \text{where } 0 < \beta < 1,$$

is that the spectral density $f(\lambda)$ permit a representation of the form

$$f(\lambda) = |P(e^{i\lambda})|^2 w(\lambda), \tag{6.1}$$

where $P(z)$ is a polynomial with zeros on $|z| = 1$ and where the function $w(\lambda)$ is strictly positive, $w(\lambda) \geqslant m > 0$, and is r times differentiable with the rth derivative satisfying a Hölder condition of order β.

Proof. By Theorem 3 or Theorem 5 we can represent the spectral density $f(\lambda)$ as (3.1) where $w(\lambda)$ satisfies (3.2). According to Theorem 2 and Lemma 11

$$\begin{aligned} \rho(n; f) &\leqslant \rho(n - k; w), \\ \rho(n; w) &\leqslant \rho(n; f), \end{aligned} \tag{6.2}$$

where k is the degree of a polynomial P. Hence it suffices to restrict oneself to the case where $f \equiv w$, that is, when in (3.1) the polynomial P is missing and the spectral density $f(\lambda)$ itself satisfies (3.2). We assume further $P \equiv 1$ unless otherwise specified.

Sufficiency of the conditions of the theorem follows immediately from Theorem 1. In fact, by this theorem,

$$\rho(\tau; f) = \rho(\tau) \leqslant \frac{1}{m} E_{\tau-1}(f). \tag{6.3}$$

On the other hand, as we saw in Jackson's theorem on the order of best approximations,†

$$E_{\tau-1}(f) \leqslant \frac{C}{\tau^{r+\beta}}. \tag{6.4}$$

As to necessity, we explained above in rough terms (see Section V.1) that the spectral density $f(\lambda)$ of a process with a regularity coefficient decreasing

* See [16], Chapter 18, or [24], Chapter 4.

† See [27], p. 275.

rapidly must be sufficiently smooth. We also showed that smallness of the regularity coefficients implies smallness of the corresponding Fourier coefficients of the function $f(\lambda)$. Our considerations here are based on the idea used in the discussion above, though we find it more efficient to estimate segments of the Fourier series of the function $f(\lambda)$ instead of the Fourier coefficients of $f(\lambda)$. The results thus obtained will be used for estimating best approximation of the function $f(\lambda)$ by trigonometric polynomials. Finally, Bernstein's converse theorems from the best approximation theory enable us to conclude that the function $f(\lambda)$ is smooth if we start from the order of best approximation of this function. The general scheme of the proof is the following.

Let r_j denote the jth Fourier coefficient of the function $f(\lambda)$. By virtue of (1.5) we have inequality

$$\left|\sum_{2^k}^{2^{k+1}} r_j e^{ij\mu}\right| = \left|\int_{-\pi}^{\pi}\left(\sum_{2^k}^{2^{k+1}} e^{ij(\mu-\lambda)}\right) f(\lambda)\,d\lambda\right|$$

$$\leqslant \rho(2^k)\int_{-\pi}^{\pi}\left|\sum_{0}^{2^k} e^{ij(\mu-\lambda)}\right| f(\lambda)\,d\lambda, \qquad k \geqslant 0.$$

If we assume in addition that $\max_\lambda |f(\lambda)| \leqslant M < \infty$, we shall have

$$\left|\sum_{2^k}^{2^{k+1}} r_j e^{ij\mu}\right| \leqslant \rho(2^k) M \int_{-\pi}^{\pi} \frac{\sin 2^{k-1}}{\sin\dfrac{\lambda}{2}}\,d\lambda \leqslant C_1 \rho(2^k) k.$$

Similar inequalities can obviously be satisfied for negative k as well.

If $2^k \leqslant \tau < 2^{k+1}$, the magnitude of the best approximation of $f(\lambda)$ by trigonometric polynomials of degree τ is known to be

$$\max_\mu \left|\sum_{|j|\geqslant 2^k} r_j e^{i\mu j}\right| \leqslant \sum_{p=k}^{\infty}\left|\sum_{2^p\leqslant|j|<2^{p+1}} r_j e^{i\mu j}\right| \leqslant C_2 \sum_{p\geqslant \ln_2 \tau+1} k\rho(2^k).$$

From the inequality obtained Theorem 8 follows almost immediately. However, we cannot guarantee for $f^{(r)}(\lambda)$ any Hölder condition of order $\beta - \varepsilon$, ε being arbitrarily small (and not of order β). □

To get rid of ε we have to carry out rather cumbersome computations. In this case we estimate $f(\lambda)$ using Fejer truncated sums (rather than the Fourier sums), which are natural in this context since they approximate $f(\lambda)$, in essence, in the same way as the polynomials of best approximation do.

Lemma 17. *For all θ, $0 < \theta < 1$, we have the estimate*

$$E_n(f) \leqslant \frac{32 \max_\lambda f(\lambda)}{\theta^2} \sum_{k=0}^{\infty} \rho(2^k n(1-\theta); f). \tag{6.5}$$

Proof. It suffices to consider the case where the series in the right-hand side of (6.5) is convergent; if this series is divergent, the inequality given in (6.5) is trivial.

We shall prove first that the function $f(\lambda)$ is bounded if $\sum \rho(2^k)$ is convergent. To this end we again take the function $\gamma(N; \mu)$ from Sections V.3 and V.4 and prove that $\gamma(N; \mu) \leqslant NM$, where the constant $M < \infty$ is independent of N and μ.

While proving Lemma 8 we obtained (3.14). If we assume that $k = 3$, it is easy to verify that

$$\gamma(3N; \mu) \leqslant 3\gamma(N; \mu) + 9\sqrt{\gamma(N; \mu)\gamma(r; \mu)} + 9\gamma(N; \mu)\rho(r).$$

For any $\varepsilon > 0$, we have for large N uniformly in μ (see Section V.3)

$$N^{1-\varepsilon} \leqslant \gamma(N; \mu) \leqslant N^{1+\varepsilon}.$$

Taking $\varepsilon = \ln 3 - \ln 2/2 \ln 6$, we obtain for large N uniformly in μ

$$\frac{\gamma(3N; \mu)}{3\gamma(N; \mu)} \leqslant 1 + 3\rho(r) + 3\frac{r^{(1+\varepsilon)/2}}{N^{(1-\varepsilon)/2}}.$$

Finally, assuming $N - 3^s$, $s = 1, 2, \ldots$, $r = r(N) = 2^s$, we have for large s uniformly in μ

$$\frac{\gamma(3^{s+1}; \mu)}{3\gamma(3^s; \mu)} \leqslant 1 + 3\rho(2^s) + 3\left(\frac{2}{3}\right)^{s/4}.$$

Therefore,

$$\lim_{s\to\infty} \frac{\gamma(3^s; \mu)}{3^s} \leqslant C \prod_1^\infty \left(1 + 3\rho(2^s) + 3\left(\frac{2}{3}\right)^{s/4}\right) = M < \infty$$

since

$$\sum_s \left[\rho(2^s) + \left(\frac{2}{3}\right)^{s/4}\right] < \infty.$$

We have thus proved the existence of a constant M such that

$$3^{-s}\gamma(3^s; \mu) \leqslant M. \tag{6.6}$$

By Lebesgue's theorem we have for almost all $\mu \in [-\pi, \pi]$

$$\lim_{N\to\infty} \frac{1}{N} \gamma(N; \mu) - \lim_N \frac{1}{N} \int_{-\pi}^{\pi} \frac{\sin^2 N \dfrac{\lambda - \mu}{2}}{\sin^2 \dfrac{\lambda - \mu}{2}} f(\lambda)\, d\lambda = 2\pi f(\mu),$$

and by (6.6) we have $f(\mu) \leqslant M/2\pi$ for all μ Therefore, we can assume that $\sup_\mu f(\mu) \leqslant M$.

Further on, we shall assume that $M = \sup_\mu f(\mu)$. Let r_j be the jth Fourier coefficient of the function $f(\lambda)$. We denote by $S_n(\lambda) = \sum_{-n}^{n} r_j e^{ij\lambda}$ the nth partial sum of the Fourier series of the function $f(\lambda)$. Finally, let us take a number θ, $0 < \theta < 1$, and let us introduce Fejer truncated sums for the function $f(\lambda)$:

$$\sigma_n(f; [n\theta]; \lambda) = \frac{1}{[n\theta]+1} \sum_{\nu=0}^{[n\theta]} S_{n-\nu}(\lambda).$$

It is seen that

$$E_n(f) \leqslant \max_{\lambda} |g(\lambda) - \sigma_n(f; [n\theta]; \lambda)|.$$

We shall show that

$$\max_{\lambda} |g(\lambda) - \sigma_n(f; [n\theta]; \lambda)| \leqslant C(\theta) \sum_{k=0}^{\infty} \rho([2^k n(1-\theta)]),$$

where $C(\theta)$ is a constant dependent on θ. Let us consider the differences

$$\Delta_k(\lambda) = \sigma_{2^{k+1}n}(f; [2^{k+1}n\theta]; \lambda) - \sigma_{2^k n}(f; [2^k n\theta]; \lambda),$$

which can, as easily seen, be expressed as

$$\Delta_k(\lambda) = \Delta_k^+(\lambda) + \Delta_k^-(\lambda),$$

where the sums Δ_k^+ and Δ_k^- are the following:

$$\Delta_k^+(\lambda) = \sum_{[2^k n\theta]+1}^{2^{k+1}n} a_j r_j e^{ij\lambda},$$

$$\Delta_k^-(\lambda) = \sum_{-2^{k+1}n}^{-[2^k n\theta]-1} a_j r_j e^{ij\lambda},$$

the coefficients a_j being independent of the function $f(\lambda)$ (the explicit form of these coefficients is not very essential).

Simple calculations enable us to rewrite Δ_k^+ and Δ_k^- as

$$\Delta_k^+(\lambda) = \int_{-\pi}^{\pi} r_k^+(\lambda - \mu) f(\mu)\, d\mu,$$

$$\Delta_k^-(\lambda) = \int_{-\pi}^{\pi} r_k^-(\lambda - \mu) f(\mu)\, d\mu,$$

where

$$r_k^+(\lambda) = \frac{1}{[2^{k+1}n\theta]+1} \sum_{\nu=0}^{[2^{k+1}n\theta]} \frac{1}{2\pi} \sum_{j=1}^{2^{k+1}n-\nu} e^{ij\lambda} - \frac{1}{[2^k n\theta]+1} \sum_{\nu=0}^{[2^k n\theta]} \frac{1}{2\pi} \sum_{j=1}^{2^k n-\nu} e^{ij\lambda},$$

$$r_k^-(\lambda) = \overline{r_k^+(\lambda)}.$$

The trigonometric polynomial $r_k^+(\lambda - \mu)$ considered as a function of μ contains only negative powers of $e^{i\mu}$ starting from $\exp\{-(2^k n - [2^k n\theta] + 1)\mu\}$. Hence (see (1.5))

$$\begin{aligned} |\Delta_k^+(\lambda)| &\leqslant \rho(2^k n - [2^k n\theta] + 1) \int_{-\pi}^{\pi} |r_k^+(\lambda - \mu)| f(\mu)\, d\mu \\ &\leqslant M\rho(2^k n - [2^k n\theta] + 1)\left[\int_{-\pi}^{\pi} |\operatorname{Re} r_k^+(\lambda - \mu)|\, d\mu \right. \\ &\qquad \left. + \int_{-\pi}^{\pi} |\operatorname{Im} r_k^+(\lambda - \mu)|\, d\mu\right]. \end{aligned} \tag{6.7}$$

Similarly,

$$|\Delta_k^-(\lambda)| \leqslant M\rho(2^k n - [2^k n\theta] + 1)\left[\int_{-\pi}^{\pi} |\operatorname{Re} r_k^-(\lambda - \mu)|\, d\mu \right.$$
$$\left. + \int_{-\pi}^{\pi} |\operatorname{Im} r_k^-(\lambda - \mu)|\, d\mu \right]. \tag{6.8}$$

We show that the integrals in the right-hand sides of these inequalities are bounded.

1. Boundedness of $\int_{-\pi}^{\pi} |\operatorname{Re} r_k^+(\lambda - \mu)|\, d\mu$. For the sake of brevity we write N instead of $2^k n$, and we write p instead of θN. Then, after some simple calculations we obtain

$$\operatorname{Re} r_k^+(\lambda - \mu)$$
$$= \frac{1}{2\pi \sin \dfrac{\lambda - \mu}{2}} \left[\frac{1}{[2p] + 1} \sin \frac{4N + 1 - [2p]}{2}(\lambda - \mu) \sin \frac{[2p] + 1}{2} \right.$$
$$\left. \times (\lambda - \mu) - \frac{1}{[p] + 1} \sin \frac{2N + 1 - [p]}{2}(\lambda - \mu) \sin \frac{[p] + 1}{2}(\lambda - \mu) \right].$$

Then the Fejer integral

$$\frac{1}{2\pi s}\int_{-\pi}^{\pi} \frac{\sin^2 \dfrac{s}{2}(\lambda - \mu)}{\sin^2 \dfrac{\lambda - \mu}{2}}\, d\mu = 1;$$

hence

$$\int_{-\pi}^{\pi} |\operatorname{Re} r_k^+(\lambda - \mu)|\, d\mu \leqslant \frac{(4N + 1 - [2p])^{1/2}([2p] + 1)^{1/2}}{[2p] + 1}$$
$$+ \frac{(2N + 1 - p)^{1/2}(p + 1)^{1/2}}{([p] - 1)}$$
$$\leqslant \frac{6}{\theta}.$$

2. Boundedness of $\int_{-\pi}^{\pi} |\operatorname{Im} r_k^+(\lambda - \mu)|\, d\mu$. Retaining the notation $2^k n = N$, we have

$$\operatorname{Im} r_k^+(\lambda - \mu) - \frac{1}{2\pi([2N\theta] + 1)} \sum_{\nu=0}^{[2N\theta]} \sum_{j=1}^{2N-\nu} \sin j(\lambda - \mu)$$
$$- \frac{1}{2\pi([N\theta] + 1)} \sum_{\nu=0}^{[N\theta]} \sum_{j=1}^{N-\nu} \sin j(\lambda - \mu).$$

But

$$\sum_{j=1}^{T} \sin \lambda = \frac{\cos \dfrac{\lambda}{2} - \cos \dfrac{2T+1}{2} \lambda}{2 \sin \dfrac{\lambda}{2}},$$

$$\sum_{T=A}^{b} \cos\left(T + \frac{1}{2}\right) \lambda = \frac{\sin \dfrac{b-a+1}{2} \lambda \cos \dfrac{b+a+1}{2} \lambda}{\sin \dfrac{\lambda}{2}}.$$

Therefore

$$\begin{aligned}
&\left|\operatorname{Im} r_k^+(\lambda - \mu)\right| \\
&= \left| -\frac{1}{4\pi([2N\theta]+1)} \frac{\sin \dfrac{[2N\theta]}{2}(\lambda-\mu)}{\sin^2 \dfrac{\lambda-\mu}{2}} \cos \frac{4N - [2N\theta] + 1}{2}(\lambda - \mu) \right. \\
&\quad \left. + \frac{1}{4\pi([N\theta]+1)} \frac{\sin \dfrac{[N\theta]+1}{2}(\lambda-\mu)}{\sin^2 \dfrac{\lambda-\mu}{2}} \cos \frac{2N - [N\theta] + 1}{2}(\lambda - \mu) \right| \\
&\leqslant \frac{1}{2\pi([2N\theta]+1)} \frac{\left|\sin \dfrac{[N\theta]+1}{2}(\lambda-\mu)\right|}{\sin^2 \dfrac{\lambda-\mu}{2}} \sin^2 \frac{4N - [2N\theta] + 1}{4}(\lambda - \mu) \\
&\quad + \frac{1}{2\pi([N\theta]+1)} \frac{\left|\sin \dfrac{[N\theta]+1}{2}(\lambda-\mu)\right|}{\sin^2 \dfrac{\lambda-\mu}{2}} \sin^2 \frac{2N - [N\theta] + 1}{4}(\lambda - \mu) \\
&\quad + \left| \frac{\sin \dfrac{[2N\theta]+1}{2}(\lambda-\mu)}{4\pi([2N\theta]+1)\sin^2 \dfrac{\lambda-\mu}{2}} - \frac{\sin \dfrac{[N\theta]+1}{2}(\lambda-\mu)}{4\pi([N\theta]+1)\sin^2 \dfrac{\lambda-\mu}{2}} \right|. \qquad (6.9)
\end{aligned}$$

The integral of the two first terms in the right-hand side of (6.9) does not exceed $5/4\theta$. For the sake of brevity we denote by $b(\lambda)$ the last term in (6.9).

For $|\lambda| \leqslant 1/2$, $|\sin\lambda - \lambda| \leqslant |\lambda|^3$, and hence

$$\int_{\{|\lambda-\mu|\leqslant(1/[2N\theta]+1)} b(\lambda)\,d\lambda$$

$$\leqslant \frac{1}{2\pi}\int_{\{|\lambda-\mu|\leqslant(1/[2N\theta]+1)\}} \frac{\left(\frac{1}{2}(2N\theta+1)|\lambda-\mu|\right)^3}{[2N\theta+1]}$$

$$+\frac{\left(\frac{1}{2}(N\theta+1)|\lambda-\mu|\right)^3}{[N\theta]+1}\ \frac{d\lambda}{\sin^2\frac{\lambda-\mu}{2}}$$

$$\leqslant 1 < \frac{1}{\theta}.$$

We have in the domain $\pi \geqslant |\lambda-\mu| \geqslant 1/([2N\theta]+1)$ that

$$b(\lambda) \leqslant \frac{\left|\sin\frac{[2N\theta]+1}{2}(\lambda-\mu)\right|}{4\pi\sin^2\frac{\lambda-\mu}{2}}\left|\frac{1}{[2N\theta]+1}-\frac{1}{2([N\theta]+1)}\right|$$

$$+\frac{1}{8\pi\left(\sin^2\frac{\lambda-\mu}{2}\right)([N\theta]+1)}$$

$$\times\left|\sin\frac{[2N\theta]+1}{2}(\lambda-\mu)-\sin([N\theta]+1)(\lambda-\mu)\right|$$

$$+\frac{1}{4\pi\left(\sin^2\frac{\lambda-\mu}{2}\right)([N\theta]+1)}$$

$$\times\left|\sin\frac{[N\theta]+1}{2}(\lambda-\mu)\cos\frac{[N\theta]+1}{2}(\lambda-\mu)\right.$$

$$\left.-\sin\frac{[N\theta]+1}{2}(\lambda-\mu)\right|$$

$$\leqslant \frac{5}{2\pi\theta^2}+\frac{3}{2\pi\theta^2}+\frac{\sin^2\frac{[N\theta]+1}{4}(\lambda-\mu)}{2\pi([N\theta]+1)\sin^2\frac{\lambda-\mu}{2}};$$

hence

$$\int_{\{|\lambda-\mu|\geqslant(1/[2N\theta]+1),\ |\lambda|\leqslant\pi\}} b(\lambda)\,d\lambda \leqslant \frac{5}{\theta^2}+\frac{4}{\theta}.$$

Finally, we obtain

$$\int_{-\pi}^{\pi} |\operatorname{Im} r^{+}(\lambda-\mu)|\,d\mu \leqslant \frac{10}{\theta^2}.$$

Further, pooling all the estimates obtained, we find that

$$\begin{aligned}\int_{-\pi}^{\pi} |r^{+}(\lambda-\mu)|\,d\mu &\leqslant \frac{16}{\theta^2},\\ \int_{-\pi}^{\pi} |r^{-}(\lambda-\mu)|\,d\mu &\leqslant \frac{16}{\theta^2}.\end{aligned} \tag{6.10}$$

Returning to the estimate $\Delta_k(\lambda)$, we deduce from (6.7) and (6.10) that

$$|\Delta_k(\lambda)| \leqslant C(\theta)\rho(2^k n - [2^k n\theta] + 1) \leqslant C(\theta)\rho(2^k[n(1-\theta)]), \tag{6.11}$$

where

$$C(\theta) \leqslant \frac{32}{\theta^2} M.$$

Since the series $\sum \rho(2^k)$ is convergent, the last inequality implies that the series $\sum \Delta_k(\lambda)$, and therefore the sequence $\{\sigma_{2^k n}(f; [2^k n\theta]; \lambda)\}$, are uniformly convergent. But the sequence of Fejer truncated sums $\sigma_{2^k n}(f; [2^k n\theta)]; \lambda)$ for almost all λ is convergent to $f(\lambda)$,* and therefore $f(\lambda)$ is the uniform limit of the polynomial $\sigma_{2^k n}(f; 2^k n\theta]; \lambda)$.

Summing up the inequalities given in (6.11) we conclude that

$$E_n(f) \leqslant \max_{\lambda} |f(\lambda) - \sigma_{2^k n}(f; [2^k n\theta]; \lambda)| \leqslant \sum |\Delta_k(\lambda)| \leqslant \frac{32M}{\theta^2} \sum_{k=0}^{\infty} \rho(2^k[n(1-\theta)]),$$

thus proving Lemma 17. □

Proof of Theorem 8. This theorem is an almost immediate corollary of Lemma 17. In fact, by (6.5) and by the assumption made on the rate of decrease of $\rho(\tau)$, we can assert that

$$E_n(f) \leqslant \frac{\text{const.}}{n^{r+\beta}\theta^2(1-\theta)^{r+\beta}} \sum_{k=0}^{\infty} 2^{-k(r+\beta)} = O(n^{-r-\beta}). \tag{6.12}$$

According to the converse theorems of approximation theory,† it follows from (6.12) that the function $f(\lambda)$ is differentiable r times, and that $f^{(r)}(\lambda)$ satisfies a Hölder condition of order β.

* See, for instance, [27], p. 533.

† See, for instance, the theorem due to S. N. Bernstein in [27], Chapter VI.

It remains only to prove that the function $f(\lambda)$ is positive. (We recall that $P \equiv 1$!) Let us assume that $f(\lambda)$ vanishes at a point λ_0. The function $f(\lambda)$ satisfies a Hölder condition of order $\geqslant \beta(|f(\lambda_1) - f(\lambda_2)| \leqslant C|\lambda_1 - \lambda_2|^\beta$, so that $|f(\lambda)| \leqslant C|\lambda - \lambda_0|^\beta$). Therefore, the lower order of zero λ_0 is

$$\underline{k}(\lambda_0) = \lim_{\lambda \to \lambda_0} \frac{|\ln f(\lambda)|}{|\ln|\lambda - \lambda_0||} \geqslant \beta > 0,$$

which fact contradicts the statement made in Section V.5, according to which $\underline{k}(\lambda_0) = 0$. We have completed the proof of Theorem 8. □

Next we examine when the regularity coefficient decreases even more rapidly, namely, exponentially rapidly.

Theorem 9. *A necessary and sufficient condition for* $\overline{\lim}_{\tau \to \infty} \sqrt[\tau]{\rho(\tau)} \leqslant e^{-\delta}$, *where* $0 < \delta < \infty$, *is that the spectral density* $f(\lambda)$ *can be extended into the complex domain taking values in the strip* $-\delta < \operatorname{Im} \zeta < \delta$ *for a variable* $\zeta = \lambda + i\mu$.

Proof. Let the function $f(\lambda)$ be analytic in the strip given above. This function has a finite number of zeros on an interval $-\pi \leqslant \lambda \leqslant \pi$. Let $|P(z)|^2$ be a polynomial of degree k whose zeros coincide with real zeros of $f(\lambda)$. Then $f(\lambda) = |P(e^{i\lambda})|^2 w(\lambda)$, where the function $w(\lambda)$ is strictly positive on the real axis and permits an analytic extension into the same strip as $f(\lambda)$. It can easily be seen that this is an analytic extension; $w(\zeta)$ is bounded in any strip of the form $-\delta' \leqslant \operatorname{Im} \zeta \leqslant \delta'$, $\delta' < \delta$.

By Bernstein's theorem* we have $\overline{\lim}_{\tau \to \infty} \sqrt[\tau]{E_\tau(w)} \leqslant e^{-\delta}$. Then by (6.3) we also have

$$\overline{\lim} \sqrt[\tau]{\rho(\tau; w)} \leqslant e^{-\delta},$$

and therefore

$$\overline{\lim_{\tau}} \sqrt[\tau]{\rho(\tau; f)} \leqslant \lim_{\tau} \sqrt[\tau]{\rho(\tau - k; w)} \leqslant e^{-\delta}.$$

Conversely, let $\overline{\lim}_\tau \sqrt[\tau]{\rho(\tau; f)} \leqslant e^{-\delta}$. Then $\rho(\tau; f) \leqslant e^{-\delta\tau}(1 + \varepsilon(\tau))^\tau$, where $\lim_n \varepsilon(\tau) = 0$. By Lemma 17

$$\begin{aligned} \lim_{\tau} [E_\tau(f)]^{1/\tau} &\leqslant \lim_{\tau} \left[\frac{CM}{\theta^2} \sum_{k=0}^{\infty} \exp\{-[\tau(1-\theta)]\delta 2^{-k}\} \right. \\ &\qquad \times (1 + \varepsilon([\tau(1-\theta)]2^k))^{|\tau(1-\theta)|2^k} \Big|^{1/\tau} \\ &= e^{-(1-\theta)\delta}, \end{aligned}$$

where the number θ is arbitrary if only $0 < \theta < 1$; hence $\lim[E_\tau(f)]^{1/\tau} \leqslant e^{-\delta}$. The converse to Bernstein's theorem states that the function $f(\lambda)$ is analytically extendable to a strip $-\delta < \operatorname{Im} \zeta < \delta$ (and bounded in any strip of the form $-\delta' \leqslant \operatorname{Im} \zeta \leqslant \delta$, $\delta' < \delta$). The theorem is proved. □

* See [1], p. 235.

Similarly, we can prove:

Theorem 10. *A necessary and sufficient condition for* $\rho(\tau) = O(e^{-\tau\delta})$ *for any* $\delta > 0$ *is that the analytic extension of* $f(\lambda)$ *be an entire function of a complex variable* $\zeta = \lambda + i\mu$.

We shall note the following.

Theorem 11. *A necessary and sufficient condition for* $\rho(\tau) = 0$ *for all* $\tau \geqslant k + 1$ *is that* $f(\lambda) = |P(e^{i\lambda})|^2$, *where* $P(z)$ *is a polynomial of degree less than* k.

Proof. Sufficiency is obvious. Necessity follows from the inequality

$$\left|\int_{-\pi}^{\pi} e^{i\lambda s} f(\lambda)\, d\lambda\right| \leqslant \rho(s) \int_{-\pi}^{\pi} f(\lambda)\, d\lambda = 0,$$

if $|s| \geqslant k + 1$. □

Note 1. Theorem 8 can be refined if we permit the index β to take on the value 1 as well, and if we invoke, instead of Jackson's theorem and Bernstein's theorem, a finer result for estimating the magnitude of best approximation by the index of smoothness.* Theorem 8 thus refined can be stated as follows.

Theorem 8′. *A necessary and sufficient condition for* $\rho(\tau) = O(\tau^{-r-\beta})$ *for some* β, $0 < \beta \leqslant 1$, *is that* $f(\lambda)$ *be represented as*

$$f(\lambda) = |P(e^{i\lambda})|^2 w(\lambda),$$

where $w(\lambda)$ *is strictly positive, has* r *derivatives, and is uniformly in* μ:

$$\max_{h \leqslant \delta} |w^{(r)}(\lambda + h) + w^{(r)}(\lambda - h) - 2w^{(r)}(\lambda)| = O(\delta^{\beta}).$$

Note 2. Lemma 17 implies the following useful assertion: If

$$\sum_{k=0}^{\infty} \rho(2^k) < \infty,$$

the spectral density $f(\lambda)$ will be continuous.

* See [27], p. 275.

CHAPTER VI

Complete Regularity and Processes with Continuous Time

VI.1 Introduction

We shall investigate in this section spectral characteristics of completely regular stationary processes $\xi(t)$ with continuous time t, $-\infty < t < \infty$. In Section IV.1 we obtained a characterization of the spectrum of Gaussian stationary processes satisfying a strong mixing condition. We note in advance that the results regarding the behavior of spectral densities $f(\lambda)$ of completely regular processes with continuous time on any finite interval of variation of λ are completely analogous to the results obtained in Chapter V for processes with discrete time. Specific difficulties arise only in the case of $f(\lambda)$ as $\lambda \to \infty$; unfortunately, the investigation of this case is less complete, although Theorems 5 and 6 do give some idea about phenomena arising here (see Sections VI.5 and VI.6).

Each completely regular process is (linearly) regular, as we have seen in Section IV. 1, and

(a) it has spectral density $f(\lambda)$ representable as $f = |g|^2$ where g is an exterior function of class $\mathscr{H}^2$ in the upper half-plane:

(b) $\int_{-\infty}^{\infty} \ln f(\lambda)/(1 + \lambda^2)\, d\lambda < \infty$.

From the above and Theorem 2 in Chapter II it follows immediately that the regularity coefficient $\rho(\tau)$ (by definition, equal to $\sup|\mathsf{M}\eta_1\bar{\eta}_2|$, where $\eta_1 \in H(-\infty, 0)$, $\eta_2 \in H(0, \infty)$, and $\mathsf{M}|\eta_1|^2 = \mathsf{M}|\eta_2|^2 = 1$) can be expressed as follows:

$$\rho(\tau) = \sup_{\varphi,\,\psi} \left| \int_{-\infty}^{\infty} e^{i\tau\lambda} \varphi(\lambda)\psi(\lambda) f(\lambda)\, d\lambda \right|, \tag{1.1}$$

where the supremum is taken over all φ, ψ in the unit sphere $E^+(f)$ of the space $L^+(f) = (1/g)\mathscr{H}^2$. (As in the previous chapter, we deal here with absolutely continuous spectral functions $F, G, \ldots$; hence, instead of $L^+(F)$,

$L^+(G)$, $\|\cdot\|_F$, $(\cdot,\cdot)_G$, etc., we prefer to write $L^+(f)$, $L^+(g)$, $\|\cdot\|_f$, $(\cdot,\cdot)_g$, where $f = F'$, $g = G'$, ...).

It can be seen that the value of $\rho(\tau)$ will not change if the supremum is taken in (1.1) over a dense subset of the sphere $E^+(f)$ rather than over the whole sphere $E^+(f)$ itself, for example, over $\mathscr{H}^2 \cap E^+(f)$, over functions of the form $\sum a_j e^{i\lambda t_j}$, $t_j \geqslant 0$, etc.

Similarly to (1.5) in Chapter V, we can deduce from (1.1) the equality

$$\rho(\tau) = \sup_{\theta} \left| \int_{-\infty}^{\infty} e^{i\lambda\tau}\theta(\lambda) f(\lambda)\, d\lambda \right|, \tag{1.2}$$

where the supremum is taken over all $\theta \in \mathscr{H}^1$, $\|\theta\|_f^{(1)} \leqslant 1$.

Finally, using the equality $L^+(f) = (1/g)\mathscr{H}^2$ and replacing φ, ψ in (1.1) with φ_1/g, ψ_1/g, where $\varphi, \psi \in \mathscr{H}^2$, we find that

$$\rho(\tau) = \sup_{\varphi_1, \psi_1} \left| \int_{-\infty}^{\infty} e^{i\lambda\tau}\varphi_1(\lambda)\psi(\lambda) \frac{\overline{g(\lambda)}}{g(\lambda)}\, d\lambda \right|, \tag{1.3}$$

where φ_1 and ψ_1 run through the unit sphere $\mathscr{H}^2$. (1.3) is similar to (1.6) in Chapter V. It also follows (compare with the deduction of (1.7) from (1.6) in Chapter V) that

$$\rho(\tau) = \sup_{\theta} \left| \int_{-\infty}^{\infty} \theta(\lambda) e^{i\lambda\tau} \frac{\overline{g(\lambda)}}{g(\lambda)}\, d\lambda \right|, \qquad \theta \in \mathscr{H}^1,\ \|\theta\|^{(1)} \leqslant 1. \tag{1.4}$$

Further, considering the integral in the right-hand side of (1.4) as a functional $l(\theta)$ on $\mathscr{L}^1$, we can arrive (as in Section V.2) at the equality

$$\rho(\tau) = \inf_{A} \left\| \frac{\bar{g}}{g} e^{i\lambda\tau} - A \right\|,$$

where inf is taken over all $A \in \mathscr{H}^\infty$ in the upper half-plane. Starting from this equality, we can proceed in the same way as in Section V.2; however, we shall encounter difficulties preventing us from obtaining as complete a result as Theorem 3 in Chapter V, since instead of polynomials we shall have to deal with entire functions of finite degree, whose independence of ε will have to be proved. We restrict ourselves as in Theorem 1 in Chapter V to spectral densities of completely regular processes.

Theorem 1. *Let the spectral density $f(\lambda)$ of the stationary process $\xi(t)$ be*

$$f(\lambda) = |\Gamma(\lambda)|^2 w(\lambda), \tag{1.5}$$

where $\Gamma(\lambda)$ is a square-summable entire function of finite degree and where the function w permits for all $\varepsilon > 0$ the representation

$$w = \exp\{r_\varepsilon + t_\varepsilon + \tilde{s}_\varepsilon\}, \tag{1.6}$$

where r_ε is a bounded real function uniformly continuous on $(-\infty, \infty)$, t_ε is a real function with $\|t_\varepsilon\|^{(\infty)} \leqslant \varepsilon$, and $\tilde{s}_\varepsilon$ is a (harmonic) adjoint of the function s_ε with $\|s_\varepsilon\|^\infty \leqslant \varepsilon$. In this case the process $\xi(t)$ is completely regular.

We explain first what we understand as the adjoint of a bounded function. Let us denote by W^2 the class of measurable functions $a(\lambda)$ for which $a/(1+|\lambda|) \in \mathscr{L}^2(-\infty, \infty)$. In particular, $\mathscr{L}^\infty \subset W^2$. Let

$$a(\lambda) = (\lambda - i)\frac{1}{\sqrt{2\pi}}\int_{-\infty}^{\infty} \alpha(u)e^{i\lambda u}\,du,$$

where $\alpha(u)$ is the Fourier transform of the function $a(\lambda)/(\lambda - i) \in \mathscr{L}^2$. Let us define the adjoint function $\tilde{a}$ by the equality

$$\tilde{a}(\lambda) = (\lambda - i)\frac{1}{\sqrt{2\pi}}\int_{-\infty}^{\infty} \alpha(u) i \operatorname{sign} u \cdot e^{i\lambda u}\,du \tag{1.7}$$

(for details see [1], p. 171). It follows from (1.7) that

$$\int_{-\infty}^{\infty} \frac{|\tilde{a}(\lambda)|}{1+\lambda^2}\,d\lambda \leqslant \frac{1}{\sqrt{2\pi}}\|(\lambda + i)^{-1}\|^{(2)} \cdot \|\alpha\|^{(2)}. \tag{1.8}$$

In proving Theorem 1 and further on we shall deal with nonnegative functions f that may turn out to be nonsummable with the value $\rho(\tau) \to 0$ being defined still as (1.1). In this case we shall need the class of functions we shall call *completely regular functions.*

Definition. A nonnegative measurable function $f(\lambda)$ is said to be a *completely regular function* if

1. $\int_{-\infty}^{\infty} \ln f(\lambda)/(1+\lambda^2)\,d\lambda < \infty$;
2. $\rho(\tau) = \sup_{\varphi, \psi} \left|\int_{-\infty}^{\infty} \varphi(\lambda)\psi(\lambda)e^{i\lambda\tau}f(\lambda)\,d\lambda\right| \xrightarrow[\tau\to\infty]{} 0$, the supremum being taken over all $\varphi, \psi \in \mathscr{H}^{2+}$ with $\|\varphi\|_f = \|\psi\|_f = 1$.

Each completely regular summable function is the spectral density of a completely regular process and, conversely, the spectral density of a completely regular process is a completely regular function.

By virtue of (1) each completely regular function is factorizable: $f(\lambda) = |g(\lambda)|^2$ where

$$g(z) = \exp\left\{\frac{1}{2\pi i}\int_{-\infty}^{\infty} \frac{1+\lambda z}{\lambda - z}\frac{\ln f(\lambda)}{1+\lambda^2}\,d\lambda\right\}$$

is the outer function in the upper half-plane. From this it follows immediately (compare with Section II.2)* that the closure $L^+(f)$ in $L(f)$ of the set $\mathscr{H}^2 \cap L(f)$ is equal to $(1/g)\mathscr{H}^2$.

Returning to Theorem 1 we prove first that the function w permitting the representation given by (1.6) is completely regular. A uniformly continuous bounded function e^{r_ε} can be adequately approximated by bounded

* The proof is as follows. The function $w^+ = \max(1, w)$ is factorizable: $w^+ = |g^+|^2$ where g^+ is an outer function. If φ is an exterior function from $\mathscr{H}^2$, $\varphi/g^+ \subset \mathscr{H}^2$ and $\varphi/g^+ \in L(w)$, so that $\mathscr{H}^2 \cap L(w)$ is nonempty. Further, φg is an exterior from $\mathscr{H}^2$, and by Lax's theorem (see Section II.1) a set dense in $\mathscr{H}^2$ is contained in $g\mathscr{H}^2$ as well as in $g\varphi\mathscr{H}^2$. Therefore, $(1/g)\mathscr{H}^2 \subset L^+(w)$. Conversely, if $\psi \in L^+(w)$, the function ψg, being the limit in $\mathscr{L}^2$ of functions $\varphi_n g$, $\varphi_n \in \mathscr{H}^2$, $\varphi_n g \in \mathscr{H}^2$, is a function from $\mathscr{H}^2$, that is, $L^+(w) \subset (1/g)\mathscr{H}^2$.

entire functions of finite degree (see [27]). Let Q_T be a bounded entire function of degree $\leqslant T$ for which

$$e^{r_\varepsilon} = Q_T(1 + \theta), \qquad |\theta| \leqslant \varepsilon.$$

Then

$$w(\lambda) = Q_T e^{\tilde{v}_\varepsilon - iv_\varepsilon}(1 + \theta_\varepsilon), \qquad \|\theta_\varepsilon\|^{(\infty)} \leqslant 7\varepsilon.$$

Let $w_\varepsilon = |Q_T|\,|e^{\tilde{v}_\varepsilon - iv_\varepsilon}|$. If ε is sufficiently small, then for all $\varphi \in L(w)$

$$\frac{1}{2}\|\varphi\|_{w_\varepsilon} \leqslant \|\varphi\|_w \leqslant 2\|\varphi\|_{w_\varepsilon}.$$

Further, by (1.8), $\tilde{v}_\varepsilon/(1 + \lambda^2) \in \mathscr{L}^1$, so that $L^+(\omega) = L^+(e^{\tilde{v}_\varepsilon}) = (1/\gamma)\mathscr{H}^2$, where the exterior function $\gamma(z)$ is defined by the equality $|\gamma(\lambda)|^2 = e^{\tilde{v}_\varepsilon}$. Obviously, $\gamma = \exp\{1/2(\tilde{v}_\varepsilon - iv)\}$.

If $\varphi, \psi \in L^+(w)$, then $\varphi = \varphi_1/\gamma$ and $\psi = \psi_1/\gamma$, where $\varphi_1, \psi_1 \in \mathscr{H}^2$. In addition, for example, for $\tau > T e^{i\lambda\tau}Q_T \in \mathscr{H}^\infty$, $e^{i\lambda\tau}\varphi_1 Q_T \in \mathscr{H}^2$. Therefore, for all $\tau > T$ and all $\varphi, \psi \in L^+(w)$

$$\int_{-\infty}^{\infty} \varphi(\lambda)\psi(\lambda)e^{i\lambda\tau}Q_T(\lambda)e^{\tilde{v}_\varepsilon - iv_\varepsilon}\,d\lambda = \int_{-\infty}^{\infty} [\varphi_1(\lambda)e^{i\lambda\tau}Q_T(\lambda)]\psi_1(\lambda)\,d\lambda = 0.$$

Hence for all $\tau > T$

$$\begin{aligned}\rho(\tau, w) &= \sup_{\varphi,\psi}\left|\int_{-\infty}^{\infty} \varphi(\lambda)\psi(\lambda)e^{i\lambda\tau}w(\lambda)\,d\lambda\right| \\ &\leqslant \sup_{\varphi,\psi}\int_{-\infty}^{\infty} |\varphi(\lambda)||\psi(\lambda)||w_\varepsilon(\lambda)||\theta_\varepsilon(\lambda)|\,d\lambda \leqslant 28\varepsilon.\end{aligned}$$

The assertion of the theorem follows from the following lemma being an analog of Theorem 2 in Chapter V.

Lemma 1. *If a function w is completely regular, Γ is a bounded entire function of finite degree $\leqslant T$, and $f = |\Gamma|^2 w$, then the function f will be completely regular, with*

$$\rho(\tau: f) \leqslant \rho(\tau - 2T; w). \tag{1.9}$$

Proof of Lemma 1 and Theorem 1. $\ln|\Gamma|/(1 + \lambda^2) \in \mathscr{L}^1$ and for $\tau \geqslant T$ both Γ and $\bar{\Gamma}$ are such that $e^{i\lambda\tau}\Gamma$, $e^{\lambda i\tau}\bar{\Gamma} \in \mathscr{H}^\infty$.* Hence, if φ and ψ belong to the unit sphere $L^+(f)$, $e^{i\lambda T}\Gamma\varphi$ and $e^{i\lambda T}\bar{\Gamma}\psi$ belong to the unit sphere of the space $L^+(w)$, and therefore

$$\begin{aligned}\rho(\tau; f) &= \sup_{\varphi,\psi}\left|\int_{-\infty}^{\infty} \varphi(\lambda)\psi(\lambda)e^{i\lambda\tau}w(\lambda)|\Gamma(\lambda)|^2\,d\lambda\right| \\ &= \sup_{\varphi,\psi}\left|\int_{-\infty}^{\infty} [\varphi e^{i\lambda T}\Gamma][\psi e^{i\lambda T}\bar{\Gamma}]e^{i\lambda(\tau - 2T)}w\,d\lambda\right| \leqslant \rho(\tau - 2T; w).\end{aligned}$$

The above proves Lemma 1 as well as Theorem 1. □

* Recall that $\bar{\Gamma}(z) = \overline{\Gamma(\bar{z})}$.

Corollary *If a spectral density is of the form* $f(\lambda) = |\Gamma(\lambda)|^2 w(\lambda)$ *where* Γ *is a square-summable function of degree* $\leqslant T$, *and if the function* w *is bounded from above and below* $(0 < m \leqslant w \leqslant M < \infty)$ *and uniformly continuous on* $(-\infty, \infty)$, *the process* $\xi(t)$ *will be completely regular, with*

$$\rho(\tau) \leqslant \frac{1}{m} A_{\tau - 2T}(\omega). \tag{1.10}$$

Here and further on we shall denote by $A_\sigma(h)$ the value of the best uniform approximation of the function h by entire functions of finite degree $\leqslant \sigma$. In fact, $w = e^{\ln w}$ and under the assumptions made in the corollary the function $\ln w$ is bounded and uniformly continuous. The inequality given by (1.10) follows readily from the calculations used in proving Theorem 1.

VI.2 The Investigation of a Particular Function, $\gamma(T; \mu)$

In Sections VI.2, VI.3, and VI.4 we shall extend the results from Sections V.3 and V.4 to the case of continuous time. We investigate first the behavior of the Fejer integral

$$\gamma(T; \mu) = \int_{-\infty}^{\infty} \frac{\sin^2 \dfrac{T\lambda}{2}}{\lambda^2} f(\lambda + \mu)\, d\lambda \tag{2.1}$$

as $T \to \infty$, where $f(\lambda)$ is the spectral density of a completely regular process $\xi(t)$. The function $\gamma(T; \mu)$ will play from now on the same role as that played by the function $\gamma(N; \mu)$ in Chapter V. We note that

$$\gamma(T; \mu) = \mathsf{M} \left| \int_0^T e^{-i\mu t} \xi(t)\, dt \right|^2, \tag{2.2}$$

to prove which we take the spectral representation of the process

$$\xi(t) = \int_{-\infty}^{\infty} e^{it\lambda} \Phi(d\lambda),$$

where $\Phi(d\lambda)$ is an orthogonal random measure ($\mathsf{M}|\Phi(d\lambda)|^2 = f(\lambda)\, d\lambda$). We write

$$\begin{aligned} \mathsf{M} \left| \int_0^T e^{-i\mu t} \xi(t)\, dt \right|^2 &= \mathsf{M} \left| \int_{-\infty}^{\infty} \frac{e^{i(\lambda - \mu)T} - 1}{i(\lambda - \mu)} \Phi(d\lambda) \right|^2 \\ &= \int_{-\infty}^{\infty} \frac{\sin^2 T \dfrac{\lambda - \mu}{2}}{(\lambda - \mu)^2} f(\lambda)\, d\lambda \\ &= \int_{-\infty}^{\infty} \frac{\sin^2 \dfrac{T\lambda}{2}}{\lambda^2} f(\lambda + \mu)\, d\lambda. \end{aligned} \tag{2.3}$$

Lemma 2. *If the stationary process $\xi(t)$ has the spectral density $f(\lambda)$, the value*

$$\mathsf{M}\left|\int_0^T \xi(t)\,dt\right|^2 = \int_{-\infty}^{\infty} \frac{\sin^2 \dfrac{T\lambda}{2}}{\lambda^2} f(\lambda)\,d\lambda$$

either tends to ∞ or is bounded as $T \to \infty$. It is bounded if and only if $\int_{-\infty}^{\infty} (f(\lambda)/\lambda^2)\,d\lambda < \infty$.

The proofs of Lemma 2 and other two lemmas, which follow, are similar to analogous assertions (Lemmas 4–6) in Chapter V. Hence we shall give a detailed proof of Lemma 4 only.

Lemma 3. *For each μ, as $T \to \infty$, the function $\gamma(T)$ either tends to ∞ or is bounded. The function $\gamma(T)$ is bounded if and only if*

$$\int_{-\infty}^{\infty} \frac{f(\lambda)}{(\lambda - \mu)^2}\,d\lambda < \infty.$$

Lemma 4. *Let $0 < a < \infty$. As $T \to \infty$ either $\inf_{|\mu| \leqslant a} \gamma(T; \mu) \to \infty$ or there exists a point $\theta \in [-a, a]$ such that $\int_{-\infty}^{\infty} (f(\lambda)/(\lambda - \theta)^2)\,d\lambda < \infty$.*

Proof of Lemma 4. The proof is similar to a great extent to the proof in the case of discrete time (Lemma 6, Chapter V). The equality $U^\tau \xi(t) = \xi(t + \tau)$ defines the group of unitary operators $\{U^\tau\}$ in a Hilbert space $H(-\infty, \infty)$. If

$$\lim_{T \to \infty} \inf_{|\mu| \leq a} \gamma(T; \mu) < \infty,$$

then, arguing in the same way as in the proof of Lemma 6 in Chapter V, we can define the sequence

$$\zeta_k \in H(-\infty, \infty), \qquad \zeta_k = \int_0^{T_k} e^{-it\theta_k} U^t \xi(0)\,dt,$$

weakly convergent to a value $\zeta \in H(-\infty, \infty)$. In this case the numerical sequence θ_k is convergent to a number $\theta \in [-a, a]$. Therefore, (weak) $\lim e^{-i\tau\theta_k}\zeta_k = e^{-i\tau\theta}\zeta$, τ denoting an arbitrary real number.

By the Riemann–Lebesgue lemma

$$B(s) = \mathsf{M}\xi(t + s)\overline{\xi(t)} = (\xi(t + s), \xi(t)) = \int_{-\infty}^{\infty} e^{i\lambda s} f(\lambda)\,d\lambda \xrightarrow[|s| \to \infty]{} 0.$$

Hence for all $h \in H(-\infty, \infty)$

$$\lim_{s \to \infty} (\xi(s), h) = 0, \qquad \text{(weak)} \lim_{s \to \infty} \xi(s) = 0,$$

and therefore,

$$\begin{aligned}\zeta - e^{-i\theta\tau} U^\tau \zeta &= \text{(weak)} \lim(\zeta_k - e^{-i\theta_k\tau} U^\tau \zeta_k) \\ &= \text{(weak)} \lim\left(\int_0^\tau e^{-i\theta_k t} \xi(t)\,dt + \int_{T_k}^{T_k + \tau} e^{-i\theta_k t} \xi(t)\,dt\right) \\ &= \int_0^\tau e^{-i\theta t} \xi(t)\,dt. \qquad (2.4)\end{aligned}$$

Let us form a stationary process $\eta(t) = e^{-i\theta t}U^t\zeta$. By (2.4) we have

$$\eta(t) - \eta(t+\tau) = e^{-i\theta t}U^t(\zeta - e^{-i\theta\tau}U^\tau\zeta) = \int_t^{t+\tau} e^{-i\theta s}\xi(s)\,ds, \tag{2.5}$$

which implies that the process $\eta(t)$ is differentiable in the mean square and that

$$\eta'(t) = -e^{-i\theta t}\xi(t). \tag{2.6}$$

Let $f_\eta(\lambda)$, $f_\xi(\lambda)$ denote spectral densities of the processes $\eta(t)$, $\xi(t)$. From (2.6) we have the following relationship between f_ξ and f_η:

$$\lambda^2 f_\eta(\lambda) = f_\xi(\lambda + \theta).$$

Therefore,

$$\int_{-\infty}^{\infty} \frac{f(\lambda)\,d\lambda}{(\lambda-\theta)^2} = \int_{-\infty}^{\infty} f_\eta(\lambda)\,d\lambda < \infty.$$

Conversely, if $\int_{-\infty}^{\infty} (f(\lambda)/(\lambda-\theta)^2)\,d\lambda < \infty$, by Lemma 3 we have

$$\overline{\lim_{T\to\infty}}\ \gamma(T;\theta) < \infty,$$

thus proving Lemma 4.

Lemma 5. *Let the process $\xi(t)$ be completely regular. Then there exists a square-integrable entire function of finite degree $\Gamma(\lambda)$ such that*

$$\int_{-\infty}^{\infty} \frac{|\Gamma(\lambda)|^2}{f(\lambda)}\,d\lambda < \infty. \tag{2.7}$$

The proof of Lemma 5 follows word by word the first part of the proof of Lemma 7, Chapter V; now we need, however, to apply results on non-interpolatability of processes with continuous time.*

Lemma 6. *Let the process $\xi(t)$ be completely regular. For any number a, $0 < a < \infty$, the spectral density $f(\lambda)$ of the process $\xi(t)$ can be expressed as*

$$f(\lambda) = w(\lambda)|P_a(\lambda)|^2, \tag{2.8}$$

where $P_a(\lambda)$ is an algebraic polynomial with real roots (located inside the interval $[-a, a]$), and $w(\lambda)$ has the property that

$$\lim_T \inf_{|\mu|\leqslant a} \int_{-\infty}^{\infty} w(\lambda)\frac{\sin^2 T\dfrac{\lambda-\mu}{2}}{(\lambda-\mu)^2}\,d\lambda = \infty. \tag{2.9}$$

Proof. Let us denote by $\mathscr{P}$ the aggregate of all polynomials $Q(\lambda)$ with roots inside $[-a, a]$ for which

$$\int_{-a}^{a} \frac{|Q(\lambda)|^2}{f(\lambda)}\,d\lambda < \infty.$$

* See, for example, [24], p. 183.

According to the previous lemma the set $\mathscr{P}$ is nonempty. (We can take as $Q(\lambda)$ a polynomial whose roots coincide with the roots of $\Gamma(\lambda)$ lying on $[-a, a]$.) Let $Q_a(\lambda)$ be one of the polynomials $Q(\lambda) \in \mathscr{P}$ of the smallest degree and coefficient one for the highest degree term. Among polynomials $P(\lambda)$ with roots inside $[-a, a]$ for which $\int_{-\infty}^{\infty} (f(\lambda)/|P(\lambda)|^2)\, d\lambda < \infty$, there exists a polynomial $P_a(\lambda)$ of maximal (finite) degree. In fact, from the inequality

$$\left(\int_{-a}^{a} \left|\frac{Q_a(\lambda)}{P(\lambda)}\right| d\lambda\right)^2 \leqslant \int_{-a}^{a} \frac{|Q_a(\lambda)|^2}{f(\lambda)}\, d\lambda \cdot \int_{a}^{a} \frac{f(\lambda)}{|P(\lambda)|^2}\, d\lambda$$

it follows that all polynomials $P(\lambda)$ divide $Q_a(\lambda)$. Next let $f(\lambda) = |P_a(\lambda)|^2 w(\lambda)$. Obviously,

$$\int_{-\infty}^{\infty} w(\lambda)\, d\lambda < \infty;$$

therefore, $w(\lambda)$ is the spectral density of a certain stationary process. Further considerations follow those used in proving Lemma 7 in Chapter V, and hence they will be omitted here. The lemma is proved. □

Lemma 7. *Let the process $\xi(t)$ be completely regular. At the points μ where $\lim_T \gamma(T; \mu) = \infty$, the function $\gamma(T; \mu)$ can be represented as $\gamma(T; \mu) = Th(T; \mu)$ where $h(T; \mu)$ is a slowly varying function T, that is, for all $x > 0$*

$$\lim_{T \to \infty} \frac{h(Tx; \mu)}{h(T; \mu)} = 1.$$

Proof of Lemma 7. The proof is essentially similar to analogous assertions made in Chapter V (Lemmas 8–10); hence we shall omit some details. We need to show that for all $x > 0$

$$\lim_T \frac{\gamma(Tx)}{\gamma(T)} = x \tag{2.10}$$

(here and below we omit the argument μ for the sake of brevity). As in Chapter V, we discuss first the case involving integral x; second, we establish some properties of the function $\gamma(T)$; and next we prove (2.10) for all x.

1. x is integral, $x = k$. Let us write $\gamma(kT)$ as

$$\gamma(kT) = \mathsf{M}|z_1 + y_1 + \ldots + z_k + y_k|^2,$$

where

$$z_j = \int_{(j-1)T+(j-1)r}^{jT+(j-1)r} e^{-i\mu t}\xi(t)\, dt, \qquad j = \overline{1, k},$$

$$y_j = \int_{jT+(j-1)r}^{jT+jr} e^{-i\mu t}\xi(t)\, dt, \qquad j = \overline{1, k-1},$$

$$y_k = -\int_{kT}^{kT+(k-1)r} e^{-i\mu t}\xi(t)\, dt,$$

and where $r = r(T)$ is such that* $r(T) \xrightarrow[T\to\infty]{} \infty$ but

$$\frac{\gamma(r)}{\gamma(T)} \xrightarrow[T\to\infty]{} 0, \qquad \max_{1\leqslant s\leqslant k} \frac{\gamma(sr)}{\gamma(T)} \xrightarrow[T\to 0]{} 0.$$

It is easy to calculate that

$$\mathsf{M}|z_j|^2 = \gamma(T), \qquad \mathsf{M}|y_i|^2 = \gamma(r), \qquad j \leqslant k-1;$$
$$\mathsf{M}|y_k|^2 = \gamma((k-1)r).$$

Hence the same arguments as those in Section V.3 enable us to assert

$$\gamma(kT) = k\gamma(T)\left(1 + o\left(k\rho(r) + k\left(\frac{\gamma(r)}{\gamma(T)}\right)^{1/2} + k\left(\frac{\gamma((k+1)r)}{\gamma(T)}\right)^{1/2}\right)\right)$$
$$= k\gamma(T)(1 + o(1)).$$

We have completed the case of integral x.

2. *x is rational*, $x = p/q$. We have

$$\lim_T \frac{\gamma\left(\frac{p}{q}T\right)}{\gamma(T)} = \lim_T \frac{\gamma\left(p\frac{T}{q}\right)}{\gamma\left(\frac{T}{q}\right)} \frac{\gamma\left(\frac{T}{q}\right)}{\gamma\left(q\frac{T}{q}\right)} = \frac{p}{q}.$$

3. *x is arbitrary*. As in Section V.3, we deduce that

$$\ln h(T) = o(\ln T),$$

and therefore for all $\varepsilon > 0$

$$\lim_T T^{\varepsilon} h(T) = \infty, \qquad \lim_T T^{-\varepsilon} h(T) = 0. \tag{2.11}$$

Relying on (2.11) and arguing in the same way as in Section V.3, we can easily show that the functions $\psi_1(x)$, $\psi_2(x)$, defined by the relations

$$\psi_1(x) = \underline{\lim}_T \frac{\gamma(Tx)}{\gamma(T)}, \qquad \psi_2(x) = \overline{\lim}_T \frac{\gamma(Tx)}{\gamma(T)},$$

are continuous. Since for rational x we have $\psi_1(x) = \psi_2(x) = x$, for all real x we shall have in turn

$$\lim_T \frac{\gamma(Tx)}{\gamma(T)} = x.$$

We have proved the lemma. □

Lemma 8. *If*

$$\gamma_a(T) = \inf_{|\mu|\leqslant a} \gamma(T; \mu) \xrightarrow[T\to\infty]{} \infty,$$

* Since $\gamma(T) = \int_{-\infty}^{\infty} (\sin^2(\frac{1}{2}T\lambda)/\lambda^2) f(\lambda + \mu)\, d\lambda \leqslant T^2 \int_{-\infty}^{\infty} f(\lambda)\, d\lambda$, we can put $r = \ln\gamma(T)$, whereas $\gamma(sr) \leqslant k^2 \ln^2 \gamma(T) \int_{-\infty}^{\infty} f(\lambda)\, d\lambda$.

the relation

$$\lim_{T} \frac{h(Tx)}{h(T)} = 1 \tag{2.12}$$

holds uniformly in all μ *and* x *such that* $|\mu| \leqslant a$, $0 < x_0 < x < x_1 < \infty$.

Lemma 8 can be proved exactly in the same way as Lemma 10 in Chapter V, and the proof will therefore be omitted.

VI.3 The Proof of the Basic Theorem on Necessity

The objective of this section is to prove Theorem 2, which is analogous to Theorem 5 in Chapter V.

Theorem 2. *Let* $\xi(t)$ *be a completely regular process with spectral density* $f(\lambda)$. *For any number* a, $0 < a < \infty$, *the function* $f(\lambda)$ *can be expressed as*

$$f(\lambda) = |P_a(\lambda)|^2 w_a(\lambda),$$

where $P_a(\lambda)$ *is an algebraic polynomial with roots on* $[-a, a]$ *and the primitive* $W_a(\lambda)$ *of the function* $w_a(\lambda)$ *satisfies the condition*

$$\omega_W(\delta) = \sup_{|\lambda| \leqslant a} \sup_{|t| \leqslant \delta} \frac{|W_a(\lambda + t) + W_a(\lambda - t) - 2W_a(\lambda)|}{|W_a(\lambda + t) - W_a(\lambda - t)|} \xrightarrow[\delta \to 0]{} 0. \tag{3.1}$$

The condition given by (3.1) is a complete analog of the condition given by (3.2) in Chapter V; (3.1) imposes on the smoothness and order of zeros of the spectral density $f(\lambda)$ the same restrictions as the condition given by (3.2) in Chapter V does in the case of discrete time. In fact, using the same techniques as those described in Section V.5, we deduce from (3.1) the following.

Corollary 1. *The spectral density of a completely regular process has no discontinuities of the first kind.*

We shall need some additional considerations to prove:

Corollary 2. *The lower order of any zero* λ_0 *of the spectral density* $f(\lambda)$ *of a completely regular process is equal either to zero or to an even integer. Therefore, the real order of any zero of* $f(\lambda)$ *can be only integral and even.*

Corollary 3. *For all* $\varepsilon > 0$

$$\varliminf_{\lambda \to \lambda_0} f(\lambda)|\lambda - \lambda_0|^{\varepsilon} = 0.$$

Proof of Corollaries 1 and 2. First we prove the following assertion: if the condition given by (3.1) is satisfied, the functions

$$h_\mu(x) = h(x) = \frac{1}{x}\int_0^x w_a(\lambda + \mu)\, d\lambda$$

vary slowly as $x \to 0$, $\lim_{n \to 0} (h(zx)/h(x)) = 1$ uniformly over all $\mu \in [-a, a]$, and $z \in [s, S]$, $0 < s < S < \infty$.

In fact, if z is integral, it can be seen that (3.1) yields

$$\frac{\dfrac{1}{zx}\, h(zx)}{\dfrac{1}{x}\, h(x)} = 1 + o(1).$$

From this it follows readily that $\lim(h(zx)/h(x)) = 1$ for all rational z. The final part of the proof coincides with the final part of the proof of Lemmas 7 and 8; in this case we should also use the following obvious inequality: for $\varepsilon < 1/q$

$$\int_0^{\varepsilon x} w_a(\lambda + \mu)\, d\lambda \leqslant \int_0^{x/q} w_a(\lambda + \mu)\, d\lambda.$$

Next we prove that the lower order $(\underline{k}(\lambda_0))$ for any zero λ_0 of the function $w_a(\lambda)$ from Theorem 2 is equal to zero. If $\underline{k}(\lambda_0) > \varepsilon > 0$, then $w_a(\lambda) < |\lambda - \lambda_0|^{\varepsilon/2}$ in some neighborhood of the point λ_0, and therefore

$$h_{\lambda_0}(x) = \frac{1}{x}\int_0^x w_a(\lambda + \lambda_0)\, dx < x^{\varepsilon/2}.$$

This is impossible, however, since the function $h(x)$ varies slowly and $\lim_{x \to 0} h(x)x^{-\varepsilon} = \infty$ for all $\varepsilon > 0$.

We have proved Corollary 2. We can prove Corollary 3 in a similar way. □

The proof of Theorem 2 is based on the same arguments as the proof of Theorem 5 in Chapter V. As noted, Theorem 2 is an assertion of the Tauberian type on the local behavior of $W_a(\lambda)$ based on the properties of the function

$$\gamma(T; \mu) = \int_{-\infty}^{\omega} \frac{\sin^2 T \dfrac{\lambda - \mu}{2}}{(\lambda - \mu)^2}\, dW_a(\lambda).$$

As a further preliminary to the proof of Theorem 2 we shall get rid of zeros of $f(\lambda)$ using Lemma 6 and Lemma 9, which follows.

Lemma 9. *If $f(\lambda)$ is the spectral density of a completely regular process $\xi(t)$, all poles and zeros of the rational function $R(\lambda) = P(\lambda)/Q(\lambda)$ are real, and* *the function* $w = |R|^2 f \in \mathscr{L}^1(-\infty, \infty)$, then $w(\lambda)$ *is the spectral density of a stationary process* $\eta(t)$, *with*

$$\rho(\tau; \omega) \leqslant \rho(\tau; f). \qquad \square$$

Proof. If functions φ, ψ belong to the unit sphere of $L^+(w)$, then φ/R, $\psi/\bar{R}$ lie in the unit sphere in $L^+(f)$. Hence

$$\rho(\tau; w) = \sup_{\varphi, \psi} \left| \int_{-\infty}^{\infty} \varphi(\lambda)\psi(\lambda)e^{i\lambda\tau}w(\lambda)\, d\lambda \right|$$

$$= \sup_{\varphi, \psi} \left| \int_{-\infty}^{\infty} \frac{\varphi(\lambda)}{R(\lambda)} \frac{\psi(\lambda)}{\overline{R(\lambda)}} e^{i\lambda\tau} f(\lambda)\, d\lambda \right| \leqslant \rho(\tau; f). \qquad \square$$

Proceeding as in Section V.4 we shall assume (starting here and concluding with Lemma 12) that $f(\lambda)$ is the spectral density of a completely regular sequence for which

$$\lim_{T} \inf_{|\mu| \leqslant a} \gamma(T; \mu) = \infty.$$

Lemma 10. *Let* $a(\lambda)$ *be an even function with the bounded third derivative vanishing outside of the interval* $[-1, 1]$. *Then, as* $T \to \infty$, *uniformly in* $|\mu| \leqslant a$

$$\frac{1}{T} \int_{-\infty}^{\infty} \frac{\sin^2 \dfrac{T\lambda}{2}}{\lambda^2} f(\lambda + \mu) a(T\lambda)\, d\lambda = \frac{1}{\pi} \int_0^1 \frac{\sin^2 \dfrac{\lambda}{2}}{\left(\dfrac{\lambda}{2}\right)^2} a(\lambda)\, d\lambda \cdot h(T)(1 + o(1)), \tag{3.2}$$

where, for brievity, $h(T)$ *denotes* $h(T; \mu)$.

Proof. The proof of this lemma coincides in essence with the proof of Lemma 12 in Chapter V. Let us introduce the Fourier transform $A(z)$ of the function $a(\lambda)$. Given $\varepsilon > 0$ let us also define s, δ, and the set B as in Section V.4. Further, let us rewrite the left-hand side of (3.2) as

$$\frac{1}{\sqrt{2\pi}} \left[\int_{1+\delta}^{s} z[h(T(z + 1)) + h(T(z - 1) - 2h(Tz)]A(z)\, dz \right.$$

$$+ \int_{1+\delta}^{s} [h(T(z + 1)) - h(T(z - 1))]A(z)\, dz$$

$$+ \int_{\delta}^{1-\delta} z[h(T(z + 1)) - h(T(z - 1)) - 2h(Tz)]A(z)\, dz$$

$$+ \int_{\delta}^{1-\delta} [h(T(z + 1)) + h(T(z - 1))]A(z)\, dz + R(T),$$

where $|R(T)| \leqslant \varepsilon h(T)$.

From this and Lemma 8 we deduce that, as $T \to \infty$, uniformly in $|\mu| \leqslant a$

$$\frac{1}{T}\int_{-\infty}^{\infty} \frac{\sin^2 \frac{T\lambda}{2}}{\lambda^2} f(\lambda + \mu)a(T\lambda)\, d\lambda$$

$$= \sqrt{\frac{2}{\pi}} \int_0^1 (1 - z)A(z)\, dz \cdot h(T)(1 + o(1))$$

$$= \frac{1}{\pi}\int_0^1 \frac{\sin^2 \frac{\lambda}{2}}{\left(\frac{\lambda}{2}\right)^2} a(\lambda)\, d\lambda\, (1 + o(1)), \tag{3.3}$$

thus proving the lemma. □

Lemma 11. *Let $a(\lambda)$ be an odd thrice differentiable function with the bounded third derivative vanishing outside of the interval $[-1, 1]$. Then as $T \to \infty$, uniformly in $|\mu| \leqslant a$*

$$\frac{1}{T}\int_{-\infty}^{\infty} \frac{\sin^2 \frac{T\lambda}{2}}{\lambda^2} f(\lambda + \mu)a(T\lambda)\, d\lambda = o(h(T)), \tag{3.4}$$

where $h(T) = h(T; \mu)$ as above.

Proof. The proof of this lemma is even simpler than the proof of its discrete analog (Lemma 13 in Chapter V). We denote by $A(z)$ the Fourier transform of the function $a(\lambda)$ and rewrite the integral to be estimated as follows:

$$\frac{1}{T}\int_B A(z)\, dz \int_{-\infty}^{\infty} \frac{\sin^2 \frac{T\lambda}{2}}{\lambda^2} \sin T\lambda z f(\lambda + \mu)\, d\lambda + R(T),$$

where $|R(T)| \leqslant \varepsilon h(T)$ and B is the set $\{z\colon \delta < z < 1 - \delta,\ 1 + \delta < z < s\}$. It is easy to calculate that

$$\frac{\sin^2 \frac{T\lambda}{2}}{\lambda^2} = \frac{1}{4}\left|\int_0^T e^{i\lambda\tau}\, d\tau\right|^2.$$

Further,

$$\sin T\lambda z = \frac{1}{2i}(e^{iT\lambda z} - e^{-iT\lambda z});$$

therefore, for z for which $Tz > T + T_0$, the products of $\exp\{iT\lambda z\}$, $\exp\{-iT\lambda z\}$, and $\sin^2(T\lambda/2)/\lambda^2$ can be written respectively as

$$e^{iT_0\lambda}\varphi_z^+(\lambda), \qquad e^{-iT_0\lambda}\varphi_z^-(\lambda),$$

where $\varphi_z^+(\lambda)$, $\varphi_z^-(\lambda)$ belong to spaces $\mathscr{H}^1$ in the upper half-plane and lower half-plane, respectively. By virtue of the property of complete regularity*

$$\frac{1}{T}\left|\int_{B\cap\{Tz>T+T_0\}} A(z)\,dz \int_{-\infty}^{\infty} \frac{\sin^2 \dfrac{T\lambda}{2}}{\lambda^2} \sin T\lambda z f(\lambda+\mu)\,d\lambda\right|$$

$$= \frac{1}{T}\left|\int_{B\cap\{Tz>T+T_0\}} A(z)\,dz \int_{-\infty}^{\infty} [e^{iT_0\lambda}\varphi_z^+(\lambda) + e^{-iT_0\lambda}\varphi_z^-(\lambda)] \times f(\lambda+\mu)\,d\lambda\right|$$

$$\leqslant \rho(T_0)\int_{-\infty}^{\infty} |A(z)|\,dz \cdot \frac{1}{T}\int_{-\infty}^{\infty} (|\varphi_z^+(\lambda)| + |\varphi_z^-(\lambda)|) f(\lambda+\mu)\,d\lambda \leqslant \varepsilon h(T) \tag{3.5}$$

for T_0 sufficiently large. Therefore, it remains only to study the behavior of the integral considered in the case where integration is carried out over the domain $\{\delta < z < 1-\delta\}$. To this end we change slightly the integrand in (3.5). We have

$$\begin{aligned}\frac{\sin^2 \dfrac{T\lambda}{2}}{\lambda^2} e^{iT\lambda z} &= \frac{1}{4} e^{iT\lambda z}\int_0^T e^{i\tau\lambda}\,d\tau \int_0^T e^{-i\tau\lambda}\,d\tau \\ &= \frac{1}{4}\int_0^T e^{i\tau\lambda}\,d\tau \int_{-Tz}^{T(1-z)} e^{-i\tau\lambda}\,d\tau \\ &= \frac{1}{4}\left[\int_0^T e^{i\lambda\tau}\,d\tau \int_0^{Tz} e^{i\lambda\tau}\,d\tau + \left|\int_0^{T(1-z)} e^{i\tau\lambda}\,d\tau\right|^2\right. \\ &\quad \left. + \int_{T(1-z)}^{T} e^{i\tau\lambda}\,d\tau \int_0^{T(1-z)} e^{-i\tau\lambda}\,d\tau\right]\end{aligned} \tag{3.6}$$

(we recall that now $\delta < z < 1-\delta$). Similarly,

$$\begin{aligned}&\frac{\sin^2 \dfrac{T\lambda}{2}}{\lambda^2} e^{-iT\lambda z} \\ &= \frac{1}{4}\left[\int_0^T e^{-i\tau\lambda}\,d\tau \int_0^{Tz} e^{-i\tau\lambda}\,d\tau + \left|\int_0^{T(1-z)} e^{i\tau\lambda}\,d\tau\right|^2\right. \\ &\quad \left. + \int_{T(1-z)}^{T} e^{-i\tau\lambda}\,d\tau \int_0^{T(1-z)} e^{i\tau\lambda}\,d\tau\right].\end{aligned} \tag{3.7}$$

Substracting (3.7) from (3.6) we obtain finally

$$\frac{\sin^2 \dfrac{T\lambda}{2}}{\lambda^2} \sin T\lambda z = [\Phi_1(\lambda; z) - \Phi_1(-\lambda; z)] + [\Phi_2(\lambda; z) + \Phi_2(-\lambda; z)],$$

* In addition to $f(\lambda)$, for all μ $f(\mu+\lambda) = f_\mu(\lambda)$ is the spectral density of a completely regular process with the same coefficient of mixing.

where

$$\Phi_1(\lambda; z) = \frac{1}{4}\int_0^T e^{i\lambda\tau}\,d\tau \int_0^{Tz} e^{i\lambda\tau}\,d\tau,$$

$$\Phi_2(\lambda; z) = \frac{1}{4}\int_{T(1-z)}^{T} e^{i\lambda\tau}\,d\tau \int_0^{T(1-z)} e^{-i\lambda\tau}\,d\tau.$$

Further, let us estimate the integrals $(1/T)\int_{-\infty}^{\infty} \Phi_i(\pm\lambda; z) \times f(\lambda + \mu)\,d\lambda$. The calculations are even simpler in this case than in the case with discrete time.

1. *Estimation of* $(1/T)\int_{-\infty}^{\infty} \Phi_1(\lambda; z) f(\lambda + \mu)\,d\lambda$. Let us express $\Phi_1(\lambda; z)$ as the sum $\Phi_{11}(\lambda; z) + \Phi_{12}(\lambda; z)$, where

$$\Phi_{11}(\lambda; z) = \frac{1}{4}\int_{T(1-z)}^{T(1-z)+\sqrt{T}} e^{i\lambda\tau}\,d\tau \int_0^{T(1-z)} e^{i\lambda\tau}\,d\tau,$$

$$\Phi_{12}(\lambda; z) = \frac{1}{4}\int_{T(1-z)+\sqrt{T}}^{T} e^{i\lambda\tau}\,d\tau \int_0^{T(1-z)} e^{i\lambda\tau}\,d\tau.$$

It is seen that

$$|\Phi_{11}(\lambda; z)| = \left|\left(\frac{\sin\dfrac{\lambda\sqrt{T}}{2}}{\lambda}\right)\left(\frac{\sin\dfrac{T(1-z)\lambda}{2}}{\lambda}\right)\right|. \tag{3.8}$$

The function $\Phi_{12}(\lambda; z)$ can be rewritten as

$$\Phi_{12}(\lambda; z) = e^{i\lambda\sqrt{T}}\varphi_1^+(\lambda)\varphi_2^+(\lambda), \tag{3.9}$$

where

$$|\varphi_1^+(\lambda)| = \left|\frac{\sin(Tz - \sqrt{T})\dfrac{\lambda}{2}}{\lambda}\right|, \qquad |\varphi_2^+(\lambda)| = \left|\frac{\sin T(1-z)\dfrac{\lambda}{2}}{\lambda}\right|, \qquad \varphi_1^+, \varphi_2^+ \in \mathscr{H}_2^+.$$

Next, using the property given by (2.12) of the functions $h(T)$, the equality given by (3.8), and the Schwarz inequality, we have

$$\frac{1}{T}\int_{-\infty}^{\infty} |\Phi_{11}(\lambda; z)| f(\lambda + \mu)\,d\lambda \leqslant T^{-1/4}[h(\sqrt{T})]^{1/2}[h(T)]^{1/2} = o(h(T)), \tag{3.10}$$

this estimate being independent of z ($\delta < z < 1 - \delta$).

The condition of complete regularity yields, if we take into account (3.9),

$$\left|\int_{-\infty}^{\infty} \Phi_{12}(\lambda; z) f(\lambda + \mu)\,d\lambda\right| \leqslant \rho(\sqrt{T})[h(Tz - \sqrt{T})h(T(1-z))]^{1/2}. \tag{3.11}$$

By Lemma 7

$$\lim_{T\to\infty} \frac{h(Tz - \sqrt{T})}{h(T)} = 1, \qquad \lim_{T\to\infty} \frac{h(T(1-z))}{h(T)} = 1$$

uniformly in z, $\delta < z < 1 - \delta$; therefore, uniformly in μ and z, $\delta \leqslant z \leqslant 1 - \delta$,

$$\left|\frac{1}{T}\int_{-\infty}^{\infty} \Phi_{12}(\lambda; z)f(\lambda + \mu)\,d\lambda\right| = o(h(T)). \tag{3.12}$$

The estimates given by (3.10) and (3.12) enable us to assert that uniformly in $|\mu| \leqslant a$, z, $\delta \leqslant z \leqslant 1 - \delta$,

$$\frac{1}{T}\int_{-\infty}^{\infty} \Phi_{1}(\lambda; z)f(\lambda + \mu)\,d\lambda = o(h(T)). \tag{3.13}$$

2. *Estimation of* $(1/T)\int_{-\infty}^{\infty} \Phi_2(\lambda; z)f(\lambda + \mu)\,d\lambda$. Again let us represent Φ_2 as the sum

$$\Phi_2(\lambda; z) = \Phi_{21}(\lambda; z) + \Phi_{22}(\lambda; z),$$

where we have now

$$\begin{aligned}\Phi_{21}(\lambda; z) &= \frac{1}{4}\int_{T(1-z)}^{T(1-z)+\sqrt{T}} e^{i\lambda\tau}\,d\tau \int_0^{T(1-z)} e^{-i\lambda\tau}\,d\tau,\\ \Phi_{22}(\lambda; z) &= \int_{T(1-z)+\sqrt{T}}^{T} e^{i\lambda\tau}\,d\tau \int_0^{T(1-z)} e^{-i\lambda\tau}\,d\tau = e^{i\lambda\sqrt{T}}\varphi_2^+(\lambda)\psi_2^+(\lambda),\end{aligned} \tag{3.14}$$

and $\varphi_2^+, \psi_2^+ \in \mathscr{H}^2$ in the upper half-plane. From (3.14) we obtain similarly to (3.10) and (3.12)

$$\frac{1}{T}\int_{-\infty}^{\infty} |\Phi_{21}(\lambda; z)|\,f(\lambda + \mu)\,d\lambda \leqslant T^{-1/4}[h(\sqrt{T})h(T(1 - z))]^{1/2} = o(h(T)), \tag{3.15}$$

$$\frac{1}{T}\left|\int_{-\infty}^{\infty} \Phi_{22}(\lambda; z)f(\lambda + \mu)\,d\lambda\right| \leqslant \rho(\sqrt{T})[h(Tz - \sqrt{T})h(T(1 - z))]^{1/2} = o(h(T)); \tag{3.16}$$

therefore, uniformly in $|\mu| \leqslant a$, $\delta \leqslant z \leqslant 1 - \delta$,

$$\frac{1}{T}\int_{-\infty}^{\infty} \Phi_{2}(\lambda; z)f(\lambda + \mu)\,d\lambda = o(h(T)). \tag{3.17}$$

Similarly, uniformly in μ and z

$$\begin{aligned}\frac{1}{T}\int_{-\infty}^{\infty} \Phi_{1}(-\lambda; z)f(\lambda + \mu)\,d\lambda &= o(h(T)),\\ \frac{1}{T}\int_{-\infty}^{\infty} \Phi_{2}(-\lambda; z)f(\lambda + \mu)\,d\lambda &= o(h(T)).\end{aligned} \tag{3.18}$$

These estimates can be proved in the same way as in the proof of (3.13) and (3.17), but the decompositions given by (3.9) and (3.14) must be replaced by decompositions of the form $e^{-i\lambda\sqrt{T}}\varphi^-(\lambda)\psi^-(\lambda)$, where we have now φ^-, $\psi^- \in \mathscr{H}^2$ in the lower half-plane. Piecing together the estimates obtained we finally conclude that, in fact, uniformly in $|\mu| \leqslant a$

$$\frac{1}{T}\int_{-\infty}^{\infty} \frac{\sin^2 \dfrac{T\lambda}{2}}{\lambda^2} f(\lambda + \mu)a(T\lambda)\,d\lambda = o(h(T)),$$

thus proving the lemma. □

Lemma 12. *Let $a(\lambda)$ be a function of bounded variation vanishing outside of the interval $[-1, 1]$. Then, as $T \to \infty$, uniformly in $|\mu| \leqslant a$*

$$\frac{1}{T}\int_{-\infty}^{\infty} \frac{\sin^2 \dfrac{T\lambda}{2}}{\lambda^2} f(\lambda + \mu)a(T\lambda)\,d\lambda = \frac{h(T)}{2\pi}\int_{-1}^{1} \frac{\sin^2 \dfrac{\lambda}{2}}{\left(\dfrac{\lambda}{2}\right)^2} a(\lambda)\,d\lambda(1 + o(1)). \tag{3.19}$$

Proof of Lemma 12 and Theorem 2. The proof of Lemma 12 follows completely the proof of Lemma 14 in Chapter V.

Let $f(\lambda)$ be the spectral density of a completely regular process. Using Lemma 6 we write $f(\lambda)$ as $w_a(\lambda)|P_a(\lambda)|^2$, where

$$\int_{-\infty}^{\infty} w_a(\lambda)\,d\lambda < \infty \qquad \text{and} \qquad \lim_{T} \inf_{|\mu| \leqslant a} \gamma(T; \mu; w_a) = \infty.$$

By Lemma 9, $w_a(\lambda)$ is the spectral density of a completely regular process. Further, using Lemma 12 we have that uniformly in $|\mu| \leqslant a$ (compare with Section V.4)

$$\frac{1}{T}\int_{0}^{1/T} w_a(\lambda + \mu)\,d\lambda - \frac{1}{T}\int_{-1/T}^{0} w_a(\lambda + \mu)\,d\lambda = o\left(\frac{1}{T}\int_{-1/T}^{1/T} w_a(\lambda + \mu)\,d\lambda\right), \tag{3.20}$$

which is equivalent to (3.1), thus completing proving Theorem 2. □

VI.4 The Behavior of the Spectral Density on the Entire Line

The theorem proved in the previous section gives no idea about the behavior of spectral density $f(\lambda)$ of a completely regular process as $\lambda \to \infty$. It is clear, for instance, that $f(\lambda)$ cannot decrease too rapidly. Since a completely regular process is linearly regular, we infer that

$$\int_{-\infty}^{\infty} \frac{|\ln f(\lambda)|}{1 + \lambda^2}\,d\lambda < \infty \tag{4.1}$$

On the other hand, the above is, in essence, the only restriction imposed on the rate of decay of $f(\lambda)$ as $\lambda \to \infty$. In fact, if $\Gamma(\lambda)$ is a square-integrable function of finite degree, the process $\xi(t)$ with spectral density $f(\lambda) = |\Gamma(\lambda)|^2$ is completely regular (this follows from Theorem 1).

However, if the factor $\Gamma(\lambda)$ of the kind indicated is such that the remainder $f(\lambda)(\Gamma(\lambda))^{-1} = w(\lambda)$ in $f(\lambda)$ is bounded (for large λ) from above and from below ($m \leqslant w \leqslant M$), we can draw certain conclusions about the behavior of $w(\lambda)$ at infinity (more precisely, about the uniform behavior of $w(\lambda)$ on the entire line).

To formulate the result in precise terms we need to introduce a class of entire functions that is to be defined as follows.

The class A is the class of entire functions (of finite degree) $\Gamma(z)$, $z = \lambda + i\mu$, whose zeros z_i satisfy the inequality

$$\sum_j \left| \operatorname{Im} \frac{1}{z_j} \right| < \infty \tag{4.2}$$

plays an essential role in the theory of entire functions.

We shall need entire functions whose zeros satisfy a stronger condition than that given by (4.2).

Let us denote by A^* the class of entire functions $\Gamma(z)$, $z = \lambda + i\mu$, of finite degree whose zeros z_j satisfy the inequality

$$\sup_{-\infty < \lambda < \infty} \sum_j \left| \operatorname{Im} \frac{1}{z_j - \lambda} \right| < \infty. \tag{4.3}$$

Here the summation is carried out over all nonreal zeros z_j of the function Γ.

Theorem 3. *Let the spectral density $f(\lambda)$ of the stationary completely regular process $\xi(t)$ permit a representation of the form*

$$f(\lambda) = |\Gamma(\lambda)|^2 w(\lambda),$$

where Γ is a square-summable entire function of the class A^* and let the function w satisfy the inequalities

$$0 < m \leqslant \inf_\lambda w(\lambda) \leqslant \sup_\lambda w(\lambda) \leqslant M < \infty.$$

Then the condition

$$\omega_W(\delta) = \sup_{-\infty < \lambda < \infty} \sup_{|t| \leqslant \delta} \frac{|W(\lambda + t) + W(\lambda - t) - 2W(\lambda)|}{|W(\lambda + t) - W(\lambda - t)|} \xrightarrow[\delta \to 0]{} 0. \tag{4.4}$$

is satisfied for the primitive $W(\lambda)$ of the function $w(\lambda)$.

To prove Theorem 3 we proceed as follows. First we prove that after the class A^* is divided into functions the completely regular function still remains completely regular (for the terminology, see Section VI.1). Next we investigate a completely regular function w for which, obviously, $\inf \gamma(T; \mu; w) \xrightarrow[T \to \infty]{} \infty$, using the methods applied in Section VI.3.

Let us note that we can consider without loss of generality all zeros of the function Γ to lie in the lower closed half-plane. In fact, $|\Gamma|^2$ denotes a value of a nonnegative entire function of the class A on a real line. By Akhiezer's theorem* such a function can be written as $|\varphi(\lambda)|^2$ where $\varphi(z)$ is an entire function of the class A with the roots in the lower half-plane. It is clear that $\varphi \in A^*$ and, if necessary, we can replace Γ by φ without changing the spectral density f. It is also seen that the function $\Gamma(z)$ can be regarded as the exterior function in the upper half-plane.

Further, let us assume $\Gamma(z) = \Gamma(\bar{z})$ and let us introduce a meromorphic function $\chi(z) = \bar{\Gamma}(z)/\Gamma(z)$. We write the function $\Gamma(z)$ as the infinite product

$$\Gamma(z) = z^p e^{az+b} \prod_j \left(1 - \frac{z}{z_j}\right) e^{z/z_j}.$$

It follows from (4.3) that

$$\chi(z) = \alpha e^{i\beta z} \prod_j \left(1 - \frac{z}{\bar{z}_j}\right)\left(1 - \frac{z}{z_j}\right)^{-1}, \tag{4.5}$$

where β is a real number, $|\alpha| = 1$.

Lemma 13. *Let the entire function $\Gamma(z) \in A^*$ and let $\ln|\Gamma|/(1 + \lambda^2) \in \mathscr{L}^1$. Then the meromorphic function $\chi(z)$ is analytic in a certain strip $|\operatorname{Im} z| < \delta$, $\delta > 0$, and is bounded in any strip $|\operatorname{Im} z| < \delta' < \delta$.*

Proof. We can consider without loss of generality that $\alpha = 1$, $\beta = 0$. Let $z_j = \alpha_j + i\beta_j$; by (4.3)

$$\sup_\lambda \sum \frac{|\beta_j|}{(\lambda - \alpha_j)^2 + \beta_j^2} = \sup_\lambda \sum \left|\operatorname{Im} \frac{1}{\lambda - z_j}\right| < \infty,$$

and therefore $\delta = \inf|\beta_j| > 0$, where inf is taken over all nonreal zeros z_j (if all zeros of $\Gamma(z)$ are real we assume $\delta = \infty$ by definition).

It is seen that the function $\chi(z)$ is analytic in the strip $|\operatorname{Im} z| < \delta$. Let $\delta' < \delta$; we prove that $|\chi(z)|$ is bounded in $|\operatorname{Im} z| < \delta'$. By virtue of (4.3) we have in the strip indicated that

$$\begin{aligned}
|\chi(z)|^2 &\leqslant \prod_j \left(1 + \frac{4|\beta_j \mu|}{(\alpha_j - \lambda)^2 + (\beta_j - \mu)^2}\right) \\
&\leqslant \exp\left\{4|\mu| \sum_j \frac{|\beta_j|}{(\alpha_j - \lambda)^2 + (\beta_j - \mu)^2}\right\} \\
&\leqslant \exp\left\{4\delta \left(1 - \frac{\delta'}{\delta}\right)^2 \sup_\lambda \sum_j \frac{|\beta_j|}{(\alpha_j - \lambda)^2 + \beta_j^2}\right\} = M_{\delta'} < \infty. \qquad \square
\end{aligned}$$

* See [19], p. 567.

Lemma 14. *By the hypothesis of Theorem 3 the function $w(\lambda)$ is completely regular.*

Proof. From Lemma 13, on the basis of theorems on approximation of bounded analytic functions* there follows the existence of entire functions $\Phi_\sigma(\lambda)$ of finite degree $\leqslant \sigma$ for which

$$\|\chi - \Phi_\sigma\|^{(\infty)} = O(e^{-\sigma\delta'}), \qquad \delta' < \delta.$$

For large σ all the functions Φ_σ are uniformly bounded; for example, $|\Phi_\sigma(\lambda)| \leqslant 2$. It can be easily seen that $e^{i\sigma\lambda}\Phi_\sigma \in \mathscr{H}^\infty$ in the upper half-plane.

Let φ, ψ be arbitrary functions from the unit sphere of a space $L^+(w)$; then $\varphi, \psi \in \mathscr{H}^2$, and therefore the functions $\varphi_1 = \varphi/\Gamma$, $\psi_1 = \psi/\Gamma$ will belong to the unit sphere of $L^+(f)$. Finally, $e^{i\sigma\lambda}\Phi_\sigma\varphi_1 \in L^+(f)$. Hence for $\sigma = \tau/2$

$$\begin{aligned}
&\left|\int_{-\infty}^{\infty} e^{i\lambda\tau}\varphi(\lambda)\psi(\lambda)w(\lambda)\,d\lambda\right| \\
&= \left|\int_{-\infty}^{\infty} e^{i\lambda\tau}\varphi_1(\lambda)\psi_1(\lambda)\chi(\lambda)f(\lambda)\,d\lambda\right| \\
&\leqslant \left|\int_{-\infty}^{\infty} e^{i\lambda\tau/2}[e^{i\lambda\tau/2}\Phi_\sigma(\lambda)\varphi_1(\lambda)]\psi_1(\lambda)f(\lambda)\,d\lambda\right| \\
&\quad + \int_{-\infty}^{\infty} |\varphi_1(\lambda)\psi_1(\lambda)|\,|\chi(\lambda) - \Phi_\sigma(\lambda)|f(\lambda)\,d\lambda \\
&\leqslant 2\rho(\tau/2) + \|\chi - \Phi_\sigma\|^{(\infty)},
\end{aligned}$$

from which it follows in turn that

$$\rho(\tau, w) \leqslant 2\rho(\tau/2) + O(e^{-\tau\delta'/2}), \qquad \delta' < \delta,$$

thus proving the lemma. □

Proof of Theorem 3. The remaining part of the proof of Theorem 3 is equivalent to the proof of Theorem 2. In fact, by the hypothesis of the theorem

$$\inf \gamma(T; x; w) \geqslant \frac{m\pi}{2} T \xrightarrow[T\to\infty]{} \infty.$$

In addition, functions of the form $\int_0^\alpha e^{it\lambda}\,dt$ are elements in $L^+(w)$. Therefore, the proofs of Lemmas 7, 8, 10–12 still hold, in which we can assume $a = \infty$. □

The condition given by (4.4) implies in general, the boundedness of $|w'(\lambda)|/|w(\lambda)|$ or, even more generally, uniform continuity on the entire axis of $\ln w(\lambda)$ (compare with the next section). This condition can be violated even for completely smooth functions. Let us consider the following

* See [27], p. 317.

Example. Let the process $\xi(t)$ have as its spectral density the function

$$f(\lambda) = (\sin^2 \lambda^2 + 1)\left(\frac{\sin \lambda}{\lambda}\right)^{2p},$$

where p is an arbitrary entire positive number. It is easily seen that $\int_{-\infty}^{\infty} (\ln f(\lambda)/(1 + \lambda^2))\, d\lambda < \infty$, and therefore the process $\xi(t)$ is regular. However, it is not completely regular, although the spectral density of this process is analytic in the whole plane of the complex variable $z = \lambda + \varepsilon\mu$ and has zeros coinciding on the real line with zeros of an entire function of finite degree $\Gamma(z) = (\sin z/z)^{2p}$.

In fact, the function $\Gamma(z) \in A^*$: it has real zeros only. By Theorem 3 the function $w(\lambda) = \sin^2 \lambda^2 + 1$ must satisfy (4.4). However, we have

$$\begin{aligned} &\frac{|W(\lambda + t) + W(\lambda - t) - 2W(\lambda)|}{W(\lambda + t) - W(\lambda - t)} \\ &\quad \geqslant \frac{1}{2t}\left|\int_{\lambda}^{\lambda+t} [\sin^2 s^2 - \sin^2(s - t)^2]\, ds\right| \\ &\quad = \frac{1}{2t}\left|\int_{\lambda}^{\lambda+t} \sin t(2s - t) \sin[s^2 + (s - t)^2]\, ds\right|. \end{aligned} \tag{4.6}$$

Let us take $\lambda = \pi/4t$ in (4.6). Then for $\lambda \leqslant s \leqslant \lambda + t$, as $t \nrightarrow 0$

$$\sin t(2s - t) = 1 + O(t^2),$$
$$\sin(s^2 + (s - t)^2) = -\cos 2s^2 + O(t^2).$$

Therefore,

$$\frac{|W(\lambda + t) + W(\lambda - t) - 2W(\lambda)|}{W(\lambda + t) - W(\lambda - t)} \geqslant \frac{1}{2t}\left|\int_{\lambda}^{\lambda+t} \cos 2s^2\, ds\right| + O(t^2).$$

Further,

$$\begin{aligned} \int_{\lambda}^{\lambda+t} \cos(2s^2)\, ds &= \int_{\lambda}^{\lambda+t} \frac{d \sin(2s^2)}{s} \\ &= \frac{\sin 2(\lambda + t)^2}{\lambda + t} - \frac{\sin 2\lambda^2}{\lambda} + \int_{\lambda}^{\lambda+t} \frac{\sin 2s^2}{s^2}\, ds \\ &= -\frac{2}{\lambda} \sin \frac{\pi^2}{8t^2} + O(|t|^3). \end{aligned}$$

Therefore, we have

$$\sup_{\lambda} \frac{|W(\lambda + t) + W(\lambda - t) - 2W(\lambda)|}{W(\lambda + t) - W(\lambda - t)} \geqslant \frac{4}{\pi}\left|\sin \frac{\pi^2}{8t^2}\right| + O(t^2) \nrightarrow 0,$$

thus proving that the process $\xi(t)$ is not completely regular.

VI.5 Sufficiency

The theorem that follows is a converse of Theorem 3, and yields one of the complete regularity criteria.

Theorem 4. *Let the spectral density $f(\lambda)$ of a stationary process $\xi(t)$ permit a representation of the form*

$$f(\lambda) = |\Gamma(\lambda)|^2 w(\lambda),$$

where Γ is a square-summable entire function of finite degree $\leqslant \sigma$ and the function w has the following properties:

1. $0 < m \leqslant w(\lambda) \leqslant M < \infty$;
2. $\sum_n \omega_W^2(2^{-n}) < \infty$; *here, as above,*

$$\omega_W(\delta) = \sup_{\lambda} \sup_{|t| \leqslant \delta} \frac{|W(\lambda + t) + W(\lambda - t) - 2W(\lambda)|}{|W(\lambda + t) - W(\lambda - t)|},$$

W being the primitive for w.

Then the process $\xi(t)$ is completely regular; in this case

$$\rho(\tau) \leqslant C \left(\frac{M}{m}\right)^3 \left(\sum_1^\infty \omega_W^2 \left(\frac{1}{\tau - 2\sigma} 2^{-(n-1)}\right)\right)^{1/2},$$

where C is an absolute constant.

We shall omit the proof of Theorem 4 here since it differs only in technicalities from the proof of the corresponding discrete analog—Theorem 6 in Chapter V.

Theorem 5. *Let the spectral density $f(\lambda)$ of the process $\xi(t)$ permit a representation of the form $f(\lambda) = |\Gamma(\lambda)|^2 w(\lambda)$ where:*

1. *Γ is a bounded entire function of finite degree $\leqslant \sigma$;*
2. *the function $\ln w(\lambda)$ is uniformly continuous on $(-\infty, \infty)$, that is,*

$$\sup_{\lambda, |h| \leqslant s} \left| \ln \frac{w(\lambda + h)}{w(\lambda)} \right| = \omega(s) \xrightarrow[s \to 0]{} 0;$$

3. $\int_1^\infty (\omega(s)/s^2)\, ds < \infty$.

Then the process $\xi(t)$ is completely regular, with

$$\rho(\tau) \leqslant C\omega\left(\frac{1}{\tau - 2\sigma}\right), \qquad \tau > 2\sigma \tag{5.1}$$

(where the constant C is dependent of w).

By Lemma 1 it suffices to prove that the function $w(\lambda)$ is completely regular. From the inequality $|\ln w(\lambda)| \leqslant \ln w(0)| + \omega(|\lambda|)$ and from (3) it follows that $\ln w/(1+\lambda^2) \in \mathscr{L}^1$. It remains only to estimate the value $\rho(\tau; w)$, relying on the following basic lemma.

Lemma 15. *By the hypothesis of Theorem 5, for any $r > 0$ there is a non-negative function $\Phi_r(\lambda)$ such that:*

1. *for all $\varphi, \psi \in L^+(w)$ the functions*

$$\varphi\Phi_r^{1/2}, \psi\Phi_r^{1/2} \in L^2(-\infty, \infty),$$

and if, in addition, $\tau > r$, then

$$\int_{-\infty}^{\infty} \varphi(\lambda)\psi(\lambda)e^{i\lambda\tau}\Phi_r(\lambda)\,d\lambda = 0; \tag{5.2}$$

2. *for all λ, $-\infty < \lambda < \infty$,*

$$|w(\lambda) - \Phi_r(\lambda)| \leqslant C_1\omega(1/r)w(\lambda). \tag{5.3}$$

We shall prove the lemma later but now we show how the inequality given by (5.1) can be deduced from this lemma. Let $\varphi(\lambda)$, $\psi(\lambda)$ be arbitrary functions from the unit sphere of $L^+(w)$. By this lemma, for all $\tau > r$

$$\left|\int_{-\infty}^{\infty} e^{i\lambda\tau}\varphi(\lambda)\psi(\lambda)w(\lambda)\,d\lambda\right|$$
$$= \left|\int_{-\infty}^{\infty} e^{i\lambda\tau}\varphi(\lambda)\psi(\lambda)[w(\lambda) - \Phi_r(\lambda)]\,d\lambda\right|$$
$$\leqslant C_1\omega(1/r)\int_{-\infty}^{\infty} |\varphi(\lambda)|\,|\psi(\lambda)|w(\lambda)\,d\lambda \leqslant C_1\omega(1/r).$$

Therefore, $\rho(\tau) \leqslant C_1\omega(1/r)$, and invoking Lemma 1 we prove (5.1).

Returning to Lemma 15 we define α by the equality

$$\alpha = \sum_{n=1}^{\infty} \frac{\omega(n)}{n^2}.$$

By virtue of (3) of Theorem 5 the series that defines α is convergent. Next let $a_k = \omega(n)/2\alpha n^2$, and let us define

$$k(x) = \frac{\sin^4 \dfrac{x}{8}}{\left(\dfrac{x}{8}\right)^r} \prod_1^{\infty} \frac{\sin^2 \dfrac{a_k x}{2}}{\left(\dfrac{a_k x}{2}\right)^2}$$

and also "Marchenko's kernel"*

$$K(x) = k(x)\left(\int_{-\infty}^{\infty} k(x)\,dx\right)^{-1}.$$

* Such kernels were introduced by V. A. Marchenko, "Some problems of approximating continuous functions on the whole axis," *III Soobtshenija Kharkov. Matem. Obtshestva* **22** (1950).

Then the sought functions $\Phi_r(\lambda)$ are to be defined by the equality

$$\Phi_r(\lambda) = r \int_{-\infty}^{\infty} K(rx) w(\lambda - x)\, dx.$$

Before proving the fact that the Φ_r satisfy all the conditions of Lemma 15 we investigate the properties of the kernels $K(x)$.

Lemma 16. *The kernels $K(x)$ have the following properties:*

1. *$K(x)$ is a square-integrable entire function of finite degree ($\leqslant 1/2 + \sum a_k \leqslant 1$);*
2. $\int_{-\infty}^{\infty} K(x)\, dx = 1$;
3. *for any $r > 0$*

$$|K(rx)| \leqslant M_r e^{-\omega(x)} \frac{\sin^4 \dfrac{rx}{8}}{\left(\dfrac{rx}{8}\right)^4}, \tag{5.4}$$

where the constant M_r depends on r only.

Proof. The properties given by (1) and (2) are obvious. To prove (3) we note first that $\lim_{x\to\infty} (\omega(x)/x) = 0$. From the fact that the integral $\int_1^\infty (\omega(x)/x^2)\, dx$ is convergent, it follows that $\underline{\lim}_{x\to\infty} (\omega(x)/x) < \infty$. Integrating by parts we arrive at the equality

$$\int_1^x \frac{\omega(y)}{y^2}\, dy = \int_1^x \frac{d\omega(y)}{y} - \frac{\omega(x)}{x} + \omega(1),$$

from which it follows that $\underline{\lim}_{x} \int_1^x (d\omega(y)/y) < \infty$. Since the function $\omega(x)$ does not decrease the last condition implies that the integral $\int_1^\infty (d\omega(y)/y)$ is convergent. Therefore, $\lim(\omega(x)/x)$ exists and is equal to zero because the integral $\int_1^\infty (\omega(x)/x^2)\, dx$ is convergent.

We can consider without loss of generality that $\lim_{x\to\infty} \omega(x) = \infty$ (otherwise, the inequality given by (5.4) is trivial). Given $x > 0$ we choose an integer N such that $N \leqslant \omega(x) < N + 1$. Then

$$\prod_1^\infty \frac{\sin^2 \dfrac{a_k x r}{2}}{\left(\dfrac{a_k x r}{2}\right)^2} \leqslant \prod_1^N \left(\frac{2}{a_k x}\right)^2 = \frac{2^{2N}(N!)^4}{x^{2N} \prod_1^N \omega^2(k)} \frac{(2\alpha)^{2N}}{r^{2N}}.$$

Using Stirling's formula we have

$$\frac{(N!)^4}{x^{2N}} \leqslant N^{4N} x^{-2N} \leqslant e^{4\omega(x)\ln(1/\omega(x)} e^{-2(\omega(x)-1)\ln x}$$

for large x, which does not exceed $e^{-\omega(x)}$ (since $\omega(x)/x \to 0$). Furthermore, by virtue of the equality $\lim_{k\to\infty} \omega(k) = \infty$,

$$\lim \frac{(2\alpha)^{2N} 2^{2N} r^{-2N}}{\omega^2(1) \ldots \omega^2(N)} = 0,$$

thus proving (5.4). □

Taking again the functions Φ_r, we note that by virtue of (3) of Lemma 16

$$\begin{aligned} \Phi_r(\lambda) &\leqslant rw(\lambda) \int_{-\infty}^{\infty} K(rx) \sup_{\lambda} \frac{w(\lambda - x)}{w(\lambda)}\, dx \\ &\leqslant rw(\lambda) \int_{-\infty}^{\infty} K(rx) e^{\omega(|x|)}\, dx \leqslant C_2 M_r w(\lambda); \end{aligned} \tag{5.5}$$

therefore, for all φ for which $\int |\varphi(\lambda)| w(\lambda)\, d\lambda < \infty$ we necessarily have $\varphi\Phi_r \in \mathscr{L}^1$. Further, let

$$\Phi_{rT}(\lambda) = r \int_{-T}^{T} K(r(x - \lambda)) w(x)\, dx, \qquad T > 0.$$

It is seen that for all T

$$\Phi_{rT}(\lambda) \leqslant \Phi_r(\lambda) \leqslant \pi M_1 w(\lambda). \tag{5.6}$$

Next, if $|\lambda| \leqslant T/2$ then similarly to (5.5)

$$\begin{aligned} &\Phi_r(\lambda) - \Phi_{rT}(\lambda) \\ &= r\left(\int_{-\infty}^{-T-\lambda} K(rx) w(x - \lambda)\, dx + \int_{T-\lambda}^{\infty} K(rx) w(x - \lambda)\, dx\right) \\ &\leqslant rw(\lambda)\left[\int_{-\infty}^{-T/2} K(rx) e^{\omega(x)}\, dx + \int_{T/2}^{\infty} K(rx) e^{\omega(x)}\, dx\right] \\ &\leqslant \frac{C_3 w(\lambda)}{T} M_r, \end{aligned}$$

from which and from (5.6) it follows that for any fixed functions $\varphi, \psi \in L(w)$ and for any fixed r

$$\lim_{T\to\infty} \int_{-\infty}^{\infty} \varphi(\lambda)\psi(\lambda) e^{i\lambda\tau} \Phi_{rT}(\lambda)\, d\lambda = \int_{-\infty}^{\infty} \varphi(\lambda)\psi(\lambda) e^{i\lambda\tau} \Phi_r(\lambda)\, d\lambda.$$

The last equality as well as the definition of the space $L^+(w)$ enable us to restrict ourselves while proving (5.2) to the investigation of integrals of the form

$$\int_{-\infty}^{\infty} \varphi(\lambda)\psi(\lambda) e^{i\lambda\tau} \Phi_{rT}(\lambda)\, d\lambda, \tag{5.7}$$

where $\varphi, \psi \in \mathscr{H}^2 \cap L(w)$, $\tau > r$, and T is a positive number.

Proof of Lemma 15 and Theorem 5. By the Paley–Wiener theorem the Fourier transform $\chi_r(x)$ of the kernel $rK(rx)$ vanishes outside of the interval $[-r, r]$; therefore

$$\Phi_{rT}(\lambda) = \int_{-T}^{T} w(x)\,dx \int_{-r}^{r} e^{i(\lambda - x)s}\chi_r(s)\,ds = \int_{-r}^{r} e^{i\lambda s}b(s)\,ds, \qquad b \in \mathcal{L}^2.$$

Therefore, by the above theorem all $\Phi_{rT}(\lambda)$ are square-summable entire functions of finite degree $\leqslant r$. Hence for all $\tau > r$ the functions $e^{i\lambda\tau}\varphi\psi\Phi_{rT} \in \mathcal{H}^1$ in the upper half-plane and all the integrals given by (5.7) are equal to zero. We have proved (5.2).

To prove the second part of Lemma 15, that is, the inequality given by (5.3), we need to note first that by virtue of the property given by (2) of the kernels $K(x)$

$$\begin{aligned}|w(\lambda) - \Phi_r(\lambda)| &\leqslant \int_{-\infty}^{\infty} K(x) \left| w(\lambda) - w\left(\lambda - \frac{x}{r}\right)\right| dx \\ &\leqslant w(\lambda) \int_{-\infty}^{\infty} K(x) \sup_{\lambda} \left| \frac{w(\lambda) - w\left(\lambda - \frac{x}{r}\right)}{w(\lambda)} \right| dx. \end{aligned} \tag{5.8}$$

For all x, $|e^x - 1| \leqslant |x|e^{|x|}$. Hence

$$\left| \frac{w(\lambda) - w\left(\lambda - \frac{x}{r}\right)}{w(\lambda)} \right| = \left| 1 - \exp\left\{ \ln \frac{w\left(\lambda - \frac{x}{r}\right)}{w(\lambda)} \right\} \right| \leqslant \omega\left(\frac{|x|}{r}\right) e^{\omega(|x|/r)}.$$

Let us substitute the inequality obtained above into (5.8). Since* $\omega(|x|/r) \leqslant (1 + |x|)\omega(1/r)$ we have (noting the property given by (5.4) of the kernels $K(x)$), that for all $r \geqslant 1$

$$\begin{aligned}|w(\lambda) - \Phi_r(\lambda)| &\leqslant w(\lambda)w(1/r) \int_{-\infty}^{\infty} K(x)(1 + |x|)e^{\omega(|x|)}\,dx \\ &\leqslant M_1\omega(1/r) \int_{-\infty}^{\infty} (1 + |x|) \frac{\sin^4 \frac{x}{8}}{(x/8)^4}\,dx \cdot w(\lambda) = C_1\omega(1/r)w(\lambda).\end{aligned}$$

We have proved completely Lemma 15 as well as Theorem 5. □

Note. As can be seen from the proof, the constant C_1 can be written as C_4M_1, where C_4 is a fixed constant and M_1 is a characteristic of the kernel K, is defined by the function $\omega(x)$. Hence, though the constant M_1 is not fixed it can be taken to be the same for the whole class of functions $w(\lambda)$, for instance, for all $w(\lambda)$ for which

$$\sup_{\lambda} \left| \ln \frac{g(\lambda + x)}{g(\lambda)} \right| \leqslant \omega(x).$$

* Each index of continuity $\omega(\delta)$ satisfies the inequality $\omega(\gamma\delta) \leqslant (1 + \gamma)\omega(\delta)$, $\gamma > 0$ (see, for instance, [27], p. 213).

Example. Let us consider a stationary process $\xi_\alpha(t)$ with spectral density $f_\alpha(\lambda) = e^{-|\lambda|^\alpha}$, $\alpha > 0$. If $\alpha \geqslant 1$, then $f_\alpha/(1+\lambda^2) \notin \mathscr{L}^1$; therefore, the process $\xi_\alpha(t)$ is not even regular. If $\alpha < 1$, then

$$\sup_\lambda \ln \frac{f_\alpha(\lambda + x)}{f_\alpha(\lambda)} = \sup_\lambda (|\lambda + x|^\alpha - |\lambda|^\alpha) \leqslant \alpha |x|^\alpha,$$

which follows from the inequality

$$(1+u)^\alpha - 1 - \alpha u^\alpha \leqslant 0, \qquad u > 0,$$

which is easily verified.

Therefore, for $\alpha < 1$ the process $\xi_\alpha(t)$ is completely regular, and $\rho(\tau) = O(\tau^{-\alpha})$. (Naturally, the constant in the notation $O(\cdot)$ is dependent of α.)

VI.6 A Special Class of Stationary Processes

We shall describe in this section regularity conditions for stationary processes $\xi(t)$ with spectral density of the form $|\Gamma(\lambda)|^{-2}$ where $\Gamma(\lambda)$ is an entire function of finite degree. Why are the processes of this special type of interest?

Stationary processes with spectral density equal to $|P(\lambda)|^{-2}$, P a polynomial, should obviously be considered the simplest. Among Gaussian stationary processes the processes of the type indicated are projections of Markov vector processes (continuous analogs of the mth Markov chains). It immediately follows from Theorem 5 that the stationary process with density $|P(\lambda)|^{-2}$ is completely regular. Simple calculations yield that the pertinent regularity coefficient satisfies the condition*

$$\rho(\tau) = O(e^{-\tau(\delta - \varepsilon)}), \tag{6.1}$$

where 2δ is the width of the analyticity strip ($|\operatorname{Im} z| < \delta$) of the function $1/P(z)$ and ε is an arbitrary positive number.

Entire functions of finite degree constitute the next complicated class of analytic functions after the class of polynomials. One might think that a simple and final criterion for complete regularity can still be found. If we succeed in finding this criterion, we solve thereby the problem (which naturally arises) of describing the class of entire functions of finite degree, which, when used as factors, retain complete regularity (see Sections VI.2–4; compare also with Lemma 11 in Chapter V).

Theorem 6. *The stationary process $\xi(t)$ with spectral density of the form $|\Gamma(\lambda)|^{-2}$, where Γ is an entire function of finite degree, is completely regular if and only if:*

1. $\dfrac{\ln|\Gamma|}{1+\lambda^2} \in \mathscr{L}^1(-\infty, \infty)$;

* It was proved in A. M. Yaglom's paper, cited in Section IV.2, that in this case $\rho(\tau)$ is the maximal root of a determinant equation; other roots of this equation coincide with eigenvalues of the operator B_τ.

2. *The function Γ is a function of class* A^*, that is,* $\sup_\lambda \sum|\mathrm{Im}(1/(\lambda - z_j))| < \infty$, *where the summation is carried over all nonreal zeros z_j of the function Γ.*

In this case necessarily $\inf|\mathrm{Im}\, z_j| = \delta > 0$ *and for all* $\varepsilon > 0$

$$\rho(\tau) = O(e^{-\tau(\delta - \varepsilon)}). \tag{6.2}$$

Proof. Let the conditions given by (1) and (2) be satisfied. We can consider without loss of generality that all zeros of the function $\Gamma(z)$ lie in the lower half-plane and that $\Gamma(z)$ and $1/\Gamma(z)$ are exterior functions in the upper half-plane (see Section VI.4). Let us introduce the meromorphic function $\chi(z) = \Gamma(z)/\bar{\Gamma}(z)$, where $\bar{\Gamma}(z) = \overline{\Gamma(\bar{z})}$ as before.

By virtue of (1.4) we have

$$\rho(\tau) = \sup_\theta \left| \int_{-\infty}^{\infty} e^{i\lambda\tau}\theta(\lambda)\chi(\lambda)\, d\lambda \right|, \tag{6.3}$$

where θ runs through the unit sphere of the space $\mathscr{H}^1$. By Lemma 13 the function $\chi(z)$ is analytic in the strip of nonzero width $|\mathrm{Im}\, z| < \delta$, $\delta = \inf|\mathrm{Im}\, z_j| > 0$, and bounded in any strip $|\mathrm{Im}\, z| \leqslant \delta' < \delta$. By S. N. Bernstein's theorems on approximations of analytic functions* there are bounded entire functions $\Phi_r(\lambda)$ of finite degree $\leqslant r$ for which

$$\sup_\lambda |\chi(\lambda) - \Phi_r(\lambda)| = O(e^{-r(\delta - \varepsilon)}), \qquad \varepsilon > 0.$$

For all $r < \tau$

$$\int_{-\infty}^{\infty} e^{i\lambda\tau}\theta(\lambda)\Phi_r(\lambda)\, d\lambda = 0.$$

Hence for $r < \tau$

$$\rho(\tau) = \sup_\theta \left| \int_{-\infty}^{\infty} e^{i\lambda\tau}\theta(\lambda)[\chi(\lambda) - \Phi_r(\lambda)]\, d\lambda \right| = O(e^{-r(\delta - \varepsilon)}),$$

which in fact proves (6.2).

The necessity of (1) follows from the fact that the process $\xi(t)$ is regular. We can consider again that Γ and $1/\Gamma$ are exterior functions in the upper half-plane. □

We prove next, by several lemmas, the necessity of the condition $\Gamma \in A^*$. If we denote by z_j the zeros of the function $\Gamma(z)$, we can write this function $\Gamma(z)$ as

$$\Gamma(z) = e^{az+b} \prod_j \left(1 - \frac{z}{z_j}\right) e^{\tau/z_j}. \tag{6.4}$$

* We recall that the class A^* was defined in Section VI.4.

* See [27], p. 317.

From the condition given by (1) proved above it follows that*

$$\sum \left| \operatorname{Im} \frac{1}{z_j} \right| < \infty.$$

Hence

$$\chi(z) = \frac{\Gamma(z)}{\bar{\Gamma}(z)} = \alpha e^{i\beta z} \prod \left(1 - \frac{z}{z_j}\right)\left(1 - \frac{z}{\bar{z}_j}\right)^{-1},$$

where β is a real number and $|\alpha| = 1$. In the upper half-plane $z = \lambda + i\mu$, $\mu > 0$,

$$|\bar{\chi}(z)e^{i\beta z}| = \left|\frac{e^{i\beta z}}{\chi(z)}\right| = \prod \left|1 - \frac{z}{\bar{z}_j}\right|\left|1 - \frac{z}{z_j}\right|^{-1}$$
$$= \prod \left[\frac{(\operatorname{Re} z_j - \lambda)^2 + (\operatorname{Im} z_j + \mu)^2}{(\operatorname{Re} z_j - \lambda)^2 + (\operatorname{Im} z_j - \mu)^2}\right]^{1/2} \leqslant 1.$$

Therefore, for all $\tau \geqslant \beta$ the function $e^{i\tau z}\bar{\chi}(z)$ is an inner function of the class $\mathscr{H}^\infty$. Hence, each time when $\theta \in \mathscr{H}^1$, the function $e^{i\tau\lambda}\theta\bar{\chi}$ also belongs to $\mathscr{H}^1$ in the upper half-plane; therefore

$$\int_{-\infty}^{\infty} e^{i\tau\lambda}\theta(\lambda)\bar{\chi}(\lambda)\,d\lambda = 0. \tag{6.5}$$

At the same time (by virtue of (1.4))

$$\sup_{\Theta \in \mathscr{H}^1} \left|\int_{-\infty}^{\infty} e^{i\lambda\tau}\theta(\lambda)\chi(\lambda)\,d\lambda\right| = \rho(\tau). \tag{6.6}$$

Let us note that if $\theta \in \mathscr{H}^1$ in the upper half-plane, then $\bar{\theta} \in \mathscr{H}^1$ in the lower half-plane. Hence (6.5) and (6.6) enable us to assert that for all θ^+ that belong to the unit sphere of $\mathscr{H}^{1+}$, for all θ^- which belong to the unit sphere of $\mathscr{H}^{1-}$, and for all $\tau \geqslant \beta$

$$\left|\int_{-\infty}^{\infty} e^{i\lambda\tau}\theta^+(\lambda)\chi(\lambda)\,d\lambda\right| \leqslant \rho(\tau),$$
$$\left|\int_{-\infty}^{\infty} e^{-i\lambda\tau}\theta^-(\lambda)\chi(\lambda)\,d\lambda\right| \leqslant \rho(\tau). \tag{6.7}$$

The inequalities given by (6.7) allow us to prove the following lemma.

Lemma 17. *Uniformly in* λ, $-\infty < \lambda < \infty$, *as* $t \to 0$

$$\int_{\lambda - t}^{\lambda} \chi(s)\,ds - \int_{\lambda}^{\lambda + t} \chi(s)\,ds = o(t). \tag{6.8}$$

Proof. Let $a(\lambda)$ be an odd thrice differentiable function vanishing outside of the interval $[-1, 1]$. We prove first that uniformly in x, $-\infty < x < \infty$, as

* See [19], p. 314.

$T \to \infty$

$$\frac{1}{T}\int_{-\infty}^{\infty} \frac{\sin^2 \frac{T\lambda}{2}}{\lambda^2} \chi(\lambda + x) a(T\lambda)\, d\lambda = o(1). \tag{6.9}$$

The equality given by (6.9) is an analog of Lemma 11. By virtue of (6.7) we have the inequalities

$$\begin{gathered} \left| \int_{-\infty}^{\infty} e^{i\lambda\tau} \left(\int_0^T e^{i\lambda u}\, du \right)^2 \chi(\lambda + x)\, d\lambda \right| \leqslant \rho(\tau) \int_{-\infty}^{\infty} \frac{\sin^2 \frac{\lambda T}{\lambda^2}}{\lambda^2}\, d\lambda = \rho(\tau)\frac{\pi T}{2}, \\ \left| \int_{-\infty}^{\infty} e^{-i\lambda\tau} \left(\int_{-T}^{T} e^{i\lambda u}\, du \right) \chi(\lambda + x)\, d\lambda \right| \leqslant \rho(\tau)\frac{\pi T}{2}, \end{gathered} \tag{6.10}$$

if only $\tau \geqslant \beta$. Now we note that inequalities of the type given by (6.10) lie at the basis of the proof of Lemma 11. Exactly the same arguments that prove Lemma 11 enable us to deduce (6.9) from (6.10); hence we shall omit the deduction.

Further, let us define the odd function $a_0(\lambda)$ by the equalities

$$a_0(\lambda) = \begin{cases} \dfrac{\lambda^2}{\sin^2 \frac{\lambda}{2}}, & 0 < \lambda < 1, \\ 0, & \lambda > 1, \end{cases}$$

$$a_0(\lambda) = -a_0(-\lambda).$$

Let $a_\varepsilon(\lambda)$, $\varepsilon > 0$, be odd thrice differentiable functions coinciding with $a_0(\lambda)$ outside of intervals $[-\varepsilon, \varepsilon]$, $[-1, -1 + \varepsilon]$, $[1 - \varepsilon, \varepsilon]$, the functions $a_\varepsilon(\lambda)$ being monotone inside these intervals. By (6.9) we have

$$\begin{aligned} \left| \frac{1}{T}\int_{-\infty}^{\infty} \frac{\sin^2 \frac{T\lambda}{2}}{\lambda^2} \chi(\lambda + x)[a_0(T\lambda) - a_\varepsilon(T\lambda)]\, d\lambda \right| &\leqslant \frac{2}{\sin^2 \frac{1}{2}} \int_{-\varepsilon}^{\varepsilon} \frac{\sin^2 \frac{\lambda}{2}}{\lambda^2}\, d\lambda + o(1) \\ &\leqslant 16\varepsilon + o(1). \end{aligned}$$

Therefore, (6.9) still holds if we replace $a(\lambda)$ by $a_0(\lambda)$; denoting $1/T$ by t we arrive at (6.8), thus proving the lemma. □

Lemma 18. *There exists a positive number $\delta > 0$ such that the function $\chi(z) = \Gamma(z)/\bar{\Gamma}(z)$ is analytic in the strip $|\text{Im } z| < \delta$.*

Proof. The function $\chi(z)$ is meromorphic, the nonreal zeros of the function $\bar{\Gamma}(z)$ being the poles of $\chi(z)$. Therefore, we need to prove that all nonreal zeros of the function $\bar{\Gamma}(z)$ lie outside of the strip $|\mathrm{Im}\, z| < \delta$. Let $\bar{z}_j = \alpha_j - i\beta_j$ be a zero of the function $\bar{\Gamma}$.

We define the entire functions $\gamma_j(z)$ and $\bar{\gamma}_j(z)$ by the equalities

$$\bar{\gamma}_j(z) = \left(1 - \frac{z}{\bar{z}_j}\right)^{-1} \bar{\Gamma}(z), \qquad \gamma_j(z) = \overline{\bar{\gamma}(\bar{z})}.$$

Further, let

$$\varphi_j(\lambda) = \frac{1}{2\pi T_j} \frac{e^{iT_j(\lambda - \alpha_j)} - 1}{\lambda - \alpha_j}, \qquad T_j = \frac{1}{|\beta_j|}.$$

It can be seen that $\bar{\gamma}_j/\gamma_j \in \mathscr{H}^\infty$ in the upper half-plane; hence $\varphi_j(\bar{\gamma}_j/\gamma_j) \in \mathscr{H}^1$ in the upper half-plane, and $\|\varphi_j(\bar{\gamma}_j/\gamma_j)\|^{(1)} = 1$. By virtue of (6.7)

$$\left|\int_{-\infty}^{\infty} e^{i\lambda\tau}\varphi_j(\lambda)\frac{\bar{\gamma}_j(\lambda)}{\gamma_j(\lambda)}\chi(\lambda)\,d\lambda\right| = \left|\int_{-\infty}^{\infty} e^{i\lambda\tau}\varphi_j(\lambda)\frac{z_j - \lambda}{\bar{z}_j - \lambda}\,d\lambda\right| \leqslant \rho(\tau). \qquad (6.11)$$

The function $\varphi_j(z)((z_j - z)/(\bar{z}_j - z))$ is analytic in the upper half-plane except for the pole at the point z_j, and vanishes as $|z| \to \infty$. Hence the integral in the right-hand side can be found from the residues of the integrand. It will be equal to

$$2\pi i\left(\operatorname{res} e^{i\lambda\tau}\varphi_j(z)\frac{z_j - z}{\bar{z}_j - z}\right) = -2(e^{-1} - 1)^2 e^{i\tau\bar{z}_j},$$

substituting which into (6.11) we find that

$$\rho(\tau) \geqslant \frac{2}{e^2}(e - 1)^2 e^{\beta_j\tau}. \qquad (6.12)$$

Therefore, if $\delta = \inf|\beta_j|$, where the inf is taken over all $\beta_j \neq 0$, then by virtue of (6.12)

$$\rho(\tau) \geqslant \frac{2}{e^2}(e - 1)^2 e^{-\delta\tau},$$

and since $\rho(\tau) \to 0$ as τ increases we necessarily have that $\delta > 0$, thus proving Lemma 18. □

Proof of Theorem 6. Next we shall show that from (6.8) and from the fact that the function $\chi(z)$ is analytic in the strip $|\mathrm{Im}\, z| < \delta$ we have the inequality for a derivative $\chi'(\lambda)$: $\sup_\lambda|\chi'(\lambda)| < \infty$. Since

$$|\chi'(\lambda)| = \left|\beta - \sum \mathrm{Im}\frac{1}{z_j - \lambda}\right| = \left|\beta - 2\sum\left|\mathrm{Im}\frac{1}{z_j - \lambda}\right|\right|,$$

we have proved this. Further on we shall assume β to be zero (for the sake of simplicity). We have no loss of generality since the factor $e^{\pm i\lambda\beta}$ can be expressed as $e^{i\tau\lambda}$ in the basic inequalities given by (6.7), which, at worst, leads to the substitution of $\rho(\tau)$ for $\rho(\tau - |\beta|)$.

As above, let $z_j = \alpha_j + i\beta_j$. By the Leibnitz formula we have

$$\begin{aligned}\chi'(\lambda) &= \chi(\lambda) \sum \frac{|2\beta_j|}{(\alpha_j - \lambda)^2 + \beta_j^2}, \\ \chi^{(s+1)}(\lambda) &= \sum_0^s C_s^k \chi^{(k)}(\lambda) \left(\sum_j \frac{2|\beta_j|}{(\alpha_j - \lambda)^2 + \beta_j^2} \right)^{(s-k)},\end{aligned} \tag{6.13}$$

the summation being carried out over those j for which $\beta_j \neq 0$. By Lemma 18 $|\beta_j| = -\beta_j \geqslant \delta > 0$, and hence

$$\begin{aligned}\left| \left(\frac{2\beta_j}{(\alpha_j - \lambda)^2 + \beta_j^2} \right)^{(r)} \right| &= \left| \left(\frac{1}{\bar{z}_j - \lambda} - \frac{1}{z_j - \lambda} \right)^{(r)} \right| \\ &\leqslant 2r! \frac{1}{|z_j - \lambda|^{r+1}} \leqslant \frac{2r!}{(2\delta)^p} \frac{(2\beta_j)^p}{|z_j - \lambda|^{r+1}}, \qquad p \geqslant 0.\end{aligned}$$

Therefore, for $s - k \geqslant 1$

$$\begin{aligned}\left| \left(\sum_j \frac{2|\beta_j|}{(a_j - \lambda)^2 + \beta_j^2} \right)^{(s-k)} \right| &\leqslant \frac{2 \cdot (s-k)!}{(2\delta)^{(s-k+1)/2}} \sum_j \frac{(2|\beta_j|)^{(s-k+1)/2}}{|z_j - \lambda|^{s-k+1}} \\ &\leqslant \frac{2 \cdot (s-k)!}{2\delta^{(s-k+1)/2}} \left(\sum_j \frac{2|\beta_j|}{|z_j - \lambda|^2} \right)^{(s-k+1)/2} \\ &= 2 \frac{(s-k)!}{(2\delta)^{(s-k+1)/2}} |\chi'(\lambda)|^{(s-k+1)/2}.\end{aligned} \tag{6.14}$$

Let the inequality $|\chi'(\lambda)| \geqslant 1$ be satisfied at a point λ. Furthermore, assume that at the same point, for all $k = 2, 3, \ldots, s$

$$|\chi^{(k)}(\lambda)| \leqslant L^k \cdot k! |\chi'(\lambda)|^k, \tag{6.15}$$

where L is a constant. Let us prove that in this case (6.15) holds also for $k = s + 1$. To this end we substitute (6.15) into (6.13). Using (6.14) and the agreement $|\chi'(\lambda)| \geqslant 1$ we find

$$\begin{aligned}|\chi^{(s+1)}(\lambda)| &\leqslant L^s s! |\chi'(\lambda)|^{s+1} + \sum_{k=0}^{s-1} s! \frac{2L^k}{(2\delta)^{(s-k+1)/2}} |\chi'(\lambda)|^s \\ &\leqslant s! |\chi'(\lambda)|^{s+1} \left(L^s + \sqrt{\frac{2}{\delta}} (L^s + (2\delta)^{-s/2}) \right) \\ &\leqslant L^{s+1} (s+1)! |\chi'(\lambda)|^{s+1}\end{aligned}$$

if the constant L is sufficiently large (for instance, if $L > 1 + 2\sqrt{2/\delta}$).

Further, from (6.13) and (6.14) it follows that at the point at which $|\chi'(\lambda)| \geqslant 1$, it is known that

$$|\chi''(\lambda)| \leqslant \left(1 + \frac{1}{\delta} \right) |\chi'(\lambda)|^2 < L^2 \cdot 2! |\chi'(\lambda)|^2,$$

where the constant L is large. Hence at all the points λ at which $|\chi'(\lambda)| > 1$, and for all $k \geqslant 2$, the inequalities given by (6.15) are satisfied, the constant L being independent of λ.

We shall show that the assumption $\sup|\chi'(\lambda)| = \infty$ leads us to contradict Lemma 17. Let a sequence of points λ_k be such that $M_k = |\chi'(\lambda_k)| \to \infty$ and let $t_k = 1/4M_kL$.

By (6.8), as $k \to \infty$

$$\int_{\lambda_k - t_k}^{\lambda_k} \chi(\lambda)\, d\lambda - \int_{\lambda_k}^{\lambda_k + t_k} \chi(\lambda)\, d\lambda = o(t_k). \tag{6.16}$$

For large $k|t_k| < \delta/2$ and by Lemma 18 the function $\chi(\lambda)$ analytic in the domain $|\lambda - \lambda_k| \leqslant t_k$ can be expanded as a Taylor series:

$$\chi(\lambda) = \chi(\lambda_k) + \chi'(\lambda_k)(\lambda - \lambda_k) + \sum_{s=2}^{\infty} \frac{(\lambda - \lambda_k)^s}{s!} \chi^{(s)}(\lambda_k).$$

Substituting this series into the right-hand side of (6.16) and using (6.15) we obtain the following inequality:

$$\begin{aligned}
&\left| \int_{\lambda_k - t_k}^{\lambda_k} \chi(\lambda)\, d\lambda - \int_{\lambda_k}^{\lambda_k + t_k} \chi(\lambda)\, d\lambda \right| \\
&= \left| \chi'(\lambda_k) t_k^2 + 2 \sum_{s=2}^{\infty} \frac{t_k^{2s}}{2s!} \chi^{(2s-1)}(\lambda_k) \right| \\
&\geqslant |\chi'(\lambda_k) t_k^2| \left| 1 - 2 \sum_{s=2}^{\infty} L^s t_k^s |\chi'(\lambda_k)|^s \right| \\
&\geqslant \frac{t_k^2}{3} |\chi'(\lambda_k)| = \frac{1}{12L} t_k \neq o(t_k).
\end{aligned}$$

The contradiction thus obtained proves the necessity of the condition $\Gamma \in A^*$. We have completed proving Theorem 6.

Note. At the same time we have proved the equality

$$\lim_{\tau \to 0} (\rho(\tau))^{1/\tau} = e^{-\delta}, \qquad \delta = \inf_{\beta_j = 0} |\beta_j|,$$

which follows from (6.2) and (6.12).

CHAPTER VII

Filtering and Estimation of the Mean

VII.1 Unbiased Estimates

VII.1.1 The Statement of the Problem

Let us consider a random process of the form

$$\xi(t) = \theta(t) + \Delta(t), \qquad t \in T, \tag{1.1}$$

where $\theta(t)$, $t \in T$, is an unknown deterministic function from a given class Θ and $\Delta(t)$, $t \in T$, is a Gaussian stationary process with zero mean and correlation function $B(t)$.

For the sake of exposition we can interpret $\theta = \theta(t)$ as a useful signal and $\Delta = \Delta(t)$ as random noise in the corresponding channel of communication. We are interested in *optimal filtering* of the random process $\xi = \xi(t)$, to determine the unknown signal $\theta \in \Theta$ in (1.1). We need to determine the operation on the signal ξ received that reproduces θ as precisely as possible. The random process $\hat{\theta} = \hat{\theta}(t)$ thus obtained must satisfy the following:

1. the mean value of the difference $\hat{\theta}(t) - \theta(t)$ must be zero:

$$\mathsf{M}[\hat{\theta}(t) - \theta(t)] = 0, \qquad t \in T, \tag{1.2}$$

2. the mean square of the difference $\hat{\theta}(t) - \theta(t)$ must be minimal:

$$\mathsf{M}[\hat{\theta}(t) - \theta(t)]^2 = \min. \tag{1.3}$$

The probability distribution of a random process of the type given by (1.1) depends on the function *parameter* $\theta \in \Theta$; therefore, there is a family of distributions P_θ where the parameter $\theta = \theta(t)$, $t \in T$, is the mean value of the corresponding distribution P_θ:

$$\theta(t) = \mathsf{M}_\theta \xi(t), \qquad t \in T. \tag{1.4}$$

In this case the correlation function remains the same for all θ, namely

$$\mathsf{M}_\theta[\xi(s) - \theta(s)][\xi(t) - \theta(t)] = B(t - s) \tag{1.5}$$

(here and further on $\mathsf{M}_\theta \eta$ denotes the mathematical expectation of the random variable η found from the pertinent probability distribution P_θ).

The problem of filtering coincides, in fact, with the problem of finding the best estimate $\hat{\theta} = \hat{\theta}(t)$ of the unknown mean $\theta = \theta(t)$. This problem is of particular interest if $\theta = \theta(t)$ is

$$\theta(t) = \sum_{k=1}^{N} \alpha_k \theta_k(t), \tag{1.6}$$

where $\theta_1(t), \ldots, \theta_N(t)$ are some prescribed functions and $\alpha_1, \ldots, \alpha_N$ are unknown (real) coefficients (*regression coefficients*).

Each coefficient of the above type is a linear functional of $\theta \in \Theta$. For example, if some scalar product (θ', θ'') is defined in a linear space of all functions $\theta = \theta(t)$ of the form given by (1.6), and if the functions $\theta_1, \ldots, \theta_N$ form a basis in this space, then

$$\alpha_k - (\theta_k^*, \theta), \qquad k = 1, \ldots, n, \tag{1.7}$$

where $\theta_1^*, \ldots, \theta_N^*$ is a conjugate system of functions to be defined by the conditions

$$(\theta_k^*, \theta_j) = \begin{cases} 1 & \text{for } j = k, \\ 0 & \text{for } j \neq k, \end{cases} \qquad j, k = 1, \ldots, N.$$

As we show further on, the best unbiased estimates $\hat{\theta}(t)$, $t \in T$, are representable as

$$\hat{\theta}(t) = \sum_{k=1}^{N} \hat{\alpha}_k \theta_k(t), \qquad t \in T, \tag{1.8}$$

where $\hat{\alpha}_1, \ldots, \hat{\alpha}_N$ are the best unbiased estimates for unknown coefficients $\alpha_1, \ldots, \alpha_N$.

Particular values of $\theta(t)$ are linear functionals of $\theta \in \Theta$. Further, we shall examine the best unbiased estimates for arbitrary linear functionals $\alpha = \alpha(\theta)$ of the unknown function $\theta = \theta(t)$, $t \in T$, assuming, naturally, that all probability distributions P_θ are mutually equivalent. We can consider without loss of generality that the P_θ are equivalent to a distribution $\mathsf{P} = \mathsf{P}_0$ corresponding to zero value of the parameter θ (that is, $\theta(t) \equiv 0$), and this will in fact be assumed henceforth.

VII.1.2 Necessary and Sufficient Statistics: The Complete Family of Distributions

By a *statistic* we shall mean any real random variable $\eta = \eta(\omega)$ measurable with respect to a σ-algebra $\mathfrak{A}(T)$ generated by the values $\xi(t) = \xi(\omega, t)$ of an *observable* random process (here $t \in T$ and $\omega \in \Omega$, where Ω denotes a space of elementary events). A natural question arises as to the class of statistics to which we can confine ourselves in estimating the unknown parameter $\theta \in \Theta$ without loss of the available information.

Given any family of statistics $\eta = \eta(\omega)$ we may consider the σ-algebra $\mathfrak{B}$ generated by these statistics. The σ-algebra $\mathfrak{B} \subseteq \mathfrak{A}(T)$ is said to be *sufficient* for the parameter $\theta \in \Theta$ estimated if the conditional probabilities $\mathsf{P}_\theta(A|\mathfrak{B})$ calculated for all events $A \in \mathfrak{A}(T)$ from the corresponding distributions P_θ are such that for each $\theta \in \Theta$ with probability 1

$$\mathsf{P}_\theta(A|\mathfrak{B}) = \mathsf{P}(A|\mathfrak{B}) \tag{1.9}$$

(that is, the conditional probabilities $\mathsf{P}_\theta(A|\mathfrak{B})$ are independent of the unknown parameter $\theta \in \Theta$).

The σ-algebra $\mathfrak{B}$ is said to be *necessary* if it is contained in a sufficient σ-algebra (perhaps augmented by sets of zero probability).*

Let us note that if $\mathfrak{B}$ is a sufficient σ-algebra, for any real variable (with the expectation $\mathsf{M}_\theta\eta$ for each $\theta \in \Theta$) we have the following formula: with probability 1

$$\mathsf{M}_\theta(\eta|\mathfrak{B}) = \mathsf{M}(\eta|\mathfrak{B}), \qquad \theta \in \Theta. \tag{1.10}$$

This fact follows immediately from (1.9) if we take into account that the conditional mathematical expectation $\mathsf{M}_\theta(\eta|\mathfrak{B})$ can be defined as the limit

$$\mathsf{M}_\theta(\eta|\mathfrak{B}) = \lim_{n\to\infty} \sum \frac{k}{n} \mathsf{P}_\theta\left(\frac{k-1}{n} \leqslant \eta < \frac{k}{n}\middle|\mathfrak{B}\right)$$

with convergence in the mean with respect to measure P_θ; due to equivalence of the distributions P_θ this limit with probability 1 is the same for any $\theta \in \Theta$.

Let

$$p_\theta(\omega) = \mathsf{P}_\theta(d\omega)/\mathsf{P}(d\omega) \tag{1.11}$$

be density of the distribution P_θ. We denote by $\mathfrak{B}$ the σ-algebra of events that is generated by all values $p_\theta(\omega)$ of $\omega \in \Omega$ ($\theta \in \Theta$).

Lemma 1. *The σ-algebra $\mathfrak{B}$ indicated is sufficient for the parameter $\theta \in \Theta$.*

Proof. In fact, for any $A \in \mathfrak{A}(T)$ and $B \in \mathfrak{B}$

$$\begin{aligned}\mathsf{P}_\theta(AB) &= \int_{AB} p_\theta(\omega)\mathsf{P}(d\omega)\\ &= \mathsf{M}[\mathsf{M}(\chi_A\chi_B p_\theta|\mathfrak{B})]\\ &= \mathsf{M}[\chi_B p_\theta \mathsf{M}(\chi_A|\mathfrak{B})] = \int_B \mathsf{P}(A|\mathfrak{B})p_\theta(\omega)\mathsf{P}(d\omega)\end{aligned}$$

and at the same time

$$\mathsf{P}_\theta(AB) = \int_B \mathsf{P}_\theta(A|\mathfrak{B})\mathsf{P}_\theta(d\omega) = \int_B \mathsf{P}_\theta(A|\mathfrak{B})p_\theta(\omega)\mathsf{P}(d\omega),$$

* Compare, for example, with [18], pp. 33–37.

where $\chi_A = \chi_A(\omega)$ and $\chi_B = \chi_B(\omega)$ are indicators of sets A and B. Since B is an arbitrary set from $\mathfrak{B}$, we have for integrands measurable with respect to $\mathfrak{B}$ for almost all $\omega \in \Omega$

$$\mathsf{P}_\theta(A|\mathfrak{B})p_\theta(\omega) = \mathsf{P}(A|\mathfrak{B})p_\theta(\omega).$$

But $p_\theta(\omega)$, being the density of the equivalent measures P_θ and P, is almost everywhere positive. Therefore, with probability 1 we have the equality given by (1.9). □

We note that the variables p_θ, $\theta \in \Theta$, defined by (1.11) satisfy the condition $\mathsf{M}p_\theta = 1$. Let us denote by $\boldsymbol{L}^1$ the space of all variables η that are measurable with respect to the σ-algebra $\mathfrak{B}$ and for which $\mathsf{M}|\eta| < \infty$.

The family of distributions P_θ, $\theta \in \Theta$, is said to be *boundedly complete in* $\boldsymbol{L}^1$ if for any bounded variable η (measurable with respect to $\mathfrak{B}$) the relation

$$\mathsf{M}_\theta\eta = \mathsf{M}\eta p_\theta = 0, \qquad \theta \in \Theta,$$

is equivalent to the fact that $\eta = 0$ with probability 1.

The property of bounded completeness implies* that each linear continuous functional in a Banach space $\boldsymbol{L}^1$ (with norm $\|\eta\| = \mathsf{M}|\eta|$) vanishing on elements $p_\theta \in \boldsymbol{L}^1$ is equal to zero identically. Therefore, the property of bounded completeness of the family of distributions P_θ, $\theta \in \Theta$, is equivalent to the fact that the linear hull of the variables p_θ, $\theta \in \Theta$, is everywhere dense in $\boldsymbol{L}^1$.

Lemma 2. *For a boundedly complete family of distributions* P_θ, $\theta \in \Theta$, *the* σ*-algebra* $\mathfrak{B}$ *indicated above (generated by variables of the type given by* (1.11)) *is sufficient as well as necessary.*

Proof. In fact, if $\mathfrak{B}$ is not necessary, then it contains a sufficient σ-algebra $\mathfrak{C}$ such that some set $B \in \mathfrak{B}$ does not belong to the *augmentation* $\mathfrak{C}^*$, in other words, a variable $\eta = \eta(\omega)$ of the form

$$\eta(\omega) = \begin{cases} 1 & \text{for } \omega \in B, \\ 0 & \text{for } \omega \notin B, \end{cases}$$

is nonmeasurable with respect to $\mathfrak{C}^*$. Since $\mathfrak{C}$ is a sufficient σ-algebra, the variable $\hat{\eta} = \mathsf{M}_\theta(\eta|\mathfrak{C})$ does not depend on $\theta \in \Theta$ and $\mathsf{M}_\theta(\eta - \hat{\eta}) \equiv 0$ for all $\theta \in \Theta$. Next, if $0 \leqslant \eta \leqslant 1$, then $0 \leqslant \hat{\eta} \leqslant 1$; therefore, the variable $\Delta = \eta - \hat{\eta}$ is bounded. Because the family of P_θ, $\theta \in \Theta$, is complete, the bounded (measurable with respect to $\mathfrak{B}$) variable $\Delta = \eta - \hat{\eta}$ with probability 1 is zero and $\eta = \hat{\eta}$. Therefore, the variable η is in fact measurable with respect to the σ-algebra $\mathfrak{C}^*$. The contradiction thus obtained shows that the σ-algebra $\mathfrak{B}$ is necessary. □

* See, for example, [14], p. 35.

We note that conditions for equivalence and an explicit form of the variables given by (1.11) have already been described in Chapter III for the Gaussian distributions P_θ considered. In fact, by Theorem 7, Chapter III, a necessary and sufficient condition for the Gaussian distributions P_θ and P to be equivalent is that the mean value $\theta(t)$, $t \in T$, belong to a space Y of all real functions $y(t)$, $t \in T$, representable as

$$y(t) = \int e^{-i\lambda t}\varphi(\lambda)F(d\lambda), \qquad t \in T, \tag{1.12}$$

where $\varphi(\lambda)$ is some function from the space $L_T(F)$ and $F(d\lambda)$ is the spectral measure of the random process $\xi(t)$ stationary with respect to the probability distribution P.

As already noted in Section I.6, the representation given by (1.12) is unique. Let us introduce a scalar product on the linear space Y by taking

$$\langle y_1, y_2 \rangle = \langle \varphi_1, \varphi_2 \rangle_F, \tag{1.13}$$

where $\varphi_1(\lambda)$ and $\varphi_2(\lambda)$ are elements in the Hilbert space $L_T(F)$ corresponding to the functions $y_1 = y_1(t)$ and $y_2 = y_2(t)$ according to (1.12). Therefore, for the equivalent distributions P_θ, $\theta \in \Theta$, the parametric set Θ is a set in the Hilbert space Y:

$$\Theta \subseteq Y. \tag{1.14}$$

To explain the fact that the embedding of Θ into Y is natural, we consider the case where the set T is finite, so that we have a finite number of Gaussian variables $\xi(t_1), \ldots, \xi(t_n)$ with unknown mean values $\theta(t_1), \ldots, \theta(t_n)$ and the same correlation matrix. If this matrix is nonsingular, Y is an n-dimensional vector space in which Θ is a definite set of vectors θ with coordinates $\theta(t_1), \ldots, \theta(t_n)$. If the correlation matrix is singular and has rank $m < n$, Y is an m-dimensional subspace in an n-dimensional vector space on which each singular Gaussian distribution P_θ is concentrated (probability distributions of variables $\xi(t_1), \ldots, \xi(t_n)$).

By (3.2) in Chapter III the values given by (1.11) have the form

$$p_\theta(\omega) = C_\theta \exp\{\eta_\theta(\omega)\}, \tag{1.15}$$

where

$$C_\theta = \exp\left\{-\frac{1}{2}\mathsf{M}\eta_\theta^2\right\}, \qquad \eta_\theta = \int \varphi_\theta(\lambda)\Phi(d\lambda);$$

here $\Phi(\lambda)$ denotes a stochastic spectral measure of a Gaussian process $\xi(t)$ stationary with respect to the distribution P, and $\varphi_\theta(\lambda)$ is the function from the space $L_T(F)$ that enters the representation given by (1.12) of the corresponding function $\theta \in Y$:

$$\theta(t) = \int e^{-i\lambda t}\varphi_\theta(\lambda)F(d\lambda), \qquad t \in T. \tag{1.16}$$

Lemma 3. *If a parametric set $\Theta \subseteq Y$ contains a parallelepiped from its closed linear hull $\bar{\Theta}$, the family of distributions P_θ will be boundedly complete.*

Proof. In a Hilbert subspace $\bar{\Theta} \subseteq Y$ with some orthonormal basis $\theta_1, \theta_2, \ldots$, the parallelepiped will be understood as a set of points $\theta = \sum c_k \theta_k$ having a finite number of nonzero coordinates $c_1, c_2, \ldots$, that satisfy the condition

$$|c_k - c_k^0| \leqslant \delta_k, \qquad k = 1, 2, \ldots,$$

where $c_1^0, c_2^0, \ldots$ are the coordinates of some fixed element and $\delta_1, \delta_2, \ldots$ are positive numbers. Let $\eta_k = \eta_{\theta_k}$ (see (1.15)), $k = 1, 2, \ldots$. It is seen from (1.16) that

$$\eta_\theta = \sum c_k \eta_k \qquad \text{for } \theta = \sum c_k \theta_k.$$

It can easily be seen that the values p_θ, $\theta \in \Theta$, defined by (1.15) (in which η_θ, $\theta \in \Theta$, are Gaussian variables from the space $H(T)$) belong to a Hilbert space $\boldsymbol{H}(t)$. It is seen that the system of p_θ, $\theta \in \Theta$, will be complete in the space $\boldsymbol{L}^1$ if it is complete in the Hilbert space $\boldsymbol{L}^2$ (denoted by $\boldsymbol{L}$ further on) consisting of variables $\eta \in \boldsymbol{H}(T)$ measurable with respect to the σ-algebra $\mathfrak{B}$ (we recall that $\mathfrak{B}$ is generated by the system of p_θ, $\theta \in \Theta$, or, equivalently, by the system of η_θ, $\theta \in \Theta$).

Let $\varphi = \varphi(y_1, \ldots, y_n)$ be a Borel function in an n-dimensional real space. Obviously, the aggregate of all variables $\eta \in \boldsymbol{L}$ of the form $\eta = \varphi(\eta_1, \ldots, \eta_n)$ forms an everywhere dense set in the whole space $\boldsymbol{L}$, and forms a closed space for fixed n.

Let us assume temporarily that for any n the system of variables p_θ of the form given by (1.15) (where $\theta = \sum_{k=1}^n c_k \theta_k$, $|c_k| \leqslant \delta_k$, $k = 1, \ldots, n$, and $\delta_1, \ldots, \delta_n$ are positive numbers, $\eta_\theta = \sum_{k=1}^n c_k \eta_k$) is everywhere dense in a subspace formed by all the variables $\eta = \varphi(\eta_1, \ldots, \eta_n)$ indicated above. Then, obviously, the linear hull of all variables p_θ, $\theta \in \Theta$, will be everywhere dense in the whole space $\boldsymbol{L}$. Hence it suffices to prove that if

$$\mathsf{M}\eta p_\theta = \mathsf{M}_\theta \eta = C_\theta \int_{\mathbb{R}^n} \varphi(y_1, \ldots, y_n) \exp\left\{\sum_{k=1}^n c_k y_k\right\} \mathsf{P}(dy) = 0$$

for all $\theta = \sum_{k=1}^n c_k \theta_k$ (where $\mathsf{P}(dy)$ is the distribution of the variables $\eta_1, \ldots, \eta_n$ in an n-dimensional space), the function $\varphi(y_1, \ldots, y_n)$ itself is equal to zero almost everywhere with respect to $\mathsf{P}(dy)$. But this fact is very well known* since the densities

$$\frac{\mathsf{P}_\theta(dy)}{\mathsf{P}(dy)} = C_\theta \exp\left\{\sum_{k=1}^n c_k y_k\right\},$$

where $|c_k| \leqslant \delta_k$, $k = 1, \ldots, n$, form an exponential family. □

VII.1.3 Unbiased Estimates

(1.12) and (1.13) yield the isometric correspondence between the Hilbert spaces Y, $L_T(F)$ and $H(T)$ under which

$$\begin{gathered} y \in Y \leftrightarrow \varphi \in L_T(F) \leftrightarrow \eta \in H(T) \\ \langle y_1, y_2 \rangle = \langle \varphi_1, \varphi_2 \rangle_F = \langle \eta_1, \eta_2 \rangle. \end{gathered} \tag{1.17}$$

* See, for instance, [18], p. 76.

Here $H(T)$ as before denotes the linear closed hull of values $\xi(t)$, $t \in T$, of a Gaussian process (stationary with respect to the distribution P) in which the scalar product is $\langle \eta_1, \eta_2 \rangle = M\eta_1\eta_2$, $\Phi(d\lambda)$ is the corresponding stochastic spectral measure, and the variable η in (1.17) is related to the functions $\varphi(\lambda)$ and $y(t)$ (in (1.12)) by the equality

$$\eta = \int \varphi(\lambda)\Phi(d\lambda). \tag{1.18}$$

Let us note that since

$$B(s - t) = \int e^{-i\lambda t} e^{i\lambda s} F(d\lambda),$$

for any s the function $y_s(t) = B(s - t)$ of $t \in T$ belongs to the space Y, the system of all functions y_s, $s \in T$, being complete. The fact that this system is complete follows, in particular, from the fact that the system of functions $e^{i\lambda s} \in L_T(F)$, corresponding to the functions $y_s \in Y$, where

$$y_s(t) = B(s - t) \leftrightarrow e^{i\lambda s} \leftrightarrow \xi(s),$$

is complete.

By virtue of (1.16) each functional of the function parameter $\theta = \theta(t)$, $t \in T$, defined as the value of the function $\theta = \theta(t)$ at some fixed point $t \in T$ can be expressed as

$$\theta(t) = \langle e^{i\lambda t}, \varphi_\theta(\lambda) \rangle_F = \langle y_t, \theta \rangle. \tag{1.19}$$

It is seen that the functional $\theta(t)$ of $\theta \in \Theta$ extends to a linear continuous functional on the Hilbert space Y—in this case uniquely, to the subspace $\bar{\Theta} \subseteq Y$, which is a linear closed hull of the elements $\theta \in \Theta$. We shall consider further on unbiased estimates for arbitrary functionals $\alpha = \alpha(\theta)$ of $\theta \in \Theta$ that are the restriction $\Theta \subseteq Y$ of some linear continuous functional in the Hilbert space Y representable by the general formula as

$$\alpha(\theta) = \langle y, \theta \rangle, \qquad \theta \in \Theta, \tag{1.20}$$

where $y = y(t)$, $t \in T$, is a function from the space Y.

It is useful to note that the functional $\alpha(\theta)$ of $\theta \in \Theta$ permits a linear unbiased estimate (that is, an estimate $\hat{\alpha} \in H(T)$ such that $\mathsf{M}_\theta\hat{\alpha} \equiv \alpha(\theta)$, $\theta \in \Theta$) if and only if this functional $\alpha(\theta)$ is the restriction on $\Theta \subseteq Y$ of some linear continuous functional in the Hilbert space Y. In particular, the variable $\xi(t)$ is a linear unbiased estimate for a functional of the type given by (1.19).

In fact, taking into account the unitary isomorphism described in (1.17), from (1.22) (which follows) we shall have for a functional of the type given by (1.20) that

$$\alpha(\theta) = \langle y, \theta \rangle = \langle \eta, \eta_\theta \rangle = \mathsf{M}_\theta \eta \qquad \text{for } \theta \in \Theta, \tag{1.21}$$

where $\eta \in H(T)$ is a variable of the form given by (1.18) corresponding to the function $y \in Y$. On the other hand each variable $\eta \in H(T)$ representable as $\eta = \int \varphi(\lambda)\Phi(d\lambda)$ is associated with a functional $\alpha(\theta)$ of the type given by (1.21) permitting extension from the set $\Theta \subseteq Y$ onto the whole space Y.

In this case the function $y \in Y$ determining the above functional can be defined by the formula

$$y(t) = \int e^{-i\lambda t}\varphi(\lambda)F(d\lambda), \qquad t \in T.$$

Lemma 4. *Each variable $\eta \in \boldsymbol{H}(T)$ has the finite mathematical expectation $\mathsf{M}_\theta\eta$ for any $\theta \in \Theta$. Also, we have the formula*

$$\mathsf{M}_\theta\eta = \langle \eta, \eta_\theta \rangle, \qquad \theta \in \Theta, \tag{1.22}$$

for variables η from the subspace $H(T)$ generated by $\xi(t)$, $t \in T$.

Proof. Each variable $\eta \in H(T)$, being the limit of Gaussian variables of the form $\sum c_k\xi(t_k)$, is Gaussian. Hence each variable $\eta_\theta \in H(T)$ in (1.15) is Gaussian; therefore the density $p_\theta = p_\theta(\omega)$ is square-integrable over measure P (in other words, the variable $p_\theta = p_\theta(\omega)$ belongs to the Hilbert space $\boldsymbol{H}(T)$). For any bounded variable $\eta \in \boldsymbol{H}(T)$ we have

$$\mathsf{M}_\theta\eta = \mathsf{M}\eta p_\theta = \langle \eta, p_\theta \rangle. \tag{1.23}$$

For any variable $\eta \in \boldsymbol{H}(T)$ one can take (in the Hilbert space $\boldsymbol{H}(T)$) a sequence of bounded variables $\eta_1, \ldots, \eta_2, \ldots$ convergent to η and fundamental in the sense of convergence in the mean with respect to measure P_θ:

$$M_\theta|\eta_m - \eta_n| = \langle |\eta_m - \eta_n|, p_\theta \rangle \leqslant \|\eta_m - \eta_n\| \cdot \|p_\theta\| \to 0$$

as $m, n \to \infty$. It is seen that the sequence $\eta_1, \eta_2, \ldots$ converges in the mean with respect to measure P_θ to the variable η since P_θ is equivalent to measure P. Therefore, $\mathsf{M}_\theta|\eta| < \infty$, (1.23) holding for the mathematical expectation $\mathsf{M}_\theta\eta$. Further, at the same time we have for the variables $\xi(t)$, $t \in T$, the following equalities (see (1.4), (1.15) and (1.16)):

$$\theta(t) = \mathsf{M}_\theta\xi(t) = \int e^{-i\lambda t}\varphi_\theta(\lambda)F(d\lambda) = \langle \xi(t), \eta_\theta \rangle.$$

From the equality thus obtained,

$$\mathsf{M}_\theta\eta = \langle \eta, \eta_\theta \rangle \qquad \text{for } \eta = \xi(t),\ t \in T,$$

it follows obviously that the mathematical expectation $\mathsf{M}_\theta\eta$ of $\eta \in \boldsymbol{H}(T)$, being a linear continuous functional of the form given by (1.22) on the Hilbert space $\boldsymbol{H}(T)$, can be written as

$$\mathsf{M}_\theta\eta = \langle \eta, p_\theta \rangle = \langle \eta, \eta_\theta \rangle$$

on the subspace $H(T)$ (generated by $\xi(t)$, $t \in T$).

It is also seen that the variable $\eta_\theta \in H(T)$ in (1.15) coincides with the projection of the variable $p_\theta \in \boldsymbol{H}(T)$ onto the subspace $H(T)$. We have proved the lemma. ☐

Let us denote by $\mathfrak{B}$ the σ-algebra generated by all variables $\eta_\theta = \eta_\theta(\omega)$ on Ω ($\theta \in \Theta$); let us also denote by $\boldsymbol{L}$ the subspace in the Hilbert space

$\boldsymbol{H}(T)$ formed by all variables $\eta \in \boldsymbol{H}(T)$ measurable with respect to the σ-algebra $\mathfrak{B}$, and let us denote by L the linear closed hull of all variables η_θ $(\theta \in \Theta)$.

As was noted, from (1.15) and (1.16) it follows that

$$\eta_\theta = \sum c_k \eta_{\theta_k} \qquad \text{for } \theta = \sum c_k \theta_k,$$

and therefore (see (1.22)) for any variable $\eta \in \boldsymbol{H}(T)$

$$\mathsf{M}_\theta \eta = \langle \eta, \eta_\theta \rangle = \sum_k c_k \langle \eta, \eta_{\theta_k} \rangle = \sum c_k \mathsf{M}_{\theta_k} \eta.$$

It can be seen that for any functional $\alpha(\theta)$ of the type given by (1.20) (see also (1.21)) given on an arbitrary parametric set $\Theta \subseteq Y$, its linear unbiased estimate $\eta \in H(T)$ is a linear unbiased estimate for the linear extension of $\alpha(\theta)$ to the linear closed hull $\bar{\Theta}$ of the initial set Θ. This fact implies that we can consider without loss of generality that

$$\Theta = \bar{\Theta} \tag{1.24}$$

in considering linear unbiased estimates.

We deal with the complete family of distributions P_θ, $\theta \in \Theta$ (see Lemma 3) under this assumption. The necessary and sufficient σ-algebra for the parameter $\theta \in \Theta$ is the σ-algebra $\mathfrak{B}$ generated by variables $\eta_\theta = \eta_\theta(\omega)$ from (1.15), that is, by variables of the form

$$\eta_\theta = \int \varphi_\theta(\lambda) \Phi(d\lambda), \tag{1.25}$$

where the functions $\varphi_\theta(\lambda) \in L_T(F)$ are found from the integral equations (1.16):

$$\theta(t) = \int e^{-i\lambda t} \varphi_\theta(\lambda) \Phi(d\lambda), \qquad t \in T$$

(the parameter θ runs through some complete system in Θ).

Let $\alpha(\theta)$ be a functional of the unknown parameter $\theta \in \Theta$ described by (1.20), and let $\eta \in \boldsymbol{H}(T)$ be some linear unbiased estimate (see (1.21)). Let

$$\hat{\alpha} = \mathsf{M}(\eta | \mathfrak{B}). \tag{1.26}$$

Since the σ-algebra $\mathfrak{B}$ is sufficient for the parameter $\theta \in \Theta$, then, at the same time,

$$\hat{\alpha} = \mathsf{M}_\theta(\eta | \mathfrak{B}), \qquad \theta \in \Theta,$$

and

$$\mathsf{M}_\theta \hat{\alpha} = \mathsf{M}_\theta \eta = \alpha(\theta) \qquad \text{for all } \theta \in \Theta;$$

furthermore,

$$\mathsf{M}_\theta(\hat{\alpha} - \alpha(\theta))^2 = \mathsf{M}_\theta(\eta - \alpha(\theta))^2 - \mathsf{M}_\theta(\eta - \hat{\alpha})^2 \leqslant \mathsf{M}(\eta - \alpha(\theta))^2. \tag{1.27}$$

It is seen that the variable $\hat{\alpha}$ defined by (1.26) is the "improved" unbiased estimate. This variable $\hat{\alpha} = \hat{\alpha}(\omega)$ is measurable with respect to the σ-algebra $\mathfrak{B}$ and is (because the family of distributions P_θ is complete) a unique estimate of this kind, since if $\eta = \eta(\omega)$ is a $\mathfrak{B}$-measurable unbiased estimate for $\alpha(\theta)$ then

$$\mathsf{M}_\theta(\hat{\alpha} - \eta) = \mathsf{M}_\theta\hat{\alpha} - \mathsf{M}_\theta\eta = 0$$

for all $\theta \in \Theta$, and therefore, in fact, $\hat{\alpha} = \eta$ (with probability 1 with respect to any distribution P_θ).

For the Gaussian variables $\eta \in H(T)$ and η_θ $(\theta \in \Theta)$ the variable $\hat{\alpha} = \mathsf{M}(\eta|\mathfrak{B})$ indicated is the projection of the variable η onto the subspace L which is generated by variables $\eta_\theta \in H(T)$ of the form given by (1.25) (see Section I.5).

We denote by L_0 the orthogonal complement to the subspace $L \subseteq H(T)$:

$$H(T) = L \oplus L_0.$$

As we have just shown, all linear unbiased estimates $\eta \in H(T)$ for $\alpha(\theta)$ have the same projection $\hat{\alpha}$ onto the subspace L and hence they can be represented as

$$\eta = \hat{\alpha} \oplus \Delta, \qquad \Delta \in L_0. \tag{1.28}$$

On the other hand, for any variable $\Delta \in L_0$ by (1.22) we get

$$\mathsf{M}_\theta\Delta = \langle \Delta, \eta_\theta \rangle - 0, \qquad \theta \in \Theta.$$

Therefore, each variable of the form given by (1.28) is a linear unbiased estimate of the functional $\alpha(\theta)$—in this case (see (1.27)) the estimate $\hat{\alpha} \in L$ is the best estimate of this kind.

It is useful to consider the relations given by (1.17). The variables $\eta_\theta \in H(T)$ are associated with functions $\Theta \in Y$; therefore (1.17) determines a unitary isomorphism between subspaces $L \subseteq H(T)$ and $\Theta \subseteq Y$ under which

$$\eta_\theta \in L \leftrightarrow \theta \in \Theta. \tag{1.29}$$

The linear continuous functional $\alpha(\theta)$ of θ is uniquely given on the subspace $\Theta \subseteq Y$ by the formula

$$\alpha(\theta) = \langle \theta_\alpha, \theta \rangle, \qquad \theta \in \Theta, \tag{1.30}$$

where $\theta_\alpha = \theta_\alpha(t)$, $t \in T$, is the function from Θ prescribing the functional $\alpha(\theta)$. By (1.17) (see also (1.21)) this function $\theta_\alpha \in \Theta$ is associated with the linear unbiased estimate η from subspaces L that coincides with the best estimate $\hat{\alpha}$. Therefore,

$$\hat{\alpha} = \int \varphi_\alpha(\lambda)\Phi(d\lambda), \tag{1.31}$$

where $\varphi_\alpha(\lambda) \in L_T(F)$ is the solution of the integral equation of the type given by (1.16):

$$\theta_\alpha(t) = \int e^{-i\lambda t}\varphi_\alpha(\lambda)F(d\lambda), \qquad t \in T, \tag{1.32}$$

with the left-hand side appearing in (1.30) for the functional $\alpha(\theta)$ considered of $\theta \in \Theta$.

Theorem 1. *The function $\alpha(\theta)$ of the unknown parameter $\theta \in \Theta$ permits a linear unbiased estimate $\eta \in H(T)$ if and only if $\alpha(\theta)$ is a linear continuous functional on the Hilbert space $\Theta \subseteq Y$. Let $\alpha(\theta)$ be such a functional to be given by* (1.30). *The best unbiased estimate $\hat{\alpha}$ for $\alpha(\theta)$, $\theta \in \Theta$, can be found by the formula given in* (1.31). *The aggregate of all linear unbiased estimates η for $\alpha(\theta)$, $\theta \in \Theta$, is geometrically a hyperplane in the space $H(T)$ of the form*

$$L_\alpha = \hat{\alpha} \oplus L_0 \tag{1.33}$$

(see (1.28)*), where L_0 is the orthogonal complement to the subspace $L \subseteq H(T)$ generated by variables η_θ of the form given by* (1.25). *The subspace L is the aggregate of all best unbiased estimates $\hat{\alpha}$ (for various linear functionals $\alpha(\theta)$ of $\theta \in \Theta$). The estimates $\hat{\alpha}_1, \ldots, \hat{\alpha}_n$ for $\alpha_1(\theta), \ldots, \alpha_n(\theta)$ are the best ones in the sense that their correlation matrix $\{s_{ij}\}$, $s_{ij} = \mathsf{M}_\theta(\hat{\alpha}_i - \alpha_i(\theta))(\hat{\alpha}_j - \alpha_j(\theta))$, is the smallest one among correlation matrices $\{\sigma_{ij}\}$ of all other unbiased estimates $\eta_1, \ldots, \eta_n$, $(\sigma_{ij} = \mathsf{M}_\theta(\eta_i - \alpha_i(\theta)))\eta_j - \alpha_j(\theta))$, namely*

$$\{s_{ij}\} \leqslant \{\sigma_{ij}\} \tag{1.34}$$

(that is, the matrix $\{s_{ij}\} - \{\sigma_{ij}\}$ is negative definite).

Proof. We have proved all the assertions of Theorem 1 except the inequality given by (1.34); the latter follows from the fact that a linear combination of the best estimates $\hat{\alpha} = \sum c_k \hat{\alpha}_k$ together with the estimates $\hat{\alpha}_1, \ldots, \hat{\alpha}_n$ belongs to the subspace L and is the best unbiased estimate for the pertinent functional $\alpha(\theta) = \sum c_k \alpha_k(\theta)$ of $\theta \in \Theta$. Comparing $\hat{\alpha}$ with the unbiased estimate $\eta = \sum c_k \eta_k$, using (1.27) we obtain

$$\mathsf{M}_\theta(\hat{\alpha} - \hat{\alpha}(\theta))^2 = \sum_{i,j} c_i c_j s_{ij} \leqslant \mathsf{M}_\theta(\eta - \alpha(\theta))^2 = \sum_{ij} c_i c_j \sigma_{ij},$$

from which the inequality given by (1.34) follows due to the fact that the coefficients $c_1, \ldots, c_n$ are arbitrary. The theorem is proved. □

It is useful to note that for any variables $\eta_1, \eta_2 \in H(T)$

$$\mathsf{M}_\theta(\eta_1 - \mathsf{M}_\theta \eta_1)(\eta_2 - \mathsf{M}_\theta \eta_2) = \mathsf{M}\eta_1\eta_2. \tag{1.35}$$

Next, let $\theta_1, \theta_2, \ldots$ be a basis in the subspace $\Theta \subseteq Y$, that is, a system complete in Θ such that for any real numbers $\alpha_1, \alpha_2, \ldots$

$$\left\| \sum_k \alpha_k \theta_k \right\|^2 \asymp \sum_k \alpha_k^2.$$

Let $\theta_1^*, \theta_2^*, \ldots$ be a conjugate system of elements:

$$\langle \theta_k^*, \theta_j \rangle = \begin{cases} 1 & \text{for } k = j, \\ 0 & \text{for } k \neq j. \end{cases}$$

Each element $\theta \in \Theta$ can be uniquely expressed as

$$\theta = \sum_k \alpha_k \theta_k,$$

where $\alpha_k = \langle \theta_k^*, \theta \rangle$ and $\sum_k \alpha_k^2 \asymp \|\theta\|^2$; for any real $\alpha_1, \alpha_2, \ldots$, satisfying the condition $\sum_k \alpha_k^2 < \infty$, the series $\sum_k \alpha_k \theta_k$ is strongly convergent and the sum of this series is some element of $\theta \in \Theta$. The conjugate system $\theta_1^*, \theta_2^*, \ldots$ is also a basis.*

Theorem 2. *Let $\alpha = \alpha(\theta)$, $\theta \in \Theta$, be a linear continuous functional on the subspace $\Theta \subseteq Y$ defined by (1.30). Then the function $\theta_\alpha = \theta_\alpha(t)$, $t \in T$, prescribing the above functional can be expressed as the series*

$$\theta_\alpha = \sum_k \alpha(\theta_k^*)\theta_k, \tag{1.36}$$

and the best unbiased estimate $\hat{\alpha}$ can be expressed as the appropriate series

$$\hat{\alpha} = \sum_k \alpha(\theta_k^*)\eta_k, \tag{1.37}$$

where $\eta_1, \eta_2, \ldots$, being variables from the subspace $L \subseteq H(T)$ corresponding (by (1.29)) to the elements $\theta_1, \theta_2, \ldots$ from the subspace $\Theta \subseteq Y$, are the best unbiased estimates of functionals of the form $\alpha_k^*(\theta) = \langle \theta_k, \theta \rangle$ of $\theta \in \Theta$. The series given in (1.37) converges in the mean square with respect to each of probability measures P_θ, $\theta \in \Theta$.

Proof. The elements $\theta_1, \theta_2, \ldots$ constitute a basis. The function $\theta_\alpha = \theta_\alpha(t)$, $t \in T$, is representable as the series

$$\theta_\alpha = \sum_k \langle \theta_\alpha, \theta_k^* \rangle \theta_k = \sum_k \alpha(\theta_k^*)\theta_k.$$

Further, under the correspondence given by (1.29) between $\Theta \subseteq Y$ and $L \subseteq H(T)$ the basis $\theta_1, \theta_2, \ldots$ of the subspace Θ becomes the basis $\eta_1, \eta_2, \ldots$ of the subspace L so that the best unbiased estimate $\alpha \in L$ (corresponding to the element $\theta_\alpha \in \Theta$) of the functional $\alpha = \alpha(\theta)$ is representable as a series of the form given by (1.37):

$$\alpha = \sum_k \alpha(\theta_k^*)\eta_k.$$

For $\alpha(\theta) = \langle \theta_k, \theta \rangle$ we obtain from (1.37) that $\alpha = \eta_k$, that is, η_k is the best unbiased estimate of the function $\alpha_k^*(\theta) = \langle \theta, \theta_k \rangle$, $\theta \in \Theta$ $(k = 1, 2, \ldots)$. Finally, the strong convergence of the series given by (1.37) is equivalent to convergence in the mean (with respect to probability measure P), which fact together with the convergence of the series consisting of the mean values

$$\sum_k \alpha(\theta_k^*)\mathsf{M}_\theta \eta_k = \sum_k \alpha(\theta_k^*)\langle \eta_k, \eta_\theta \rangle = \sum_k \alpha(\theta_k^*)\langle \theta_k, \theta \rangle = \langle \theta_\alpha, \theta \rangle = \alpha(\theta)$$

* See, for instance, [4]; $\{\theta_i\}$, $\{\theta_i^*\}$ are biorthogonal bases.

is equivalent to convergence in the mean square with respect to any measure P_θ, $\theta \in \Theta$. The theorem is proved. □

Let us note that if the initial basis $\theta_1, \theta_2, \ldots$ of the subspace $\Theta \subseteq Y$ is orthonomoral, we shall have the formula

$$\mathsf{M}_\theta[\alpha_1 - \alpha_1(\theta)][\alpha_2 - \alpha_2(\theta)] = \langle \alpha_1, \alpha_2 \rangle = \sum_k \alpha_1(\theta_k)\alpha_2(\theta_k) \tag{1.38}$$

for any linear functionals $\alpha_1 = \alpha_1(\theta)$ and $\alpha_2 = \alpha_2(\theta)$ of $\theta \in \Theta$. In particular, the best unbiased estimates $\hat{\theta}(t)$, $t \in T$, for functionals $\theta(t)$ of $\theta \in \Theta$ of the form given by (1.19) (where the parameter t runs through the set T) form with respect to each of the probability measures P_θ a random process with the corresponding mean value $\theta(t)$, $t \in T$, and the correlation function

$$\mathsf{M}_\theta[\hat{\theta}(s) - \theta(s)][\hat{\theta}(t) - \theta(t)] = \sum_k \theta_k(s)\theta_k(t), \tag{1.39}$$

where the functions $\theta_k = \theta_k(t)$ of $t \in T$, $k = 1, 2, \ldots$, form an orthonormal basis in the subspace Θ of the function space Y.

We shall consider the problem of estimating the regression coefficients (see Section I.1). Let the subspace Θ have a finite dimension equal to N. Then any N linearly independent elements from Θ form a basis in Θ in the same way as any N linearly independent variables from the subspace L form a basis in L. If $\theta_1, \ldots, \theta_N$ is the basis in Θ, any element $\theta \in \Theta$ can be expressed as

$$\theta = \sum_{k=1}^{N} \alpha_k \theta_k,$$

where $\alpha_k = \langle \theta_k^*, \theta \rangle$, $k = 1, \ldots, N$, are so-called *regression coefficients* in the scheme given by (1.1); $\theta_1^*, \ldots, \theta_N^*$ is a system conjugate to $\theta_1, \ldots, \theta_N$. Let $\eta_1, \ldots, \eta_N$ be variables in the subspace L corresponding, by (1.29), to the elements $\theta_1, \ldots, \theta_N$. These variables can be defined by the formula (1.25),

$$\eta_k = \int \varphi_k(\lambda)\Phi(d\lambda), \qquad k = 1, \ldots, n,$$

where $\varphi_k(\lambda) \in L_T(F)$ is the solution of an equation of the form given by (1.16):

$$\theta_k(t) = \int e^{-i\lambda t}\varphi_k(\lambda)F(d\lambda), \qquad t \in T.$$

The variables η_k are the best unbiased estimates of the functionals $\alpha_k^*(\theta) = \langle \theta_k, \theta \rangle$ of $\theta \in \Theta$. Similarly, the elements θ_k^* are associated with the best unbiased estimates $\hat{\alpha}_k$ of the coefficients $\alpha_k = \langle \theta_k^*, \theta \rangle$, $k = 1, \ldots, N$. By virtue of the isomorphism (1.29) these best estimates $\hat{\alpha}_1, \ldots, \hat{\alpha}_N$ form a system conjugate to $\eta_1, \ldots, \eta_N$:

$$\langle \hat{\alpha}_k, \eta_j \rangle = \begin{cases} 1 & \text{for } k = j, \\ 0 & \text{for } k \neq j. \end{cases}$$

The variables $\eta_1, \ldots, \eta_N$ form a basis in the subspace $L \subseteq H(T)$. If we take advantage of the representation

$$\hat{\alpha}_i = \sum_j s_{ij}\eta_j, \qquad k = 1, \ldots, N, \tag{1.40}$$

the coefficients s_{ij} $(i, j = 1, \ldots, N)$ can be found from the equations

$$\sum_j s_{ij}\langle \eta_j, \eta_k \rangle = \begin{cases} 1 & \text{for } i = k, \\ 0 & \text{for } i \neq k. \end{cases}$$

It is seen that the matrix $\{s_{ij}\}$ is the inverse of the correlation matrix $\{\langle \eta_i, \eta_k \rangle\}$, where

$$\langle \eta_i, \eta_j \rangle = \langle \varphi_i, \varphi_j \rangle_F = \int \varphi_i(\lambda)\overline{\varphi_j(\lambda)}F(d\lambda), \qquad i, j = 1, \ldots, N. \tag{1.41}$$

Taking into account (1.40), it can easily be calculated that $\{s_{ij}\}$ coincides with the correlation matrix of the best estimates $\hat{\alpha}_1, \ldots, \hat{\alpha}_N$:

$$\mathsf{M}_\theta(\hat{\alpha}_i - \alpha_i)(\hat{\alpha}_j - \alpha_j) = \langle \hat{\alpha}_i, \hat{\alpha}_j \rangle = s_{ij}, \qquad i, j = 1, \ldots, N. \tag{1.42}$$

We conclude by noting that the best unbiased estimate for a functional of the type given by (1.19)—the mean value of $\theta(t)$ at a fixed point $t \in T$—being a linear combination $\theta(t) = \sum_{k=1}^N \alpha_k\theta_k(t)$ can be represented as (1.8):

$$\theta(t) = \sum_{k=1}^N \hat{\alpha}_k\theta_k(t), \qquad t \in T.$$

VII.2 Estimation of the Mean Value and the Method of Least Squares

As noted above, the equivalence condition of all the Gaussian distributions P_θ, $\theta \in \Theta$, considered, with the mean values $\theta = \theta(t)$, $t \in T$, is equivalent to the fact that each function $\theta = \theta(t)$ permits the representation given by (1.16), implying that $\theta = \theta(t)$ is the restriction of the linear continuous functional $\langle \varphi, \varphi_\theta \rangle_F$ of $\varphi \in L_T(F)$ on the elements $e^{i\lambda t} \in L_T(F)$. It is seen that in this case $\theta = \theta(t)$ permits an analogous representation with respect to any spectral measure $G(d\lambda)$ for which the corresponding space $L_T(G)$ is contained in $L_T(F)$ and $\|\varphi\|_F \leqslant \|\varphi\|_G$ for all $\varphi(\lambda) \in L_T(G)$.

If A is the operator on the Hilbert space $L_T(G)$ into the Hilbert space $L_T(F)$ defined by $A\varphi(\lambda) = \varphi(\lambda)$, the condition indicated above implies that the operator $B = A^*A$ is bounded in $L_T(G)$ (A^* is the adjoint of A). Under this condition, as we have just noted, the function $\theta(t)$, $t \in T$, permits the representation

$$\theta(t) = \int e^{-i\lambda t}\psi_\theta(\lambda)G(d\lambda), \qquad t \in T, \tag{2.1}$$

where $\psi_\theta(\lambda) \in L_T(G)$.

Under the condition where the operator $B = A^*A$ is a nuclear operator, a representation of the type given by (2.1) holds for the Gaussian process

$\xi = \xi(t)$, $t \in T$, considered (stationary with respect to the distribution P): with probability 1

$$\xi(t) = \int e^{-i\lambda t}\eta(\lambda)G(d\lambda), \qquad t \in T \tag{2.2}$$

(for this, see Section I.6), where $\eta = \eta(\lambda)$ is a Guassian random function with trajectories in the Hilbert space $L_T(G)$. Since all initial distributions P_θ are equivalent, (2.2) is satisfied (with probability 1) with respect to any of the distributions P_θ. In this case the pertinent probability distribution of the random function $\eta(\lambda) \in L_T(G)$ is Gaussian with mean value $\psi_\theta(\lambda) \in L_T(G)$ and correlation operator coinciding with the operator $B = A^*A$ given above.

In fact,

$$\theta(t) = \mathsf{M}_\theta\xi(t) = \int e^{-i\lambda t}\mathsf{M}_\theta\eta(\lambda)G(d\lambda) = \int e^{-i\lambda t}\psi_\theta(\lambda)G(d\lambda), \qquad t \in T;$$

therefore, because of uniqueness of (2.1),

$$\mathsf{M}_\theta\eta(\lambda) = \psi_\theta(\lambda). \tag{2.3}$$

Furthermore (see Section I.6),

$$\begin{aligned} &\mathsf{M}_\theta[\langle\varphi, \eta\rangle_G - \langle\varphi, \psi_\theta\rangle_G][\langle\psi, \eta\rangle_G - \langle\psi, \psi_\theta\rangle_G] \\ &= \mathsf{M}[\langle\varphi, \eta\rangle_G\langle\psi, \eta\rangle_G] = \mathsf{M}[\eta(\varphi)\eta(\psi)] \\ &= \langle\varphi, \psi\rangle_F = \langle A\varphi, A\psi\rangle_F = \langle A^*A\varphi, \psi\rangle_G = \langle B\varphi, \psi\rangle_G, \end{aligned} \tag{2.4}$$

where (see (6.9) in Chapter I)

$$\eta(\varphi) = \int \varphi(\lambda)\Phi(d\lambda) = \langle\varphi, \eta\rangle_G, \qquad \varphi \in L_T(G).$$

The observable random process $\xi = \xi(t)$, $t \in T$, will be considered, moreover, to be an element of the Hilbert space X consisting of functions of the form

$$x(t) = \int e^{-i\lambda t}\varphi(\lambda)G(d\lambda), \qquad t \in T \tag{2.5}$$

(where $\varphi(\lambda) \in L_T(G)$), with the scalar product

$$\langle x_1, x_2\rangle = \langle\varphi_1, \varphi_2\rangle_G. \tag{2.6}$$

We consider some estimates of the unknown mean value $\theta = \theta(t)$, $t \in T$, as an element of the space X.

As shown above (see (1.24)), in considering unbiased estimates we can regard without loss of generality the parametric set Θ as a closed subspace in the Hilbert space Y, analogous to the Hilbert space X introduced above but having spectral measure $F(d\lambda)$, that is, the spectral measure of the random process $\xi = \xi(t)$ stationary with respect to P. The set Θ, being a subspace in the Hilbert space X, is not (generally speaking) closed. Moreover, the subspace $\Theta \subseteq X$ is closed if and only if it is finite dimensional.

In fact, the operator on the Hilbert space Y into the Hilbert space X determining the mapping

$$y(t) \in Y \to y(t) \in X$$

coincides with the adjoint operator A^*, more precisely, with the operator A^* in Y coinciding on functions $y \in Y$ with the values $A^*y \in Y$ that correspond by (1.17) to the values $A^*\varphi$. Also, $\varphi \in L_T(F)$ corresponds to $y \in Y$:

$$\begin{aligned} y(t) &= \int e^{-i\lambda t}\varphi(\lambda)F(d\lambda) \\ &= \langle e^{i\lambda t}, \varphi\rangle_F \\ &= \langle Ae^{i\lambda t}, \varphi\rangle_F = \langle e^{i\lambda t}, A^*\varphi\rangle_G = \int e^{-i\lambda t}[A^*\varphi(\lambda)]G(d\lambda) \end{aligned}$$

Therefore, when the subspace Θ in X is closed (when the subspace $A^*\Theta \subseteq X$ is closed) the one-to-one mapping

$$\theta(t) \in Y \xleftrightarrow{A^*} \theta(t) \in X$$

maps the closed subspace $\Theta \subseteq Y$ onto the closed subspace $A^*\Theta \subseteq X$. Therefore, by Banach's theorem, there is a bounded inverse operator $(A^*)^{-1}$. The operator $B = A^*A$, being nuclear, the bounded operator inverse to A^* exists if and only if the subspace Θ is finite dimensional.

Let us denote by $\tilde{\Theta}$ the closure of Θ in a Hilbert space X and consider an arbitrary unbiased estimate $\eta \in X$ for a parameter $\theta \in \Theta$ with values in the space X:

$$\mathsf{M}_\theta\eta = \theta, \qquad \theta \in \Theta.$$

The fact that the estimate $\eta \in X$ is unbiased implies that $M_\theta\langle\eta, x\rangle = \langle\theta, x\rangle$ for any $x \in X$. If $\theta_1, \theta_2, \ldots$ is an orthonormal basis in the subspace $\Theta \subseteq X$, and if $x_1, x_2, \ldots$ is the complement of this basis to the orthonormal basis in the whole space X, then

$$\eta = \sum_k \langle\eta, \theta_k\rangle\theta_k + \sum_k \langle\eta, x_k\rangle x_k$$

and

$$\begin{aligned} \mathsf{M}_\theta\eta &= \sum_k [\mathsf{M}_\theta\langle\eta, \theta_k\rangle]\theta_k + \sum_k [\mathsf{M}_\theta\langle\eta, x_k\rangle]x_k \\ &= \sum_k \langle\theta, \theta_k\rangle\theta_k + \sum_k \langle\theta, x_k\rangle x_k = \sum_k \langle\theta, \theta_k\rangle\theta_k = \theta. \end{aligned}$$

From this it is seen that the projection $\tilde{\eta}$ of the unbiased estimate η onto the subspace $\tilde{\Theta}$ is an unbiased estimate for the unknown mean value:

$$\tilde{\eta} = \sum_k \langle\eta, \theta_k\rangle\theta_k$$

and

$$\mathsf{M}_\theta\tilde{\eta} = \sum_k \langle\theta, \theta_k\rangle\theta_k = \theta, \qquad \theta \in \Theta.$$

Further, since the variables $\eta - \tilde{\eta}$ and $\tilde{\eta} - \theta$ are orthogonal,

$$\|\eta - \theta\|^2 = \|\tilde{\eta} - \theta\|^2 + \|\eta - \tilde{\eta}\|^2$$

and

$$\mathsf{M}_\theta\|\tilde{\eta} - \theta\|^2 = \mathsf{M}_\theta\|\eta - \theta\|^2 - \mathsf{M}_\theta\|\eta - \tilde{\eta}\|^2 \leqslant \mathsf{M}_\theta\|\eta - \theta\|^2. \tag{2.7}$$

It is seen that $\tilde{\eta} \in \tilde{\Theta}$ is the *improved* estimate as compared with the primary estimate $\eta \in X$.

The variable ξ is the simplest unbiased estimate for the parameter $\theta \in \Theta$. We denote the projection of ξ onto the subspace $\tilde{\Theta}$ by $\tilde{\xi}$:

$$\tilde{\xi} = \sum_k < \xi, \theta_k > \theta_k. \tag{2.8}$$

For the unbiased estimate $\tilde{\xi}$, called the *least-squares estimate*, we have

$$\mathsf{M}_\theta\|\tilde{\xi} - \theta\|^2 = \sum_k \mathsf{M}_\theta(\langle \xi, \theta_k \rangle - \langle \theta, \theta_k \rangle)^2 = \sum_k \langle B^*\theta_k, \theta_k \rangle < \infty, \tag{2.9}$$

since the correlation operator B^* of the Gaussian variable $\xi \in X$ is a nuclear operator (see Section I.4).

Each of the real variables $\eta_k = \langle \xi, \theta_k \rangle$, $k = 1, 2, \ldots, n$, is an unbiased estimate for the functional $\alpha_k(\theta) = \langle \theta, \theta_k \rangle$ of $\theta \in \Theta$. We described in Theorem 1 the best unbiased estimates $\hat{\alpha}_k$, $k = 1, 2, \ldots$ for which

$$\mathsf{M}_\theta(\hat{\alpha}_k - \langle \theta, \theta_k \rangle)^2 \leqslant \mathsf{M}_\theta(\langle \xi, \theta_k \rangle - \langle \theta, \theta_k \rangle)^2, \qquad k = 1, 2, \ldots . \tag{2.10}$$

It can be seen from (2.9) and (2.10) that

$$\mathsf{M} \sum_k \hat{\alpha}_k^2 = \sum_k \mathsf{M}\hat{\alpha}_k^2 \leqslant \sum_k \mathsf{M}\langle \xi, \theta_k \rangle^2 = \sum_k \langle B^*\theta_k, \theta_k \rangle < \infty;$$

hence with probability 1

$$\sum_k \hat{\alpha}_k^2 < \infty. \tag{2.11}$$

We consider the variable $\hat{\theta} \in \tilde{\Theta}$ defined with probability 1 by

$$\hat{\theta} = \sum_k \hat{\alpha}_k \theta_k, \tag{2.12}$$

where the $\hat{\alpha}_k$ are the best estimates for the coefficients $\alpha_k(\theta) = \langle \theta, \theta_k \rangle$.

It is clear that $\hat{\theta}$ is the unbiased estimate for the unknown mean value $\theta \in \Theta$. For any unbiased estimate $\eta \in X$, (1.27) yields

$$\begin{aligned}\mathsf{M}_\theta\|\hat{\theta} - \theta\|^2 &= \sum_k \mathsf{M}_\theta(\hat{\alpha}_k - \langle \theta, \theta_k \rangle)^2 \\ &\leqslant \sum_k \mathsf{M}_\theta(\langle \eta, \theta_k \rangle - \langle \theta, \theta_k \rangle)^2 \leqslant \mathsf{M}_\theta\|\eta - \theta\|^2.\end{aligned} \tag{2.13}$$

Therefore, the variable $\hat{\theta}$ is the best (in the sense of (2.13)) unbiased estimate for the parameter $\theta \in \Theta$.

Theorem 3. *There exists the best unbiased estimate $\hat{\theta}$ for any unknown parameter $\theta \in \Theta$, representable as an element from the function space X in the form*

$$\hat{\theta}(t) = \int e^{-i\lambda t}\hat{\eta}(\lambda)G(d\lambda), \qquad t \in T \tag{2.14}$$

(where $\hat{\eta}(\lambda) \in L_T(G)$). For each fixed $t \in T$ the value $\hat{\theta}(t)$ yields the best unbiased estimate for the value $\theta(t)$ of the unknown mean $\theta \in \Theta$. The random process $\hat{\theta}(t)$, $t \in T$, is obtained as a result of linear transformation of the observable random process $\xi(t)$, $t \in T$, defined by

$$\hat{\theta}(t) = \mathscr{P}\xi(t), \; t \in T, \tag{2.15}$$

where $\mathscr{P}$ is the operator of projection in the space $H(T)$ onto the subspace L generated by variables of the type (1.25). *The random process $\hat{\theta} = \hat{\theta}(t)$ permits the expansion* (*see* (1.31))

$$\hat{\theta}(t) = \int \varphi(\lambda, t)\Phi(d\lambda), \tag{2.16}$$

where $\varphi(\lambda, t)$ is the projection of $e^{i\lambda t} \in L_T(F)$ onto the subspace generated by all functions $\varphi_\theta(\lambda)$ from (1.16). *The variable $\hat{\eta}(\lambda) \in L_T(G)$ in* (2.14) *with probability* 1 *is*

$$\hat{\eta}(\lambda) = \sum_k \hat{\alpha}_k \varphi_k(\lambda), \tag{2.17}$$

where the $\hat{\alpha}_k$, $k = 1, 2, \ldots$, are the same random coefficients as those in the primary formula (2.12), *and the functions $\varphi_k(\lambda) \in L_T(G)$ correspond to $\theta_k \in X$:*

$$\theta_k(t) = \int e^{-i\lambda t}\varphi_k(\lambda)G(d\lambda), \qquad t \in T, k = 1, 2, \ldots . \tag{2.18}$$

Proof. By virtue of (2.18)

$$\theta_k(t) = \langle e^{i\lambda t}, \varphi_k(\lambda)\rangle, \qquad t \in T,$$

where φ_k, $k = 1, 2, \ldots$, is an orthonormal system in the space $L_T(G)$ since θ_k, $k = 1, 2, \ldots$, is an orthonormal system in the space X. Hence

$$\sum_k \theta_k(t)^2 \leqslant \|e^{i\lambda t}\|_G^2 < \infty$$

for each fixed $t \in T$, and since $\sum_k \hat{\alpha}_k^2 < \infty$ with probability 1 (see (2.11)) the series

$$\hat{\theta}(t) = \sum_k \hat{\alpha}_k \theta(t), \qquad t \in T, \tag{2.19}$$

is convergent with probability 1 for each fixed $t \in T$. Obviously, the variable $\hat{\theta}(t)$ (t fixed), being the limit as $n \to \infty$ of the variables $\sum_{k=1}^n \hat{\alpha}_k \theta_k(t)$ from the closed subspace $L \subseteq H(T)$, belongs to L, and is also the best unbiased estimate for the unknown value

$$\theta(t) = \mathsf{M}_\theta \hat{\theta}(t) = \sum_k [\mathsf{M}_\theta \hat{\alpha}_k]\theta_k(t), \qquad t \in T.$$

All these estimates were described in Theorem 1, the latter implying (2.15) and (2.16), thus proving the theorem. □

Thus, there exists the best unbiased estimate of $\hat{\theta} = \hat{\theta}(t)$, which reproduces best the unknown function parameter $\theta = \theta(t)$ from the subspace $\Theta \in X$. This estimate $\hat{\theta} \in X$ is contained in the closure $\tilde{\Theta}$ of the subspace Θ (see (2.12)). We have as well the following

Theorem 4. *The best unbiased estimate $\hat{\theta}$ belongs to the initial subspace Θ if and only if Θ is finite dimensional.*

Proof. By virtue of (2.16)

$$\hat{\theta}(t) = \int [\mathscr{P}e^{i\lambda t}]\Phi(d\lambda), \qquad t \in T,$$

where $\mathscr{P}$ is the operator of projection of the Hilbert space $L_T(F)$ onto the space L generated by all the functions $\varphi_\theta(\lambda)$ related to $\theta \in \Theta$ by (1.16). The subspace L is unitarily isomorphic to Θ regarded as a subspace in the Hilbert space Y (see (1.29)). If the trajectories of some equivalent random process $\hat{\theta} = \hat{\theta}(t)$, $t \in T$, enter Y with positive probability, there will exist a Gaussian function $\zeta(\lambda) \in L_T(F)$ such that with probability 1

$$\hat{\theta}(t) = \langle e^{i\lambda t}, \zeta(\lambda)\rangle_F, \qquad t \in T$$

(see Section I.6); the correlation operator of the above function can be found from the relations

$$\begin{aligned} &\mathsf{M}[\langle e^{i\lambda s}, \zeta(\lambda)\rangle_F \langle e^{i\lambda t}, \zeta(\lambda)\rangle_F] \\ &= \mathsf{M}[\hat{\theta}(s)\hat{\theta}(t)] \\ &= \langle \mathscr{P}e^{i\lambda s}, \mathscr{P}e^{i\lambda t}\rangle_F = \langle \mathscr{P}^*\mathscr{P}e^{i\lambda s}, e^{i\lambda t}\rangle_F, \qquad s, t \in T, \end{aligned}$$

which, as can be seen, is $\mathscr{P}^*\mathscr{P} = \mathscr{P}$. A correlation operator must be nuclear (see Theorem 1, Chapter I). The projection operator $\mathscr{P}$ is nuclear only if the subspace L (therefore, the unitarily isomorphic space Θ) is finite dimensional. The theorem is proved. □

Let us discuss the finite-dimensional case. The best unbiased estimate of $\hat{\theta} = \hat{\theta}(t)$ for the unknown mean value $\theta = \theta(t)$, $t \in T$, can be expressed as

$$\hat{\theta} = \sum_{k=1}^{N} \eta_k \theta_k,$$

where the $\eta_k = \hat{\alpha}_k$ are the best unbiased estimates of the coefficients $\alpha_k = \langle \theta, \theta_k \rangle$, $\theta \in \Theta$, in the expansion

$$\theta = \sum_{k=1}^{N} \alpha_k \theta_k$$

of $\theta \in \Theta$ as basis elements $\theta_1, \ldots, \theta_N$. For the orthonormal basis $\theta_1, \ldots, \theta_N$, the corresponding estimates $\eta_1, \ldots, \eta_N$ are independent Gaussian variables with the same dispersion equal to 1 (see Theorem 2).

The best unbiased estimate $\hat{\theta}$ is a maximum point of the *likelihood function* $l(\theta) = \log p_\theta$ of $\theta \in \Theta$ where $p_\theta = p_\theta(\omega)$ is the density $\mathsf{P}_\theta(d\omega)/\mathsf{P}(d\omega)$ defined by (1.15):

$$l(\theta) = \eta_\theta - \frac{1}{2}\|\eta\|^2 = \sum_k \alpha_k \eta_k - \frac{1}{2}\sum_k \alpha_k^2.$$

In fact, it is seen that the maximum of the function $l(\theta)$ can be attained at $\alpha_k = \eta_k$ $(k = 1, \ldots, N)$:

$$\max_{\theta \in \Theta} l(\theta) = \frac{1}{2}\sum_k \eta_k^2 = l(\hat{\theta}).$$

It should be noted that the *maximum likelihood method* is inapplicable in the finite-dimensional case where the Gaussian process $l(\theta)$, $\theta \in \Theta$, is unbounded. This fact follows immediately, for example, from the following the relation: with probability 1

$$\sup_{\theta \in \Theta} l(\theta) \geqslant \sup(\eta_1, \eta_2, \ldots) - \frac{1}{2} = \infty,$$

where η_k, $k = 1, 2, \ldots$, is an infinite sequence of independent Gaussian variables with dispersions equal to 1 and mean values $\alpha_k = \langle \theta, \theta_k \rangle$ such that $\sum_k \alpha_k^2 < \infty$ (here $\theta_1, \theta_2, \ldots$, is the orthonormal basis in the finite-dimensional subspace $\Theta \subseteq Y$, and $\eta_1, \eta_2, \ldots$, are the corresponding variables in the subspace $L \subseteq H(T)$—see Theorem 2).

It should be mentioned that formal extremum conditions for the likelihood function $l(\theta)$ of $\theta = \sum_k \alpha_k \theta_k$, namely

$$\frac{\partial l(\theta)}{\partial \alpha_k} = 0, \qquad k = 1, 2, \ldots,$$

lead to the best unbiased estimates $\eta_k = \hat{\alpha}_k$ for the coefficients $\alpha_k = \langle \theta, \theta_k \rangle$, $k = 1, 2, \ldots$. Since the coefficients $\hat{\alpha}_k = \eta_k$, $k = 1, 2, \ldots$, are such that the series $\sum \eta_k^2$ is divergent with probability 1, there is no element representable as $\sum \hat{\alpha}_k \theta_k$ in the Hilbert space Y. In fact, the best unbiased estimate $\hat{\theta}(t)$ for each fixed $t \in T$ can be represented as the series given by (2.19):

$$\hat{\theta}(t) = \sum_k \eta_k \theta_k(t).$$

We used above the Hilbert space $L_T(G)$ corresponding to the finite spectral measure $G(d\lambda)$, which fact enabled us to consider the spectral representation (2.2) for particular values of the primary random process $\xi(t)$, $t \in T$, coinciding on the functions $\varphi(\lambda) = e^{i\lambda t}$ with the values of the random functional

$$\eta(\varphi) = \langle \varphi, \eta \rangle_G, \qquad \varphi \in L_T(G),$$

and having with respect to the distribution P_θ the mean value $\psi_\theta(\lambda) \in L_T(G)$, where ψ_θ determines the linear functional on the Hilbert space $L_T(G)$ coinciding with the values $\theta(t)$, $t \in T$, on the functions $\varphi(\lambda) = e^{i\lambda t}$ (see (2.3), (2.4)). Instead of the random process $\xi(t)$ of the continuous parameter $t \in T$, we could consider first the generalized random process $\eta(\varphi)$ of $\varphi \in L_T(G)$ with the mean value ψ_θ, where

$$\langle \varphi, \psi_\theta \rangle_G = \mathsf{M}_\theta \langle \varphi, \eta \rangle_G, \qquad \varphi \in L_T(G),$$

and the correlation functional $B = A^*A$, where the operator A in $L_T(G)$ onto $L_T(F)$ operates according to $A\varphi = \varphi$. It should be mentioned here (see Section I.6) that the space $L_T(G)$ can be defined as the closure of the space of functions of the form

$$\varphi(\lambda) = \int_T e^{i\lambda t} c(t)\, dt,$$

where $c(t)$ is an infinitely differentiable function vanishing outside an interval in T.

In considering a random functional of the type given by (2.2), or, equivalently, a random variable η in a Hilbert space $L_T(G)$, we speak about estimates of the mean value $\psi_\theta \in L_T(G)$. The technique and results described above can be applied here: we need only go from the parametric set T over to a parametric set $L_T(G)$, and go from the values $\xi(t)$ and $\theta(t)$, $t \in T$, to values $\langle \varphi, \eta \rangle_G$ and $\langle \varphi, \psi_\theta \rangle_\theta$, $\varphi \in L_T(G)$.

$G(d\lambda) = (1/2\pi)\, d\lambda$ represents a particular case, where $L_T(G)$ coincides with the space of functions of the form

$$\varphi(\lambda) = \int_T e^{i\lambda t} c(t)\, dt,$$

where $c(t) \in \mathscr{L}^2(T)$ is the space of all square-integrable functions. By Parseval's equality we have

$$\langle \varphi_1, \varphi_2 \rangle_G = \langle c_1(t), c_2(t) \rangle = \int c_1(t) c_2(t)\, dt.$$

It is clear that considering the random function $\eta(\lambda) \in L_T(G)$ of the form

$$\eta(\lambda) = \int_T e^{i\lambda t} \xi(t)\, dt$$

is actually equivalent to considering the initial random process $\xi(t)$, $t \in T$, as a function in a Hilbert space $\mathscr{L}^2(T)$ (noting that $\psi_\theta(\lambda) = \int_T e^{i\lambda t}\theta(t)\, dt$).

VII.3 Consistent Pseudo-Best Estimates

We assume that the observable random process $\xi(t)$, $t \in T$, is an element from the Hilbert space X defined above of real functions $x = x(t)$ of $t \in T$ of the form given by (2.5). Let us consider $\tilde{\theta} \in X$, the least-squares estimate for the unknown mean value $\theta \in \Theta$, $\Theta \subseteq X$, given by (2.8).

In (2.8) there are the estimates

$$\tilde{\alpha}_k = \langle \theta_k, \xi \rangle = \langle \theta_k, \tilde{\theta} \rangle$$

for the coefficients $\alpha_k = \langle \theta, \theta_k \rangle$, $k = 1, 2, \ldots$, in the expansion of $\theta \in \Theta$ with respect to the orthonormal basis $\theta_1, \theta_2, \ldots \in \tilde{\Theta}$. These estimates as well as estimates such as

$$\tilde{\alpha} = \langle \theta_\alpha, \tilde{\theta} \rangle = \langle \theta_\alpha, \xi \rangle$$

for the functionals

$$\alpha(\theta) = \langle \theta_\alpha, \theta \rangle \tag{3.1}$$

of $\theta \in \Theta$ in the Hilbert space X (where $\theta_\alpha \in \tilde{\Theta}$) are said to be *least-squares estimates represented in the spectral form*

$$\tilde{\alpha} = \int \varphi_\alpha(\lambda) \Phi(d\lambda), \tag{3.2}$$

where $\varphi_\alpha(\lambda) \in L_T(G)$ is the solution of the integral equation

$$\theta_\alpha(t) = \int e^{-i\lambda t} \varphi_\alpha(\lambda) G(d\lambda), \qquad t \in T. \tag{3.3}$$

In fact, the results given in Section I.6 imply that

$$\tilde{\alpha} = \langle \theta_\alpha, \tilde{\theta} \rangle = \langle \theta_\alpha, \xi \rangle = \langle \varphi_\alpha, \eta \rangle = \int \varphi_\alpha(\lambda) \Phi(d\lambda),$$

where $\eta(\lambda) \in L_T(G)$ is the random function from (2.2).

We note that in the case $G(d\lambda) = (1/2\pi)\, d\lambda$ we deal with classical least-squares estimates when the observable random process $\xi = \xi(t)$, $t \in T$, is regarded as a point in a classical Hilbert space $\mathscr{L}^2(T)$ of real functions $x = x(t)$ of $t \in T$ with the scalar product

$$\langle x_1, x_2 \rangle = \langle \psi_1, \psi_2 \rangle_G = \begin{cases} \sum_t x_1(t) x_2(t) & \text{for discrete } t, \\ \int_T x_1(t) x_2(t)\, dt & \text{for continuous } t, \end{cases}$$

where $\psi_1, \psi_2 \in L_T(G)$ are related to $x_1, x_2 \in \mathscr{L}^2(T)$ by the usual Fourier transform

$$\psi(\lambda) = \begin{cases} \sum_t e^{i\lambda t} x(t) & \text{for discrete } t, \\ \int_T e^{i\lambda t} x(t)\, dt & \text{for continuous } t. \end{cases}$$

The estimates given by (3.3) for functionals $\alpha(\theta) = \langle \theta_\alpha, \theta \rangle$ of $\theta \in \Theta \subseteq \mathscr{L}^2(T)$ are

$$\tilde{\alpha} = \begin{cases} \sum_t \theta_\alpha(t) \xi(t) & \text{for discrete } t, \\ \int_T \theta_\alpha(t) \xi(t)\, dt & \text{for continuous } t. \end{cases} \tag{3.4}$$

The structure of (3.2) is the same as that of (1.31) (see Theorem 1). Therefore if $G(d\lambda)$ were the spectral measure of the stationary process $\xi(t)$, $t \in T$, the estimates $\tilde{\alpha}$ of the type given by (3.2) would be the best unbiased estimates for functionals of the type given by (3.1).

We shall consider another class of estimates generalizing least-squares estimates and including the best unbiased estimates.

(3.2) and (3.3) have meaning in the case where the conditions (given at the start of Section 2)

$$L_T(G) \subseteq L_T(F), \qquad \|\varphi\|_F \leqslant \|\varphi\|_G, \qquad \varphi \in L_T(G) \tag{3.5}$$

are satisfied as well as in the case where trajectories of the observable process $\xi = \xi(t)$, $t \in T$, are contained in the pertinent Hilbert space X.

If $G(d\lambda)$ were the spectral measure of the stationary process $\xi = \xi(t)$, (3.2) and (3.3) would yield the best unbiased estimates for functionals of the type given by (3.1). The measure $G(d\lambda)$ that is used in constructing the estimates given by (3.2) is referred to as a *pseudo-spectral measure*; the estimates given by (3.2) themselves are referred to as *pseudo-best estimates* (for functionals of the type given by (3.1)).

Then

$$\mathsf{M}_\theta \int \varphi(\lambda)\Phi(d\lambda) = \langle \varphi, \psi_\theta \rangle_G, \qquad \varphi \in L_T(G), \tag{3.6}$$

where $\psi_\theta(\lambda) \in L_T(G)$ is the function in (2.1) for $\theta \in \Theta$.

The spectral representation given by (2.1) was proved (see (2.3)) for the case where the random process $\xi = \xi(t)$, $t \in T$, is a random element in the Hilbert space X (constructed from the pseudo-spectral measure $G(d\lambda)$). For an arbitrary finite measure $G(d\lambda)$, with $\varphi(\lambda) = e^{i\lambda t}$, (3.6) coincides with (2.1) and extends to the linear closed hull of the functions $\varphi(\lambda) = e^{i\lambda t}$ ($t \in T$), that is, to the whole Hilbert space $L_T(G)$ (compare with (1.22)).

By virtue of (3.6) we have

$$\mathsf{M}_\theta \tilde{\alpha} = \mathsf{M}_\theta \int \varphi_\alpha(\lambda)\Phi(d\lambda) = \langle \varphi_\alpha, \psi_\theta \rangle_G = \langle \theta_\alpha, \theta \rangle = \alpha(\theta), \qquad \theta \in \Theta, \tag{3.7}$$

and therefore the pseudo-best estimates are unbiased.

An essential property of pseudo-best estimates consists in the fact that they are invariant with respect to multiplication of the pseudo-spectral measure $G(d\lambda)$ by any constant k.

Indeed, the function $\varphi_\alpha(\lambda)$ in (3.2) is uniquely defined from the relation

$$\alpha(\theta) = \langle \varphi_\alpha, \psi_\theta \rangle_G, \qquad \theta \in \Theta, \tag{3.8}$$

where the functions $\psi_\theta(\lambda) \in L_T(G)$ and $\theta(t) \in \Theta$ are related by (2.1). It is seen that

$$\theta(t) = \int e^{-i\lambda t}[k^{-1}\psi_\theta(\lambda)]kG(d\lambda)$$

and

$$\langle \varphi_\alpha, k^{-1}\psi_\theta \rangle_{kG} = \langle \varphi_\alpha, \psi_\theta \rangle_G = \alpha(\theta);$$

therefore, the function $\varphi_\alpha(\lambda)$ in (3.2) with respect to the pseudo-spectral measure $kG(d\lambda)$ is the same as that with respect to the primary measure $G(d\lambda)$.

Suppose we deal with a sequence of *expanding* observations in considering the random process $\xi = \xi(t)$ on sets $T = T_1, T_2, \ldots$:

$$T_1 \subseteq T_2 \subseteq \ldots \quad .$$

We speak here about pseudo-best estimates $\tilde{\alpha}_n$ of the type given by (3.2) constructed from the observations of the random process $\xi = \xi(t)$ on the pertinent set $T = T_n$.

Let $\alpha(\theta)$ be a linear functional of the unknown mean value $\theta = \theta(t)$, $t \in T^*$, where

$$T^* = \bigcup T_n.$$

This functional is continuous for $n = n_0$ with respect to the scalar product

$$\langle \theta_1, \theta_2 \rangle_n = \langle \psi^n_{\theta_1}, \psi^n_{\theta_2} \rangle_G \tag{3.9}$$

of the same type as (2.6), that is, the functions $\theta = \theta(t)$, $t \in T^*$, from a parametric space Θ and the functions $\psi^n_\theta(\lambda) \in L_{T_n}(G)$ are related by the relation of the type (2.1):

$$\theta(t) = \int e^{-i\lambda t}\psi_\theta^n(\lambda)G(d\lambda), \qquad t \in T_n. \tag{3.10}$$

Lemma 5. *The scalar products indicated above are such that*

$$\|\theta\|_1 \leqslant \|\theta\|_2 \leqslant \dots \ . \tag{3.11}$$

Proof. In fact, if (3.10) holds for some T_n then, taking the projection ψ_θ^m of the element $\psi_\theta^n \in L_{T_n}(G)$ onto the space $L_{T_m}(G) \subseteq L_{T_n}(G)$, $m \leqslant n$, we shall have

$$\theta(t) = \langle e^{i\lambda t}, \psi_\theta^n\rangle_G = \langle e^{i\lambda t}, \psi_\theta^m\rangle_G, \qquad t \in T_m,$$

that is, (3.10) holds for each $T_m \subseteq T_n$ as well, the pertinent element ψ_θ^m being the projection of the primary element $\psi_\theta^n \in L_{T_n}(G)$ onto $L_{T_m}(G)$.

It follows from the above that

$$\|\theta\|_m = \|\psi_\theta^m\|_G \leqslant \|\psi_\theta^n\|_G = \|\theta\|_n$$

for any $m \leqslant n$, which was to be proved. □

The inequalities given by (3.11) enable us to conclude that if the linear functional $\alpha(\theta)$ of $\theta \in \Theta$ for $n = n_0$ is continuous with respect to the scalar product given by (3.9), this functional will possess such a property for all $n \geqslant n_0$. Therefore, for any $n \geqslant n_0$ we have a representation of the type given by (3.1) or (3.8), namely

$$\alpha(\theta) = \langle \varphi_\alpha^n, \psi_\theta^n\rangle_G, \qquad \theta \in \Theta, \tag{3.12}$$

associated with a pseudo-best unbiased estimate of the type given by (3.2):

$$\tilde{\alpha}_n = \int \varphi_\alpha^n(\lambda)\Phi(d\lambda). \tag{3.13}$$

The estimates $\tilde{\alpha}_n$, $n \geqslant n_0$, are said to be *consistent* if

$$\mathsf{M}_\theta[\tilde{\alpha}_n - \alpha(\theta)]^2 = \|\varphi_\alpha^n\|_F^2 \to 0 \tag{3.14}$$

as $n \to \infty$.

Let us assume temporarily that $G(d\lambda)$ is the true spectral measure and that, accordingly, $\tilde{\alpha}_n$ are the best unbiased estimates constructed from the observations of $\xi(t)$, $t \in T_n$.

We recall (see Theorem 1) that here the best estimate $\tilde{\alpha}$ constructed from the observations of $\xi(t)$, $t \in T$, is the unique unbiased estimate in the space $L \subseteq H(T)$ generated by variables η_θ, $\theta \in \Theta$, of the type given by (1.25). In terms of the Hilbert space $L_T(G)$ ($G(d\lambda)$ is spectral measure) we have that among all the functions $\phi(\lambda) \in L_T(G)$ determining the unbiased estimates $\int\varphi(\lambda)\Phi(d\lambda)$ for the functional $\alpha(\theta)$ of $\theta \in \Theta$ in question, the function $\varphi_\alpha(\lambda)$ only determining the best estimate lies in the subspace $L \subseteq L_T(G)$ generated by all possible elements $\psi_\theta(\lambda) \in L_T(G)$ from (2.1). In this case $\varphi_\alpha(\lambda)$ coincides with the projection of the *spectral characteristic* $\varphi(\lambda)$ of any other unbiased estimate $\int\varphi(\lambda)\Phi(d\lambda)$ onto the space $L \subseteq L_T(G)$ indicated. According to the above, the function $\varphi_\alpha^n(\lambda)$ coincides with the projection of any function $\varphi_\alpha^m(\lambda)$ $(m \leqslant n)$ onto $L_n \subseteq L_{T_n}(G)$, where L_n denotes the subspace generated by all functions $\psi_\theta^n(\lambda) \in L_{T_n}(G)$, $\theta \in \Theta$. Consequently,

$$\|\varphi_\alpha^m - \varphi_\alpha^n\|_G^2 = \|\varphi_\alpha^m\|_G^2 - \|\varphi_\alpha^n\|_G^2 \to 0$$

as $m, n \to \infty$, and the limit

$$\varphi_\alpha^*(\lambda) = \lim_{n\to\infty} \varphi_\alpha^n(\lambda) \tag{3.15}$$

exists (in the space $L_{T^*}(G)$). Taking the space $L_{T^*}(F)$ (under the condition of (3.5) for $T = T^*$) we arrive at the fact that there exists an element $\varphi_\alpha^*(\lambda) \in L_T(F)$ such that

$$\|\varphi_\alpha^n - \varphi_\alpha^*\|_F \to 0. \tag{3.16}$$

It is seen (see (3.14)) that the consistency is valid if and only if

$$\varphi_\alpha^*(\lambda) = 0 \tag{3.17}$$

(almost everywhere with respect to $F(d\lambda)$). Thus, under the additional assumption that the measure $G(d\lambda)$ is equivalent to the true spectral measure $F(d\lambda)$ and (compare with (3.5))

$$\begin{gathered} L_{T^*}(G) \subseteq L_{T^*}(F), \\ \|\varphi\|_F \leqslant \|\varphi\|_G, \qquad \varphi \in L_{T^*}(G), \end{gathered} \tag{3.18}$$

a necessary and sufficient condition for the estimates $\tilde{\alpha}_n$ to be consistent is that the limiting function $\varphi_\alpha^*(\lambda)$ in (3.15) be zero almost everywhere with respect to $G(d\lambda)$; obviously, this condition is equivalent to the condition

$$\lim_{n\to\infty}\|\varphi_\alpha^n\|_G = 0. \tag{3.19}$$

We define $\tilde{\Theta}_n$ as the space of all real functions $\theta = \theta(t)$, $t \in T^*$, being the closure of the parametric space Θ with respect to the norm $\|\theta\| = \|\psi_\theta^n\|_G$.

Let

$$\tilde{\Theta} = \bigcap \tilde{\Theta}_n.$$

We shall consider the function

$$\theta^*(t) = \int e^{-i\lambda t}\varphi_\alpha^*(\lambda)G(d\lambda), \qquad t \in T^*, \tag{3.20}$$

related with the limiting function $\varphi_\alpha^*(\lambda)$ by a formula of the type (3.10). It can be easily seen that $\theta^* \in \tilde{\Theta}$ since $\theta^*(t)$ is the limit (in each fixed space $\tilde{\Theta}_m$) of the sequence of functions

$$\theta^n(t) = \int e^{-i\lambda t}\varphi_\alpha^n(\lambda)G(d\lambda), \qquad n \geqslant m.$$

Each linear functional $\alpha(\theta)$ of $\theta \in \Theta$ of the type considered (that is, continuous in some norm $\|\theta\|_m = \|\psi_\theta^m\|_G$) extends to the closure $\tilde{\Theta}_m$ and, all the more, to the space $\tilde{\Theta}$ with respect to the continuity. In particular, the value

$$\alpha(\theta^*) = \lim_{n\to\infty} \alpha(\theta^n) = \lim_{n\to\infty} \langle \varphi_\alpha^n, \varphi_\alpha^n \rangle_G = \|\varphi_\alpha^*\|_G^2$$

can be found from the continuity from (3.12). Therefore, a necessary and sufficient condition for the pseudo-best estimates $\tilde{\alpha}_n$ to be consistent is that

$$\alpha(\theta^*) = 0. \tag{3.21}$$

We can consider without loss of generality that $\Theta = \tilde{\Theta}$ since the class of pseudo-best unbiased estimates for the primary parametric set Θ coincides with the class of pseudo-best unbiased estimates for the linear closed hull of $\tilde{\Theta}$ (see the remark in Section VII.1). In this connection we shall assume that

$$\tilde{\Theta} = \Theta,$$

and denote by Θ^* the subspace in Θ formed by functions $\theta = \theta(t)$, $t \in T^*$ permitting a spectral representation of the type (3.10):

$$\theta(t) = \int e^{-i\lambda t}\psi_\theta(\lambda)G(d\lambda), \qquad t \in T^*. \tag{3.22}$$

For $\theta \in \Theta^*$ we obtain from (3.10) and (3.12) (holding for all $n \geqslant n_0$)

$$\alpha(\theta) = \langle \varphi_\alpha^n, \psi_\theta \rangle_G = \langle \varphi_\alpha^*, \psi_\theta \rangle_G, \tag{3.23}$$

which implies in the case of consistency of the pseudo-best estimates $\tilde{\alpha}_n$ that

$$\alpha(\theta) = 0 \qquad \text{for } \theta \in \Theta^*. \tag{3.24}$$

This fact together with the fact that the estimates in (3.21) are consistent leads to the following.

Theorem 5. *The condition of* (3.24) *is necessary and sufficient for the pseudo-best estimates $\tilde{\alpha}_n$ of a linear functional $\alpha(\theta)$ of $\theta \in \Theta$ to be consistent.*

We note that if the subspace Θ^* of all functions $\theta = \theta(t)$, $t \in T^*$, with spectral representation given by (3.22), contains the trivial element $\theta(t) \equiv 0$ only, the condition of (3.24) will be satisfied for each linear functional $\alpha(\theta)$, $\theta \in \Theta$, and by Theorem 5 all pseudo-best estimates for all linear functionals $\alpha(\theta)$ of $\theta \in \Theta$ will be consistent. Conversely, the consistency of all pseudo-best estimates implies that (see (3.24))

$$\alpha(\theta) = \langle \varphi_\alpha^n, \psi_\theta \rangle_G = 0$$

for all linear functionals (for all $\varphi_\alpha^n \in L_{T_n}(G)$, $n = 1, 2, \ldots$), and therefore $\psi_\theta = 0$, that is, the subspace Θ^* is trivial.

Let us find spectral conditions necessary and sufficient for

$$\Theta^* = 0, \tag{3.25}$$

that is, such that no function $\theta(t) \in \Theta$ satisfies (3.22).

We assume that the pseudo-spectral measure $G(d\lambda)$ is absolutely continuous and that the pseudo-spectral density $g(\lambda) = G(d\lambda)/d\lambda$ is bounded.

Under these conditions, if the function $\theta(t)$, $t \in T^*$, is representable as (3.22), it will coincide, for $t \in T$, with the square-summable function $\theta(t)$, $-\infty < t < \infty$, being the Fourier transform of a square-integrable function $\tilde{\theta}(\lambda) = \psi_\theta(\lambda)g(\lambda)$ satisfying

$$\int \frac{|\tilde{\theta}(\lambda)|^2}{g(\lambda)}\, d\lambda < \infty \tag{3.26}$$

(compare with (4.6), Chapter III). On the other hand, if there exists a function $\theta(t)$, $-\infty < t < \infty$, of the type described above, it will permit, for $t \in T^*$, a representation given by (3.22) in which $\psi_\theta(\lambda) \in L_{T^*}(G)$ is the projection of the function $\tilde{\theta}(\lambda)/g(\lambda)$ in $L_{(-\infty,\infty)}(G)$ onto the space $L_{T^*}(G)$.

Therefore, in the case of bounded pseudo-spectral density $g(\lambda)$ we have arrived at the following.

Theorem 6. *A necessary and sufficient condition for pseudo-best estimates to be consistent is that each function $\theta(t)$, $t \in T^*$, from a closed parametric space Θ have the property that either $\theta(t)$, $t \in T^*$, is not a square-summable function,* or that the Fourier transform $\tilde{\theta}(\lambda)$ satisfies*

$$\int \frac{|\tilde{\theta}(\lambda)|^2}{g(\lambda)}\, d\lambda = \infty \tag{3.27}$$

for any extension to the square-summable function $\theta(t)$, $-\infty < t < \infty$.

This theorem implies the following necessary and sufficient condition for classical least-squares estimates to be consistent (corresponding to the pseudo-spectral density $g(\lambda) = 1$):

$$\begin{aligned} &\sum_{t \in T^*} \theta(t)^2 = \infty && \text{for discrete } t; \\ &\int_{T^*} \theta(t)^2\, dt = \infty && \text{for continuous } t. \end{aligned} \tag{3.28}$$

The subspace Θ^* of functions of the type given by (3.22) is, in general, nontrivial.

Theorem 7. *For a parametric set Θ^* the pseudo-best estimates $\tilde{\alpha}_n$ converge, as $n \to \infty$, to the pseudo-best estimate $\tilde{\alpha}$ (of the same functional $\alpha(\theta)$, $\theta \in \Theta^*$), constructed from the observations of $\xi(t)$, $t \in T^*$.*

Proof. (3.23) shows that

$$\alpha(\theta) = \langle \varphi_\alpha^*, \psi_\theta \rangle_G, \qquad \theta \in \Theta^*, \tag{3.29}$$

which fact together with the limit relation

$$\tilde{\alpha}_n = \int \varphi_\alpha^n \Phi(d\lambda) \to \int \varphi_\alpha^* \Phi(d\lambda)$$

* That is, $\sum_{-\infty}^{\infty} \theta(t)^2 = \infty$ for discrete t and $\int_{-\infty}^{\infty} \theta(t)^2\, dt = \infty$ for continuous t.

proves the theorem since, by virtue of the general formula (3.2), the pseudo-best estimate for a functional of the type (3.29) is given by

$$\alpha^* = \int \varphi_\alpha^*(\lambda)\Phi(d\lambda). \qquad \square$$

VII.4 Estimation of Regression Coefficients

VII.4.1 General Remarks

We discuss the case where the parametric space Θ is finite dimensional and where the problem of estimating linear functionals $\alpha(\theta)$ of $\theta \in \Theta$ of the type (3.1) considered reduces to the problem of estimating coefficients $\alpha_1, \ldots, \alpha_N$ in the expansion of functions $\theta(t)$ from Θ with respect to some basis $\theta_1(t), \ldots, \theta_N(t)$:

$$\theta(t) = \alpha_1\theta_1(t) + \ldots + \alpha_N\theta_N(t), \qquad t \in T^*. \tag{4.1}$$

Let us distinguish a subspace Θ^* of all functions $\theta = \theta(t)$, $t \in T^*$, from Θ, permitting the spectral representation given by (3.22). Let us also choose in a subspace $\Theta^* \subset \Theta$ some basis $\theta_1, \ldots, \theta_{N^*}$, having completed this basis to the basis $\theta_1, \ldots, \theta_N$ in the whole space Θ.

Let $T_1 \subseteq T_2 \subseteq \ldots$ be a sequence of sets ($\bigcup_n T_n = T^*$) on each of which all functions $\theta_k(t)$, $k = 1, \ldots, N$, permit the spectral representation given by (3.10):

$$\theta_k(t) = \int e^{-i\lambda t}\psi_k^n(\lambda)G(d\lambda), \qquad t \in T_n. \tag{4.2}$$

Let $\tilde{\alpha}_1^n, \ldots, \tilde{\alpha}_N^n$ be the pseudo-best estimates for the coefficients $\alpha_1, \ldots, \alpha_N$:

$$\tilde{\alpha}_k^n = \int \varphi_k^n(\lambda)\Phi(d\lambda), \qquad k = 1, \ldots, N. \tag{4.3}$$

Here $\varphi_1^n, \ldots, \varphi_N^n$ are the conjugates of $\psi_1^n, \ldots, \psi_N^n$ (see (1.40)), namely

$$\varphi_k^n(\lambda) = \sum_{j=1}^{N} s_{kj}^n\psi_j^n(\lambda), \tag{4.4}$$

where the matrix $\{s_{kj}^n\}$ is the inverse of the matrix with the elements

$$\langle\psi_k^n, \psi_j^n\rangle_G = \int \psi_k^n(\lambda)\psi_j^n(\lambda)G(d\lambda), \qquad k, j = 1, \ldots, N,$$

and

$$s_{kj}^n = \langle\varphi_k^n, \varphi_j^n\rangle_G, \qquad k, j = 1, \ldots, N.$$

It follows from (4.2) and (4.4) that

$$\int e^{-i\lambda t}\varphi_k^n(\lambda)G(d\lambda) = \sum_{j=1}^{N} s_{kj}^n \int e^{-i\lambda t}\psi_j^n(\lambda)G(d\lambda)$$
$$= \sum_{j=1}^{N} s_{kj}^n\theta_j(t), \qquad t \in T_n.$$

Taking the limit as $n \to \infty$ we have

$$\sum_{j=1}^{N} s_{kj}^{*}\theta_j(t) = \int e^{-j\lambda t}\varphi_k^{*}(\lambda)G(d\lambda), \qquad t \in T^{*},$$

where

$$\varphi_k^{*}(\lambda) = \lim_{n\to\infty} \varphi_k^{n}(\lambda) \in L_{T^{*}}(G)$$

and

$$s_{kj}^{*} = \lim_{n\to\infty} s_{kj}^{n} = \lim_{n\to\infty} \langle \varphi_k^{n}, \varphi_j^{n}\rangle_G$$

(see the limit relation (3.15)). It is easily seen that linear combinations $\sum_{j=1}^{N} s_{kj}^{*}\theta_j(t)$ of basis elements $\theta_1, \ldots, \theta_N$ belong to the subspace Θ^{*} with basis elements $\theta_1, \ldots, \theta_{N^{*}}$, which fact is possible only if

$$\sum_{j=N^{*}+1}^{N} s_{kj}^{*}\theta_j(t) = 0, \qquad k = 1, \ldots, N.$$

Since $\theta_{N^{*}+1}, \ldots, \theta_N$ are linearly independent, the equalities thus obtained are feasible only if

$$s_{kk}^{*} = \|\varphi_k^{*}(\lambda)\|_G^2 = 0, \qquad k = N^{*}+1, \ldots, N.$$

These equalities imply that

$$\lim_{n\to\infty} \tilde{\alpha}_k^{n} = \alpha_k, \qquad k = N^{*}+1, \ldots, N, \tag{4.5}$$

that is, the method for obtaining the pseudo-best estimates provides exact estimation of the coefficients α_k, $k = N^{*}+1, \ldots, N$. For the remaining coefficients $\alpha_1, \ldots, \alpha_{N^{*}}$ the pseuso-best estimates $\tilde{\alpha}_1, \ldots, \tilde{\alpha}_{N^{*}}$ from the observations $\xi(t)$, $t \in T^{*}$, are to be given by the formulas indicated above, namely

$$\tilde{\alpha}_k = \int \varphi_k(\lambda)\Phi(d\lambda), \qquad k = 1, \ldots, N^{*},$$

where

$$\varphi_k(\lambda) = \sum_{j=1}^{N^{*}} s_{kj}\psi_j(\lambda),$$

the functions $\psi_k(\lambda)$ are found from the equations

$$\theta_k(t) = \int e^{-i\lambda t}\psi_k(\lambda)G(d\lambda), \qquad t \in T^{*}, k = 1, \ldots, N^{*}, \tag{4.6}$$

and

$$\{s_{kj}\} = \{\langle \varphi_k, \varphi_j\rangle_G\} = \{\langle \psi_k, \psi_j\rangle_G\}^{-1}. \tag{4.7}$$

Assume all estimates $\tilde{\alpha}_1^{n}, \ldots, \tilde{\alpha}_N^{n}$ to be consistent and, as can easily be deduced from the results obtained in Section VII.3, a necessary and sufficient condition for $\tilde{\alpha}_1^{n}, \ldots, \tilde{\alpha}_N^{n}$ to be such is that for any real $c_1, \ldots, c_N$

$$\lim_{n\to\infty} \left\| \sum_{k=1}^{N} c_k\psi_k^{n}(\lambda) \right\|_G = \infty.$$

Let us suppose that the (complex) measure $H_{kj}^n(d\lambda)$ defined by

$$H_{kj}^n(d\lambda) = \frac{\psi_k^n(\lambda)\overline{\psi_j^n(\lambda)}}{\|\psi_k^n\|_G\|\psi_j^n\|_G} G(d\lambda) \tag{4.8}$$

converges weakly to measure $H_{kj}(d\lambda)$ as $n \to \infty$, that is, for any bounded and continuous function $\varphi(\lambda)$

$$\lim_{n\to\infty} \int \varphi(\lambda) H_{kj}^n \, d\lambda = \int \varphi(\lambda) H_{kj}(d\lambda).$$

We also assume that we have weak convergence $H_{kj}^n \Rightarrow H_{kj}$ for all k, $j = 1, \ldots, N$, that is, we have weak convergence of matrix measures $H^n \Rightarrow H$:

$$H^n = \{H_{kj}^n\}; \qquad H = \{H_{kj}\}.$$

We call a function $\varphi(\lambda)$ *H-admissible* if under the condition of weak convergence $H^n \Rightarrow H$ we have the relation

$$\lim_{n\to\infty} \int \varphi(\lambda) H^n(d\lambda) = \int \varphi(\lambda) H(d\lambda). \tag{4.9}$$

Admissible functions form a linear class closed with respect to uniform convergence and containing bounded continuous functions as well as piecewise continuous functions with a finite number of discontinuities at points λ of measure zero $H(\lambda) = 0$ (and, of course, the linear uniformly closed hull of such functions).

If we introduce the diagonal matrix

$$D_n = \begin{bmatrix} \|\psi_1^n\|_G & & 0 \\ & \ddots & \\ 0 & & \|\psi_N^n\|_G \end{bmatrix}, \tag{4.10}$$

from (4.3) and (4.4) we can obtain for the correlation matrix $\{\sigma_{kj}^n\}$ of the estimates $\tilde{\alpha}_1^n, \ldots, \tilde{\alpha}_N^n$

$$\begin{aligned} \{\sigma_{kj}^n\} &= \left\{\int \varphi_k^n(\lambda)\overline{\varphi_j^n(\lambda)} F(d\lambda)\right\} \\ &= \{s_{kj}^n\} \left\{\int \psi_k^n(\lambda)\overline{\psi_j^n(\lambda)} F(d\lambda)\right\} \{s_{kj}^n\} \\ &\quad - \{s_{kj}^n\} D_n \left\{\int h(\lambda) \frac{\psi_k^n(\lambda)\overline{\psi_j^n(\lambda)}}{\|\psi_k^n\|_G \|\psi_j^n\|_G} G(d\lambda)\right\} D_n \{s_{kj}^n\} \\ &= \{s_{kj}^n\} D_n \left\{\int h(\lambda) H_{kj}^n(d\lambda)\right\} D_n \{s_{kj}^n\}, \end{aligned}$$

where

$$h(\lambda) = \frac{F(d\lambda)}{G(d\lambda)},$$

and where $\{s_{kj}^n\}$ and D_n are the matrices given by (4.7) and (4.10).

We assume that the density $h(\lambda) = F(d\lambda)/G(d\lambda)$ is admissible. By the condition for weak convergence, (4.9), we have

$$\lim_{n\to\infty} \int h(\lambda)H_{kj}^{(n)}(d\lambda) = \int h(\lambda)H_{kj}(d\lambda), \qquad k, j = 1, \ldots, N,$$

and, furthermore,

$$D_n\{s_{kj}^n\}D_n = \left\{\int \frac{\psi_k^n(\lambda)\psi_j^n(\lambda)}{\|\psi_k^n\| \cdot \|\psi_j^n\|} G(d\lambda)\right\}^{-1} = \left\{\int H_{kj}^n(d\lambda)\right\}^{-1}.$$

Therefore, under the additional condition of nonsingularity of the matrix

$$H = \left\{\int H_{kj}(d\lambda)\right\},$$

referred to below as nonsingularity of measure $H(d\lambda)$, we have

$$\lim_{n\to\infty} D_n\{s_{kj}^n\}D_n = H^{-1}.$$

As a result we obtain

$$\lim D_n\{\sigma_{kj}^n\}D_n = H^{-1}\left[\int h(\lambda)H(d\lambda)\right]H^{-1}. \tag{4.11}$$

Thus, under the condition that matrix measures $H^n(d\lambda)$ with components of the type (4.8) are weakly convergent to some measure $H(d\lambda)$, and since the matrix $H = \int H(d\lambda)$ is nonsingular and the density $h(\lambda) = F(d\lambda)/G(d\lambda)$ is admissible, we have for the correlation matrix $\{\sigma_{kj}^n\}$ of pseudo-best estimates

$$\{\sigma_{kj}^n\} \sim D_n^{-1}[H^{-1} \int h(\lambda)H(d\lambda)H^{-1}]D_n^{-1}$$

(where the diagonal matrix D_n is given by (4.10)); in particular,

$$\sigma_{kk}^n = \mathsf{M}_\theta[\tilde{\alpha}_k^n - \alpha_k]^2 = O\{\|\psi_k^n\|_G^{-2}\}, \qquad k = 1, \ldots, N.$$

VII.4.2 Asymptotics of Errors in Pseudo-Best Estimates (Discrete Time)

We shall find below explicit asymptotic formulas of the type (4.11) for the correlation matrix $\{\sigma_{kj}^n\}$ of pseudo-best estimates $\tilde{\alpha}_1^n, \ldots, \tilde{\alpha}_N^n$, constructed from the observations of a random process $\xi(t)$ on an interval $0 \leqslant t \leqslant n$ (discrete time t). As a rule we shall assume that there exists a continuous spectral density $f(\lambda)$ satisfying the condition

$$c_1 \leqslant f(\lambda) \leqslant c_2 \tag{4.12}$$

(for some positive c_1 and c_2), the functions $\theta_1(t), \ldots, \theta_N(t)$ being such that

$$\begin{gathered} \lim_{n\to\infty} \sum_0^n \theta_k(t)^2 = \infty, \\ \max_{0\leqslant t\leqslant n} |\theta_k(t)| = o\left\{\sqrt{\sum_0^n \theta_k(t)^2}\right\}, \qquad k = 1, \ldots, N \end{gathered} \tag{4.13}$$

and that for all $k, j = 1, \ldots, N$ the limit

$$\lim_{n\to\infty} \frac{\sum_0^n \theta_k(t+s)\theta_j(t)}{\sqrt{\sum_0^n \theta_k(t)^2}\sqrt{\sum_0^n \theta_j(t)^2}} = R_{kj}(s) \tag{4.14}$$

exists for each s. It is easily seen that the matrix function

$$R(t) = \{R_{kj}(t)\}, \qquad -\infty < t < \infty,$$

is positive definite in the sense that for any real $t_1, t_2, \ldots$ and for any real $c_1, c_2, \ldots$

$$\sum_{p,q} c_p c_q R_{k_p k_q}(t_p - t_q) \geqslant 0.$$

It is known* that such a matrix function permits a spectral representation

$$R(t) = \int e^{i\lambda t} H^0(d\lambda), \tag{4.15}$$

where $H^0(d\lambda) = \{H^0_{kj}(d\lambda)\}$ is a positive definite matrix measure, that is, the components $H^0_{kj}(d\lambda)$ are complex measures (of bounded variation) and the matrix $H^0(\Delta) = \{H^0_{kj}(\Delta)\}$ is positive definite for any measure set Δ.

$H^0(d\lambda)$ is said to be the spectral measure† of the vector function $\{\theta_1(t), \theta_2(t), \ldots, \theta_N(t)\}$. Let us assume that $H^0(d\lambda)$ is nonsingular, that is, the matrix $H_0 = \int H^0(d\lambda)$ is nonsingular.

By virtue of (4.13), as $n \to \infty$

$$\frac{\sum_0^n \theta_k(t+s)\theta_j(t)}{\sqrt{\sum_0^n \theta_k(t)^2}\sqrt{\sum_0^n \theta_j(t)^2}} \sim \frac{\sum_0^{n-s} \theta_k(t+s)\theta_j(t)}{\sqrt{\sum_0^n \theta_k(t)^2}\sqrt{\sum_0^n \theta_j(t)}}$$

$$= \int e^{i\lambda s} \frac{\tilde{\theta}^n_k(\lambda)\overline{\tilde{\theta}^n_j(\lambda)}}{\|\tilde{\theta}^n_k\|_{G^0}\|\tilde{\theta}^n_j\|_{G^0}} G^0(d\lambda)$$

$$= \int e^{i\lambda s} H^{0,n}_{kj}(d\lambda),$$

where

$$\tilde{\theta}^n_k(\lambda) = \sum_0^n e^{i\lambda t}\theta_k(t), \qquad k = 1, \ldots, N,$$

$$G^0(d\lambda) = \frac{1}{2\pi}\, d\lambda,$$

* See, for instance, [24], p. 30.

† This concept is successfully used by Grenander and Rosenblatt in investigating asymptotic effectiveness of least-squares estimates—see, for instance, [12]; also, compare with Theorem 10 below.

and

$$H_{kj}^{0;\,n}(d\lambda) = \frac{\tilde{\theta}_k^n(\lambda)\overline{\tilde{\theta}_j^n(\lambda)}}{\|\tilde{\theta}_k^n\|_{G^0}\|\tilde{\theta}_j^n\|_{G^0}}\, G^0(d\lambda), \qquad k, j = 1, \ldots, N.$$

Therefore for each t

$$R_{kj}(t) = \int e^{i\lambda t} H_{kj}^0(d\lambda) = \lim_{n\to\infty} \int e^{i\lambda t} H_{kj}^{0;\,n}(d\lambda),$$

which is equivalent to the fact that the sequence of the measures $H_{kj}^{0;\,n}(d\lambda)$ converges weakly to the measure $H_{kj}^0(d\lambda)$, as $n \to \infty$:

$$H_{kj}^{0;\,n} \Rightarrow H_{kj}^0, \qquad k, j = 1, \ldots, N.$$

Therefore, the spectral measure $H^0(d\lambda)$ of functions $\{\theta_1(t), \ldots, \theta_N(t)\}$ defined by (4.15) coincides with the matrix measure given in (4.11) corresponding to the pseudo-spectral measure $G^0(d\lambda) = (1/2\pi)\, d\lambda$.

As a result we have obtained an asymptotic formula of the type given by (4.11) for the correlation matrix $\{\sigma_{kj}^n\}$ of least-squares estimates—pseudo-best estimates corresponding to the pseudo-spectral measure $G^0(d\lambda) = (1/2\pi)\, d\lambda$, namely

$$\lim_{n\to\infty} D_n^0\{\sigma_{kj}^n\}D_n^0 = 2\pi R(0)^{-1} \int f(\lambda) H^0(d\lambda) R(0)^{-1} \tag{4.16}$$

for an H^0-admissible* spectral density, where $R(0) = \int H^0(d\lambda)$ and D_n^0 is a diagonal matrix of the form

$$D_n^0 = \begin{bmatrix} \sqrt{\sum_0^n \theta_1(t)^2} & & 0 \\ & \ddots & \\ 0 & & \sqrt{\sum_0^n \theta_N(t)^2} \end{bmatrix}. \tag{4.17}$$

Next we take pseudo-best estimates $\tilde{\alpha}_1^n, \ldots, \tilde{\alpha}_N^n$ corresponding to the pseudo-spectral measure $G(d\lambda)$ with density

$$g(\lambda) = \frac{1}{2\pi|Q(e^{i\lambda})|^2}, \tag{4.18}$$

where $Q(z) = \sum_{k=0}^m q_k z^k$ is a polynomial with real coefficients having no zeros in a unit circle $|z| \leqslant 1$.

Let us find the solution $\psi(\lambda) \in L_{[0,\,n]}(G)$ of the equation

$$\theta(t) = \int e^{-i\lambda t}\psi(\lambda) \frac{d\lambda}{2\pi|Q(e^{i\lambda})|^2}, \qquad 0 \leqslant t \leqslant n. \tag{4.19}$$

* See the definition in Subsection VII.4.1.

Consider the function $\theta(t)$ given by (4.19) for all t. For $0 \leqslant t \leqslant n$, $\theta(t)$ coincides with the primary function $\theta(t)$, $0 \leqslant t \leqslant n$. If we introduce the shift operator Δ, where $\Delta\theta(t) = \theta(t+1)$, and the operator $Q(\Delta) = \sum_{k=0}^{m} q_k \Delta^k$, we have the relations

$$Q(\Delta)\theta(t) = \int [Q(\Delta)e^{-i\lambda t}]\psi(\lambda)\frac{d\lambda}{2\pi|Q(e^{i\lambda})|^2} = \frac{1}{2\pi}\int e^{-i\lambda t}\frac{\psi(\lambda)}{Q(e^{i\lambda})}\,d\lambda = 0$$

for $t < 0$, since $Q(z)$ is an exterior function and $\psi(\lambda)/Q(e^{i\lambda})$ belongs to a subspace $L_{(0,\infty)}(G^0)$.

In the case $n \geqslant m$ the equation obtained above,

$$Q(\Delta)\theta(t) = \sum_{k=0}^{m} q_k\theta(t+k) = 0, \qquad t = -1, -2, \ldots,$$

enables us to define successive values of the function $\theta(t)$ in question for $t = -1, -2, \ldots$ as follows:

$$\begin{aligned}
\theta(-1) &= -\frac{1}{q_0}\sum_{k=1}^{m} q_k\theta(k-1);\\
&\ \vdots\\
\theta(-j) &= -\frac{1}{q_0}\sum_{k=1}^{m} q_k\theta(k-j);\\
&\ \vdots
\end{aligned}$$

Similarly, we can find values of $\theta(t)$ for $t = n+1, n+2, \ldots$ from the equation

$$Q(\Delta^{-1})\theta(t) = \sum_{k=0}^{m} q_k\theta(t-k) = 0, \qquad t = n+1, n+2, \ldots,$$

namely

$$\begin{aligned}
\theta(n+1) &= -\frac{1}{q_0}\sum_{k=1}^{m} q_k\theta(n+1-k),\\
&\ \vdots\\
\theta(n+j) &= -\frac{1}{q_0}\sum_{k=1}^{m} q_k\theta(n+j-k),\\
&\ \vdots
\end{aligned}$$

Further,

$$\begin{aligned}
Q(\Delta^{-1})Q(\Delta)\theta(t) &= \int [Q(\Delta^{-1})Q(\Delta)e^{-i\lambda t}]\psi(\lambda)\frac{d\lambda}{2\pi|Q(e^{i\lambda})|^2}\\
&= \frac{1}{2\pi}\int e^{-i\lambda t}\psi(\lambda)\,d\lambda, \qquad -\infty < t < \infty,
\end{aligned}$$

which implies that a solution $\psi(\lambda)$ of (4.19) is given by

$$\psi(\lambda) = \sum_0^n e^{i\lambda t} z(t), \tag{4.20}$$

where

$$z(t) = Q(\Delta^{-1})Q(\Delta)\theta(t), \qquad -\infty < t < \infty.$$

(We note that the function $z(t) = Q(\Delta^{-1})Q(\Delta)\theta(t)$ can be defined from the function $\theta(t)$ initially given for $m \leqslant t \leqslant n - m$.)

Next it is easily seen that

$$\begin{aligned}\sum_{t=0}^n e^{i\lambda t} Q(\Delta^{-1})Q(\Delta)\theta(t) &= \sum_{t=0}^n e^{i\lambda t} \sum_{k,\, j=0}^m q_k q_j \theta(t - k + j) \\ &= \sum_{k,\, j=0}^m q_k q_j e^{i\lambda(k-j)} \sum_{t=0}^n e^{i\lambda(t-k+j)}\theta(t - k + j) \\ &= |Q(e^{i\lambda})|^2 \sum_0^n e^{i\lambda s}\theta(s) + r(\lambda),\end{aligned}$$

where in the trigonometric sum

$$r(\lambda) = \sum_{-m}^m a_k e^{i\lambda k} + \sum_{n-m}^{n+m} b_k e^{i\lambda k}$$

the coefficients

$$a_k = \sum_{j=0}^{m-1} c'_{kj}\theta(j),$$

$$b_k = \sum_{j=0}^{m-1} c''_{kj}\theta(n - j)$$

are linear combinations of values $\theta(j)$, $j = 0, \ldots, m - 1$, and $\theta(k)$, $k = n - 1, \ldots, n - m + 1$. It is also seen that a solution $\psi(\lambda)$ of (4.19) is as follows:

$$\psi(\lambda) = |Q(e^{i\lambda})|^2 \tilde{\theta}(\lambda) + r(\lambda), \tag{4.21}$$

where

$$\tilde{\theta}(\lambda) = \sum_0^n e^{i\lambda t}\theta(t).$$

Let us apply the results obtained to functions $\theta_1(t), \ldots, \theta_N(t)$. It follows from (4.21) that

$$\psi_k^n(\lambda) = |Q(e^{i\lambda})|^2 \tilde{\theta}_k^n(\lambda) + r_k^n(\lambda),$$

where, under the condition of (4.13),

$$\|r_k^n\|_G = o\{\|\tilde{\theta}_k^n\|_{G^0}\}, \qquad k = 1, \ldots, N$$

(we recall that $G^0(d\lambda) = d\lambda/2\pi$ and $G(d\lambda) = 1/2\pi|Q(e^{i\lambda})|^2$). Hence, under the condition of existence of a (matrix) spectral measure $H^0(d\lambda)$ in the functions

$\{\theta_1(t), \ldots, \theta_N(t)\}$ for any bounded and continuous function $\varphi(\lambda)$ we have

$$\begin{aligned}\lim_{n\to\infty} &\int \varphi(\lambda) \frac{\psi_k^n(\lambda)\overline{\psi_j^n(\lambda)}}{\|\tilde{\theta}_k^n\|_{G^0}\|\tilde{\theta}_j^n\|_{G^0}} G(d\lambda)\\ &= \lim_{n\to\infty} \int \varphi(\lambda)|Q(e^{i\lambda})|^2 \frac{\tilde{\theta}_k^n(\lambda)}{\|\tilde{\theta}_k^n\|_{G^0}} \frac{\overline{\tilde{\theta}_j^n(\lambda)}}{\|\tilde{\theta}_j^n\|_{G^0}} G^0(d\lambda)\\ &= \int \varphi(\lambda)|Q(e^{i\lambda})|^2 H_{kj}^0(d\lambda); \end{aligned} \tag{4.22}$$

in particular,

$$\lim_{n\to\infty} \frac{\langle \psi_k^n, \psi_j^n\rangle_G}{\|\tilde{\theta}_k^n\|_{G^0}\|\tilde{\theta}_j^n\|_{G^0}} = \int |Q(e^{i\lambda})|^2 H_{kj}^0(d\lambda).$$

It is seen that the matrix measure $H(d\lambda) = \{H_{kj}(d\lambda)\}$ in the asymptotic formula (4.11) has the components

$$H_{kj}(d\lambda) = \left[\int |Q(e^{i\lambda})|^2 H_{kk}^0(d\lambda)\right]^{-1/2} |Q(e^{i\lambda})|^2 H_{kj}^0(d\lambda) \times \left[\int |Q(e^{i\lambda})|^2 H_{jj}^0(d\lambda)\right]^{-1/2}, \tag{4.23}$$

and that the diagonal matrix D_n is such that (see (4.10) and (4.17))

$$\lim_{n\to\infty} D_n^0 D_n^{-1} = \begin{bmatrix} \sqrt{\int |Q(e^{i\lambda})|^2 H_{11}^0(d\lambda)} & & 0 \\ & \ddots & \\ 0 & & \sqrt{\int |Q(e^{i\lambda})|^2 H_{NN}^0(d\lambda)} \end{bmatrix}^{-1}$$

We have arrived at the following:

Theorem 8. *Let the functions $\theta_1(t), \ldots, \theta_N(t)$ have nonsingular spectral measure $H^0(d\lambda)$ and satisfy (4.13). Let the spectral density $f(\lambda)$ be H^0-admissible. Then for any pseudo-spectral density $g(\lambda)$ of the form*

$$g(\lambda) = \frac{1}{2\pi|Q(e^{i\lambda})|^2}, \tag{4.24}$$

*for the correlation matrix $\{\sigma_{kj}^n\}$ of pseudo-best estimates $\tilde{\alpha}_1^n, \ldots, \tilde{\alpha}_N^n$ we have the following asymptotic relation:**

$$\begin{aligned}\lim_{n\to\infty} &D_n^0\{\sigma_{kj}^n\}D_n^0\\ &= 2\pi\left[\int \frac{1}{g(\lambda)} H^0(d\lambda)\right]^{-1} \left[\int h(\lambda) \frac{1}{g(\lambda)^2} H^0(d\lambda)\right]\left[\int \frac{1}{g(\lambda)} H^0(d\lambda)\right]^{-1}, \end{aligned} \tag{4.25}$$

* It should be noted that (4.25) is satisfied for any continuous positive pseudo-spectral density of $g(\lambda)$. See Yu. A. Rozanov and M. V. Kozlov, "On asymptotically efficient estimation of regression coefficients," *Doklady Akademii Nauk SSSR* **188** (1969) 1075–1078 (English translation). From the practical point of view it suffices to restrict oneself to pseudo-spectral densities of the type (4.18).

where

$$h(\lambda) = \frac{f(\lambda)}{g(\lambda)}$$

and where D_n^0 is the diagonal matrix given by (4.17).

Proof. We consider arbitrary linear unbiased estimates using the same notation,

$$\tilde{\alpha}_k^n = \int \varphi_k^n(\lambda)\Phi(d\lambda), \qquad k = 1, \ldots, N,$$

as that used for the pseudo-best estimates considered above. Let us show that under the condition given by (4.12) for the correlation matrix

$$\{\sigma_{kj}^n\} = \left\{\int \varphi_k^n(\lambda)\overline{\varphi_j^n(\lambda)} f(\lambda)\, d\lambda\right\}$$

we have the lower asymptotic bound

$$\lim_{n\to\infty} D_n^0\{\sigma_{kj}^n\}D_n^0 \geqslant 2\pi\left\{\int \frac{1}{f(\lambda)} H^0(d\lambda)\right\}^{-1}. \tag{4.26}$$

To prove Theorem 8 we shall take advantage of a general inequality: if $x_1, \ldots, x_m$ are arbitrary elements in a Hilbert space and if $y_1, \ldots, y_n$ are linearly independent, then

$$\begin{aligned} \{\langle x_i, x_j\rangle\} &\geqslant \{\langle \tilde{x}_i, \tilde{x}_j\rangle\} \\ &= \{\langle \tilde{x}_i, y_k\rangle\}\{\langle y_k, y_l\rangle\}^{-1}\{\langle \tilde{x}_j, y_l\rangle\}^* \\ &= \{\langle x_i, y_k\rangle\}\{\langle y_k, y_l\rangle\}^{-1}\{\langle x_j, y_l\rangle\}^*, \end{aligned} \tag{4.27}$$

where $\tilde{x}_1, \ldots, \tilde{x}_n$ denote the projections of the initial elements $x_1, \ldots, x_n$ onto the subspace generated by $y_1, \ldots, y_n$ and are found by the equations

$$\tilde{x}_i = \sum_{k,l} c_{kl}\langle x_i, x_l\rangle y_k, \qquad i = 1, \ldots, m,$$

$$\{c_{kl}\} = \{\langle y_k, y_l\rangle\}^{-1}.$$

Applying (4.27) to $\varphi_1^n(\lambda), \ldots, \varphi_N^n(\lambda)$ and $\psi_1^n(\lambda)/h(\lambda), \ldots, \psi_N^n(\lambda)/h(\lambda)$ from the Hilbert space $L_{(-\infty,\infty)}(F)$, we have

$$\begin{aligned} \{\sigma_{kj}^n\} &= \{\langle \varphi_i^n, \varphi_j^n\rangle_F\} \\ &\geqslant \{\langle \varphi_i^n, \psi_k^n\rangle_G\}\left\{\int \frac{2\pi g(\lambda)}{h(\lambda)}\psi_k^n(\lambda)\overline{\psi_l^n(\lambda)}G^0(d\lambda)\right\}^{-1}\{\langle \varphi_j^n, \psi_l^n\rangle_G\}^*, \end{aligned}$$

since $\psi_1^n/h, \ldots, \psi_N^n/h$ are linearly independent. Under the condition given by (4.12), $\psi_1^n, \ldots, \varphi_N^n$ belong to the Hilbert space $L_{(-\infty,\infty)}(G)$ and

$$\left\langle \varphi_i^n(\lambda), \frac{\psi_k^n(\lambda)}{h(\lambda)}\right\rangle_F = \langle \varphi_i^n, \psi_k^n\rangle_G, \qquad i, k = 1, \ldots, N,$$

$$\left\langle \frac{\psi_k^n(\lambda)}{h(\lambda)}, \frac{\psi_l^n(\lambda)}{h(\lambda)}\right\rangle_F = \int \frac{g(\lambda)}{f(\lambda)}\psi_k^n(\lambda)\overline{\psi_j^{(n)}(\lambda)}G(d\lambda), \qquad k, j = 1, \ldots, N.$$

(Let us recall that $h(\lambda) = f(\lambda)/g(\lambda)$ and $g(\lambda) = 1/2\pi|Q(e^{i\lambda})|^2$.)

Furthermore, it follows from (4.2) and from the fact that the estimates $\tilde{\alpha}_1^n, \ldots, \tilde{\alpha}_N^n$ are unbiased that

$$\begin{aligned} \alpha_i &\equiv M_\theta \tilde{\alpha}_i^n \\ &\equiv \int \varphi_i^n(\lambda) M_\theta \Phi(d\lambda) \\ &\equiv \int \varphi_i^n(\lambda) \sum_{k=1}^N \alpha_k \psi_k^n(\lambda) G(d\lambda), \qquad i = 1, \ldots, N. \end{aligned}$$

From this it follows, in turn, due to arbitrariness of the coefficients $\alpha_1, \ldots, \alpha_n$, that

$$\{\langle \varphi_i^n, \psi_k^n \rangle_G\} = E$$

(E is the unit matrix). Therefore

$$\{\sigma_{kj}^n\} \geqslant \left\{ \int \frac{g(\lambda)}{f(\lambda)} \psi_k^n(\lambda) \overline{\psi_j^n(\lambda)} G(d\lambda) \right\}^{-1}.$$

Since under the condition $1/f(\lambda) \geqslant c > 0$ given by (4.12) the inequality $\int (1/f(\lambda)) H^0(d\lambda) \geqslant c \int H^0(d\lambda)$ holds true, the matrix $\int (1/f(\lambda)) H^0(d\lambda)$ as well as the matrix $H^0 = \int H^0(d\lambda)$ are nonsingular. Hence (4.22) implies that

$$\begin{aligned} \lim_{n\to\infty} D_n^0 \{\sigma_{kj}^n\} D_n^0 &\geqslant \lim_{n\to\infty} \left\{ \int \frac{g(\lambda)}{f(\lambda)} \frac{\psi_k^n(\lambda)}{\|\tilde{\theta}_k^n\|_{G^0}} \frac{\overline{\psi_j^n(\lambda)}}{\|\tilde{\theta}_j^n\|_{G^0}} G(d\lambda) \right\}^{-1} \\ &= 2\pi \left\{ \int \frac{1}{f(\lambda)} H^0(d\lambda) \right\}^{-1}. \end{aligned}$$

Therefore, we have the inequality given by (4.26) under the condition of (4.12) for any unbiased linear estimates $\tilde{\alpha}_1^n, \ldots, \tilde{\alpha}_N^n$ with the correlation matrix $\{\sigma_{kj}^n\}$.

We shall show that the inequality in (4.26) becomes an equality for best estimates (with correlation matrix $\{s_{kj}^n\}$).

Obviously it suffices to show that

$$\lim_{n\to\infty} D_n^0 \{s_{kj}^n\} D_n^0 \leqslant 2\pi \left[\int \frac{1}{f(\lambda)} H^0(d\lambda) \right]^{-1}. \tag{4.28}$$

Let us prove (4.28). Under the condition of (4.12) the continuous positive function $1/f(\lambda)$ can be approximized arbitrarily closely by trigonometric (positive) polynomials representable as $2\pi |Q(e^{i\lambda})|^2$. Hence we can choose in (4.25) the pseudo-spectral density

$$g(\lambda) = \frac{1}{2\pi |Q(e^{i\lambda})|^2},$$

differing arbitrarily slightly from the spectral density $f(\lambda)$. Since $\{s_{kj}^n\} \leqslant \{v_{kj}^n\}$ always, it follows that

$$\lim_{n\to\infty} D_n^0 \{s_{kj}^n\} D_n^0 \leqslant 2\pi \left[\int \frac{1}{g(\lambda)} H^0(d\lambda) \right]^{-1} \left[\int \frac{f(\lambda)}{g(\lambda)^2} H^0(d\lambda) \right] \left[\int \frac{1}{g(\lambda)} H^0(d\lambda) \right]^{-1},$$

where $g(\lambda)$ can be taken arbitrarily close to $f(\lambda)$. Therefore, (4.28) holds as well. □

(4.26) and (4.28) imply the following:

Theorem 9. *Under the condition of* (4.12), *if there exists a nonsingular spectral measure* $H^0(d\lambda)$ *for the correlation matrix* $\{s_{kj}^n\}$ *of best unbiased estimates, then we have the following asymptotic formula*:

$$\lim_{n\to\infty} D_n^0\{s_{kj}^n\}D_n^0 = 2\pi\left[\int \frac{1}{f(\lambda)} H^0(d\lambda)\right]^{-1}. \tag{4.29}$$

The results obtained above enable us to deduce conditions for pseudo-best estimates to be asymptotically effective.

We recall that effectiveness of the unbiased estimate $\tilde{\alpha}$ of the parameter α yielding the best unbiased estimate $\hat{\alpha}$ with minimal dispersion can be expressed as $\mathsf{M}(\hat{\alpha}-\alpha)^2/\mathsf{M}(\tilde{\alpha}-\alpha)^2$. Hence the estimates $\tilde{\alpha}_1^n, \ldots, \tilde{\alpha}_N^n$ are referred to as *asymptotically effective* if

$$\lim_{n\to\infty} \frac{\mathsf{M}(\hat{\alpha}_k^n-\alpha_k)^2}{\mathsf{M}(\tilde{\alpha}_k^n-\alpha_k)^2} = 1, \qquad k = 1, \ldots, N.$$

Let us show that in our case the fact that the pseudo-best estimates $\tilde{\alpha}_1^n, \ldots, \tilde{\alpha}_N^n$ are asymptotically effective is equivalent to the condition

$$\lim_{n\to\infty} D_n^0\{s_{kj}^n\}D_n^0 = \lim D_n^0\{\sigma_{kj}\}D_n^0. \tag{4.30}$$

In fact, in the case considered there exist nonsingular limit matrices

$$\{a_{kj}\} = \lim D_n^0\{s_{kj}^n\}D_n^0, \qquad \{b_{kj}\} = \lim D_n^0\{\sigma_{kj}^n\}D_n^0,$$

that are positive definite, so that all the diagonal elements $a_{kk} = \lim d_k^n s_{kk}^n d_k^n$ and $b_{kk} = \lim d_k^n \sigma_{kk}^n d_k^n$ are nonzero (here the d_k^n denote the diagonal elements of the normalizing matrix D_n^0). (4.30) implies that $\{a_{kj}\} = \{b_{kj}\}$ and, in particular, that

$$\lim (d_k^n)^2 s_{kk}^n = \lim (d_k^n)^2 \sigma_{kk}^n \neq 0.$$

Therefore, under the condition given by (4.30) the condition of asymptotic effectiveness is satisfied: $\lim s_{kk}^n/\sigma_{kk}^n = 1$, $k = 1, \ldots, N$. This, in turn, implies that $\sigma_{kk}^n - s_{kk}^n = o\{\sigma_{kk}^n\}$, where $\sigma_{kk}^n \sim b_{kk}(d_k^n)^{-2}$. Since the matrix $\{\sigma_{kj}^n - s_{kj}^n\}$ is nonnegative, for any $k, j = 1, \ldots, N$

$$|\sigma_{kj}^n - s_{kj}^n| \leqslant (\sigma_{kk}^n - s_{kk}^n)^{1/2}(\sigma_{jj}^n - s_{jj}^n)^{1/2} = o\{\sqrt{\sigma_{kk}^n\sigma_{jj}^n}\}.$$

Therefore, when the estimates in question are asymptotically effective we have

$$\begin{aligned} b_{kj} &= \lim d_k^n[s_{kj}^n + (\sigma_{kj}^n - s_{kj}^n)]\, d_j^n \\ &= a_{kj} + \lim d_k^n(\sigma_{kj}^n - s_{kj}^n)\, d_j^n = a_{kj}, \qquad k, j = 1, \ldots, N, \end{aligned}$$

that is, we have (4.30).

Let us make use of the asymptotic formulas given by (4.25) and (4.29). Consider a normalized matrix measure (of the type (4.23))

$$H^*(d\lambda) = \left[\int \frac{1}{g(\lambda)} H^0(d\lambda)\right]^{-1/2} \frac{1}{g(\lambda)} H^0(d\lambda) \left[\frac{1}{g(\lambda)} H^0(d\lambda)\right]^{-1/2},$$

where $g(\lambda)$ is a pseudo-spectral density. Then if (4.25) and (4.29) are satisfied, a necessary and sufficient condition for the estimates considered to be asymptotically effective is that the equality

$$\int h(\lambda) H^*(d\lambda) = \left[\int \frac{1}{h(\lambda)} H^*(d\lambda)\right]^{-1} \tag{4.31}$$

(equivalent to (4.30)) be satisfied with $h(\lambda) = f(\lambda)/g(\lambda)$.

We shall show that (4.31) can be satisfied if and only if the scalar matrix $(f(\lambda)/g(\lambda))E$ is almost everywhere constant with respect to $H^*(d\lambda)$, that is, $(f(\lambda)/g(\lambda))E = C(C = \{c_{kj}\})$, more precisely,

$$\left(\frac{f(\lambda)}{g(\lambda)} E - C\right) H^*(d\lambda) = 0, \tag{4.32}$$

where E is the unit matrix, and $C = \{c_{kj}\}$ is the constant term of the matrix $(f(\lambda)/g(\lambda))E$.

To prove the above fact we introduce the densities

$$h_{kj}(\lambda) = \frac{H_{kj}(d\lambda)}{m(d\lambda)}, \qquad k, j = 1, \ldots, N$$

(with respect to some measure $m(d\lambda)$) and express the positive matrix $\{h_{kj}(\lambda)\}$ as the product

$$\{h_{kj}(\lambda)\} = \{\varphi_{ik}(\lambda)\}\{\varphi_{kj}(\lambda)\},$$

where $\{\varphi_{ik}\}$ is the positive square root of the matrix $\{h_{ij}\}$. The vector functions $\varphi_i(\lambda) = \{\varphi_{i1}(\lambda), \ldots, \varphi_{iN}(\lambda)\}$ can be regarded as elements of a Hilbert space in which the scalar product of $\varphi = \{\varphi_1, \ldots, \varphi_N\}$ and $\psi = \{\psi_1, \ldots, \psi_N\}$ is given by the formula

$$\langle \varphi, \psi \rangle = \int \sum_{k=1}^{N} \varphi_k(\lambda) \overline{\psi_k(\lambda)} m(d\lambda).$$

If we consider the elements

$$\sqrt{h}\varphi_i = \{\sqrt{h(\lambda)}\varphi_{i1}, \ldots, \sqrt{h(\lambda)}\varphi_{iN}\}, \qquad i = 1, \ldots, N,$$

and

$$\frac{1}{\sqrt{h}}\varphi_i = \left\{\frac{1}{\sqrt{h(\lambda)}}\varphi_{i1}, \ldots, \frac{1}{\sqrt{h(\lambda)}}\varphi_{iN}(\lambda)\right\}, \qquad i = 1, \ldots, N,$$

we shall find that

$$\{\langle\sqrt{h}\varphi_i, \sqrt{h}\varphi_j\rangle\} = \int h(\lambda)H^*(d\lambda),$$

$$\left\{\left\langle\frac{1}{\sqrt{h}}\varphi_i, \frac{1}{\sqrt{h}}\varphi_j\right\rangle\right\} = \int\frac{1}{h(\lambda)}H^*(d\lambda),$$

and

$$\left\{\left\langle\sqrt{h}\varphi_i, \frac{1}{\sqrt{h}}\varphi_j\right\rangle\right\} = \int H^*(d\lambda) = E.$$

Therefore, (4.31) can be rewritten as

$$\{\langle\sqrt{h}\varphi_i, \sqrt{h}\varphi_j\rangle\}$$
$$= \left\{\left\langle\sqrt{h}\varphi_i, \frac{1}{\sqrt{h}}\varphi_k\right\rangle\right\}\left\{\left\langle\frac{1}{\sqrt{h}}\varphi_k, \frac{1}{\sqrt{h}}\varphi_l\right\rangle\right\}^{-1} \times \left\{\left\langle\sqrt{h}\varphi_j, \frac{1}{\sqrt{h}}\varphi_l\right\rangle\right\}.$$

Generally speaking, instead of the equality sign there must be an inequality sign in the above expression (see (4.27)). We have equality only if the vectors $\sqrt{h(\lambda)}\varphi_1(\lambda), \ldots, \sqrt{h(\lambda)}\varphi_N(\lambda)$ are linear combinations of the vectors $(1/\sqrt{h(\lambda)})\varphi_1(\lambda), \ldots, (1/\sqrt{h(\lambda)})\varphi_N(\lambda)$, that is, if

$$\sqrt{h(\lambda)}\varphi_{ik}(\lambda) = \sum_{j=1}^{N} c_{ij}\frac{1}{\sqrt{h(\lambda)}}\varphi_{jk}(\lambda), \qquad k = 1, \ldots, N, i = 1, \ldots, N,$$

for some constant matrix $C = \{c_{ij}\}$ almost everywhere with respect to $m(d\lambda)$.

From this it follows that

$$h(\lambda)\sum_{k=1}^{N}\varphi_{ik}(\lambda)\varphi_{kj}(\lambda) = h(\lambda)h_{ij}(\lambda) = \sum_{k=1}^{N}\left[\sum_{l=1}^{N}c_{il}\varphi_{lk}(\lambda)\right]\varphi_{kj}(\lambda) = \sum_{k}c_{ik}h_{kj}(\lambda);$$

therefore, almost everywhere with respect to $m(d\lambda)$

$$\left(\frac{f(\lambda)}{g(\lambda)}E - \{c_{ik}\}\right)\{h_{kj}(\lambda)\} = 0,$$

or

$$\left(\frac{f(\lambda)}{g(\lambda)}E - C\right)H^*(d\lambda) = 0,$$

which was to be proved. We can deduce easily from (4.32) the following:

Theorem 10. *Under the condition of* (4.12) *a necessary and sufficient condition for the pseudo-best estimates of the coefficients* $\alpha_1, \ldots, \alpha_N$ *in the expansion given by* (4.1) (*corresponding to the pseudo-spectral density* $g(\lambda)$ *is that the matrix function* $(f(\lambda)/g(\lambda))E$ *be constant almost everywhere with respect to* $H^0(d\lambda)$.

VII.4.3 Asymptotics of Errors in Pseudo-Best Estimates (Continuous Time)

We shall consider the case of continuous t involving pseudo-best estimates $\tilde{\alpha}_1^n, \ldots, \tilde{\alpha}_N^n$ from the observations of a process $\xi(t)$ on an interval $0 \leqslant t \leqslant \tau$, where $\tau = \tau_n \to \infty$. The question arises as to what extent asymptotic formulas for the correlation matrix $\{\sigma_{kj}^n\}$ of the pseudo-best estimates we have obtained in the case of discrete time t can be generalized to the case of continuous time.

The spectral measure $H^0(d\lambda) = \{H_{kj}^0(d\lambda)\}$ of the vector function

$$\{\theta_1(t), \ldots, \theta_N(t)\}$$

is to be defined in the same way as in the case of discrete time, that is the components $H_{kj}^0(d\lambda)$ are given by the relations

$$R_{kj}(s) = \lim_{n\to\infty} \frac{\int_0^\tau \theta_k(t+s)\theta_j(t)\,dt}{\sqrt{\int_0^\tau \theta_k(t)^2\,dt}\sqrt{\int_0^\tau \theta_j(t)^2\,dt}}$$

$$= \int e^{i\lambda s} H_{kj}^0(d\lambda), \qquad k, j = 1, \ldots, N \tag{4.33}$$

(assuming, in addition, that the function $\{R_{kj}(s)\}$ is continuous at 0). Let

$$\lim_{n\to\infty} \int_0^\tau \theta_k(t)^2\,dt = \infty$$

and let

$$\max_{0\leqslant t\leqslant\tau} |\theta_k(t)| = o\left\{\sqrt{\int_0^\tau \theta_k(t)^2\,dt}\right\}, \qquad k = 1, \ldots, N.$$

Suppose that the spectral density $f(\lambda)$ is H^0-admissible,* $H^0(d\lambda)$ being nonsingular. As in the case of discrete time, for the correlation matrix $\{\sigma_{kj}^n\}$ of least-squares estimates we have (4.16)

$$\lim_{n\to\infty} D_n^0\{\sigma_{kj}^n\}D_n^0 = 2\pi R(0)^{-1} \int f(\lambda)H^0(d\lambda)R(0)^{-1},$$

where $R(0) = H^0 = \int H^0(d\lambda)$ and D_n^0 is a diagonal matrix of the form

$$D_n^0 = \begin{pmatrix} \sqrt{\int_0^\tau \theta_1(t)^2\,dt} & & 0 \\ & \ddots & \\ 0 & & \sqrt{\int_0^\tau \theta_N(t)^2\,dt} \end{pmatrix}. \tag{4.34}$$

* See the definition in Subsection VII.4.1.

Considering pseudo-best estimates corresponding to a pseudo-spectral density of the form

$$g(\lambda) = \frac{1}{2\pi|Q(i\lambda)|^2}, \tag{4.35}$$

where $Q(z) = \sum_{k=0}^{m} q_k z^k$ is a polynomial with real coefficients, immediately raises the question as to whether the passage to the limit in a relation of the type (4.22) is justified since the function $Q(i\lambda)$ is unbounded for $-\infty < \lambda < \infty$. The same question arises as to a formula of the type (4.29) since the spectral density $f(\lambda)$ is integrable and the function $1/f(\lambda)$ is not bounded for $-\infty < \lambda < \infty$.

Let us assume that the functions $\theta_1(t), \ldots, \theta_N(t)$ have $2m$ continuous derivatives and that for functions $\{y_1(t), \ldots, y_N(t)\}$, where

$$y_k(t) = Q\left(\frac{d}{dt}\right)\theta(t), \qquad k = 1, \ldots, N, \tag{4.36}$$

there exists a nonsingular spectral measure denoted $H(d\lambda) = \{H_{kj}(d\lambda)\}$. Furthermore, we assume that

$$\max_{0 \leqslant t \leqslant \tau} \left| \frac{d^k}{dt^k} \theta_j(t) \right| = o\left\{ \sqrt{\int_0^\tau y_j(t)^2 \, dt} \right\}, \qquad k = 0, \ldots, 2m; j = 1, \ldots, N. \tag{4.37}$$

The fact that the spectral measure $H(d\lambda)$ exists under the condition of (4.37) is equivalent to the fact that the sequence of matrix measures $M^n(d\lambda) = \{M_{kj}^n(d\lambda)\}$ with the components

$$M_{kj}^n(d\lambda) = \frac{\tilde{y}_k^n(\lambda)\ \overline{\tilde{y}_j^n(\lambda)}}{\|\tilde{y}_k^n\|_{G^0}\ \|\tilde{y}_j^n\|_{G^0}} G^0(d\lambda), \quad k, j = 1, \ldots, N, \tag{4.38}$$

where

$$\tilde{y}_k^n(\lambda) = \int_0^\tau e^{i\lambda t} y_k(t)\, dt, \qquad k = 1, \ldots, N$$

($\tau = \tau_n \to \infty$) and $G^0(d\lambda) = (1/2\pi)\, d\lambda$, is weakly convergent (to $H(d\lambda)$), that is, for any bounded and continuous function $\varphi(\lambda)$

$$\lim_{n \to \infty} \int \varphi(\lambda) M^n(d\lambda) = \int \varphi(\lambda) H(d\lambda).$$

Next let us take (4.3) describing the pseudo-best estimates constructed from a pseudo-spectral density $g(\lambda)$ of the form (4.35). We consider the pertinent integral equation

$$\theta(t) = \int e^{-i\lambda t} \psi(\lambda) \frac{d\lambda}{2\pi|Q(i\lambda)|^2}, \qquad 0 \leqslant t \leqslant \tau \tag{4.39}$$

(of the type (4.6)), with respect to the function $\psi(\lambda) \in L_{[0,\tau]}(G)$.

Let the function $\theta(t)$, $-\infty < t < \infty$, be defined for all t as the left-hand side in (4.39). Since the polynomial $Q(z)$ in (4.35) is an exterior function (for the lower half-plane), the relation $\psi(\lambda)/Q(-i\lambda)$ will be a function from $L_{[0,\infty)}(G^0)$ for each function $\psi(\lambda)$ from a subspace $L_{[0,\infty)}(G)$ (see Section II.2). Hence, for the solution $\psi(\lambda) \in L_{[0,\tau]}(F)$ and the appropriate function $\theta(t)$ we shall have

$$Q\left(-\frac{d}{dt}\right)\theta(t) = \int e^{-i\lambda t}\frac{\psi(\lambda)}{Q(-i\lambda)}\frac{d\lambda}{2\pi} = 0 \qquad \text{for } t < 0.$$

For analogous reasons we have

$$Q\left(\frac{d}{dt}\right)\theta(t) = \int e^{-i\lambda t}\frac{\psi(\lambda)}{Q(i\lambda)}\frac{d\lambda}{2\pi} = 0 \qquad \text{for } t > \tau$$

(since $\psi(\lambda)/Q(i\lambda) \in L_{(-\infty,\tau]}(G^0)$).

It is seen that the solution $\psi(\lambda) \in L_{[0,\tau]}(G)$ of the integral equation (4.39) is the generalized Fourier transform* of the generalized function

$$z(t) = Q\left(\frac{d}{dt}\right)Q\left(-\frac{d}{dt}\right)\theta(t) = Q\left(-\frac{d}{dt}\right)Q\left(\frac{d}{dt}\right)\theta(t).$$

For $0 < t < \tau$ the left hand side of (4.39) coincides with the function $\theta(t)$ initially given having $2m$ usual derivatives. For $-\infty < t < \tau$ the usual function $Q(-d/dt)\theta(t)$, as we have shown above, vanishes for $-\infty < t < 0$. Therefore the generalized function $z(t) = Q(d/dt)\,[Q(-d/dt)\theta(t)]$ for $-\infty < t < \tau$ is obtained by applying the differential operator $Q(d/dt)$ to a usual piecewise smooth function $Q(-d/dt)\theta(t)$, equal to zero for $-\infty < t < 0$ and having $2m$ derivatives for $0 < t < \tau$.

A similar situation is observed for $z(t)$ if $0 < t < \infty$ so that we arrive finally at the following formula for the generalized function $z(t)$:*

$$z(t) = z_0(t) + \sum_{k=0}^{m-1} a_k'\delta^{(k)}(t) + \sum_{k=0}^{m-1} b_k'\delta^{(k)}(t-\tau),$$

where the coefficients a_k' and b_k' are linear combinations of derivatives, not higher than $m-1$, of the functions $Q(-d/dt)\theta t$ and $Q(d/dt)\theta(t)$ at boundary points $t = 0$ and $t = \tau$. The generalized functions $\delta^{(k)}(t)$ and $\delta^{(k)}(t-\tau)$ are the kth derivatives of δ functions at the same points $t = 0$ and $t = \tau$, and $z_0(t)$ is a usual function of the form

$$z_0(t) = \begin{cases} Q\left(-\frac{d}{dt}\right)Q\left(\frac{d}{dt}\right)\theta(t) & \text{for } 0 < t < \tau, \\ 0 & \text{for the remaining } t. \end{cases}$$

Next, if we introduce the function

$$y(t) = Q\left(\frac{d}{dt}\right)\theta(t), \qquad 0 \leqslant t \leqslant \tau,$$

* See, for instance, [8].

* See, for instance, [24], p. 191.

the solution $\psi(\lambda)$ of (4.39) can be expressed as

$$\psi(\lambda) = \int_0^\tau e^{i\lambda t}\left[Q\left(-\frac{d}{dt}\right)y(t)\right]dt + \sum_{k=0}^{m-1} a'_k(i\lambda)^k + \sum_{k=0}^{m-1} b'_k e^{i\lambda\tau}(i\lambda)^k.$$

Integrating by parts we obtain

$$\int_0^\tau e^{i\lambda t}\left[Q\left(-\frac{d}{dt}\right)y(t)\right]dt = Q(i\lambda)\int_0^\tau e^{i\lambda t}y(t)\,dt + \sum_{k=0}^{m-1} a''_k(i\lambda)^k + \sum_{k=0}^{m-1} b''_k e^{i\lambda\tau}(i\lambda)^k,$$

where the coefficients a''_k and b''_k are linear combinations of derivatives, not higher than $m-1$, of a function $y(t)$ at the boundary points $t = 0$ and $t = \tau$. Using the equality thus obtained we can express the solution $\psi(\lambda)$ of (4.39) as (compare with (4.21))

$$\psi(\lambda) = Q(i\lambda)\tilde{y}(\lambda) + r(\lambda), \tag{4.40}$$

where

$$\tilde{y}(\lambda) = \int_0^\tau e^{i\lambda t}y(t)\,dt.$$

The coefficients a_k and b_k in the expression

$$r(\lambda) = \sum_0^{m-1} a_k(i\lambda)^k + \sum_0^{m-1} b_k e^{i\lambda\tau}(i\lambda)^k$$

are linear combinations of derivatives, not higher than $2m-1$, of the primary function $\theta(t)$, $0 \leqslant t \leqslant \tau$ at the boundary points $t = 0$ and $t = \tau$:

$$a_k = \sum_{j=0}^{2m-1} c'_{kj}\theta^{(j)}(0), \qquad b_k = \sum_{j=0}^{2m-1} c''_{kj}\theta^{(j)}(\tau).$$

We apply the results obtained to the functions $\psi_1^n(\lambda), \ldots, \psi_N^n(\lambda)$, with the aid of which the pseudo-best estimates $\tilde{\alpha}_1^n, \ldots, \tilde{\alpha}_N^n$ can be constructed (see (4.6)). By virtue of (4.40)

$$\psi_k^n(\lambda) = Q(i\lambda)\tilde{y}_k^n(\lambda) + r_k^n(\lambda), \qquad k = 1, \ldots, N, \tag{4.41}$$

where under the condition of (4.37) the residual $r_k^n(\lambda)$ satisfies the relation

$$|r_k^n(\lambda)| = o\{|\lambda|^{m-1}\|\tilde{y}_k^n\|_{G^0}\}.$$

Hence for the pseudo-spectral measure $G(d\lambda) = d\lambda/2\pi|Q(i\lambda)|^2$ (where $Q(z)$ is a polynomial of degree $2m$)

$$\|r_k^n\|_G = o\{\|\tilde{y}_k^n\|_{G^0}\}.$$

We can easily deduce from (4.41) that

$$\lim_{n\to\infty}\frac{\|\psi_k^n\|_G}{\|\tilde{y}_k^n\|_{G^0}} = 1.$$

Under the condition for weak convergence of the measures $M^n_{kj}(d\lambda)$ introduced above (see (4.38)), we have

$$\lim_{n\to\infty} \int \varphi(\lambda) \frac{\psi^n_k(\lambda)\overline{\psi^n_j(\lambda)}}{\|\psi^n_k\|_G \|\psi^n_j\|_G} G(d\lambda)$$

$$= \lim_{n\to\infty} \int \varphi(\lambda) \frac{y^n_k(\lambda)}{\|y^n_k\|_{G^0}} \frac{\overline{y^n_j(\lambda)}}{\|y^n_j\|_G} G^0(d\lambda)$$

$$= \int \varphi(\lambda) H_{kj}(d\lambda), \qquad k, j = 1, \ldots, N \tag{4.42}$$

for any bounded and continuous function $\varphi(\lambda)$. It is seen that the matrix measures $H^n(d\lambda) = \{H^n_{kj}(d\lambda)\}$ with the components

$$H^n_{kj}(d\lambda) = \frac{\psi^n_k(\lambda)\overline{\psi^n_j(\lambda)}}{\|\psi^n_k\|_G \|\psi^n_j\|_G} G(d\lambda), \qquad k, j = 1, \ldots, N,$$

are weakly convergent to the spectral measure $H(d\lambda) = \{H_{kj}(d\lambda)\}$ of a vector function $\{y_1(t), \ldots, y_N(t)\}$.

Therefore, for the correlation matrix $\{\sigma^n_{kj}\}$ of pseudo-best estimates the asymptotic formula given by (4.11), in which $H(d\lambda)$ is the spectral measure of the functions given by (4.36), is satisfied, that is,

$$\lim_{n\to\infty} D_n\{\sigma^n_{kj}\}D_n = H^{-1}\left[\int \frac{f(\lambda)}{g(\lambda)} H(d\lambda)\right] H^{-1}, \tag{4.43}$$

where $H = \int H(d\lambda)$, D_n is a diagonal matrix of the form

$$D_n = \begin{pmatrix} \|\tilde{y}^n_1\|_{G^0} & & 0 \\ & \ddots & \\ 0 & & \|\tilde{y}^n_N\|_{G^0} \end{pmatrix},$$

and the spectral density $f(\lambda)$ and the pseudo-spectral density $g(\lambda) = 1/2\pi|Q(i\lambda)|^2$ are such that the relation $f(\lambda)/g(\lambda)$ is H-admissible.*

Let us find out how the spectral measure $H(d\lambda)$ is related to the spectral measure $H^0(d\lambda)$.

Integrating by parts we can arrive at the equality analogous to (4.41),

$$\tilde{y}^n_k(\lambda) = \int_0^\tau e^{i\lambda t}\left[Q\left(\frac{d}{dt}\right)\theta(t)\right] dt = Q(-i\lambda)\tilde{\theta}^n_k(\lambda) + r^n_k(\lambda),$$

where (with the previous notation)

$$\tilde{\theta}^n_k(\lambda) = \int_0^\tau e^{i\lambda t}\theta_k(t)$$

and the residual $r^n_k(\lambda)$ satisfies (under the condition of (4.37)) the relation

$$\|r^n_k\|_G = o\{\|\tilde{y}^n_k\|_{G^0}\}.$$

* Compare with (3.5) to be satisfied if the relation $f(\lambda)/g(\lambda)$ is bounded.

Hence (under the condition for weak convergence of the sequence $M^n(d\lambda)$)

$$\lim_{n\to\infty}\int \frac{1}{|Q(i\lambda)|^2}\frac{|\tilde{y}_k^n(\lambda)|^2}{\|\tilde{y}_k^n\|_{G^0}} G^0(d\lambda) = \lim_{n\to\infty}\int \frac{1}{|Q(i\lambda)|^2} M_{kk}^n(d\lambda) = \int \frac{1}{|Q(i\lambda)|^2} H_{kk}(d\lambda)$$

and simultaneously

$$\lim_{n\to\infty}\int \frac{1}{|Q(i\lambda)|^2}\frac{|\tilde{y}_k^n(\lambda)|^2}{\|\tilde{y}_k^n\|_{G^0}} G^0(d\lambda) = \lim_{n\to\infty}\frac{\|\tilde{\theta}_k^n\|_{G^0}}{\|\tilde{y}_k^n\|_{G^0}}.$$

It is seen that there exists a finite nonzero limit

$$d_k = \lim_{n\to\infty}\frac{\|\tilde{\theta}_k^n\|_{G^0}}{\|\tilde{y}_k^n\|_{G^0}}, \qquad k = 1, \ldots, N.$$

Furthermore, as can easily be seen,

$$\begin{aligned}\int \frac{\varphi(\lambda)}{|Q(i\lambda)|^2} H_{kj}(d\lambda) &= \lim_{n\to\infty}\int \frac{\varphi(\lambda)}{|Q(i\lambda)|^2}\frac{\tilde{y}_k^n(\lambda)\overline{\tilde{y}_j^n(\lambda)}}{\|\tilde{y}_k^n\|_{G^0}\|\tilde{y}_j^n\|_{G^0}} G^0(d\lambda)\\ &= d_k\left[\lim_{n\to\infty}\int \varphi(\lambda)\frac{\tilde{\theta}_k^n(\lambda)\tilde{\theta}_j^n(\lambda)}{\|\tilde{\theta}_k^n\|_{G^0}\|\tilde{\theta}_j^n\|_{G^0}} G^0(d\lambda)\right] d_j\\ &= d_k\left[\int \varphi(\lambda)H^0(d\lambda)\right] d_j\end{aligned}$$

for any bounded and continuous function $\varphi(\lambda)$.

This fact implies that under the condition of existence of a spectral measure $H(d\lambda)$ in functions $y_k(t) = Q(d/dt)\theta_k(t)$, $k = 1, \ldots, N$, there exists a spectral measure $H^0(d\lambda)$ in the primary functions $\theta(t)$, $k = 1, \ldots, N$,

$$H^0(d\lambda) = d^{-1}\left[\frac{1}{|Q(i\lambda)|^2} H(d\lambda)\right] d^{-1}, \tag{4.44}$$

where the diagonal matrix

$$d = \begin{pmatrix} d_1 & & 0 \\ & \ddots & \\ 0 & & d_N \end{pmatrix}$$

has the elements

$$d_k = \lim_{n\to\infty}\frac{\|\tilde{\theta}_k^n\|_{G^0}}{\|\tilde{y}_k^n\|_{G^0}}, \qquad k = 1, \ldots, N,$$

on the diagonal.

Next, if using (4.44) we go from $H(d\lambda)$ over to $H^0(d\lambda)$ in (4.42), then, having replaced the normalizing matrix D_n by the matrix

$$D_n^0 = \begin{pmatrix} \|\tilde{\theta}_1^n\|_{G^0} & & 0 \\ & \ddots & \\ 0 & & \|\tilde{\theta}_N^n\|_{G^0} \end{pmatrix} = dD_n,$$

we shall arrive, in particular, at the following.

Theorem 11. *Let the pseudo-spectral density $g(\lambda) = 1/2\pi|Q_0(i\lambda)|^2$ be such that the function $h(\lambda) = f(\lambda)/g(\lambda)$ is bounded, and let the spectral density $f(\lambda)$ be continuous,* assuming also that there exists a nonsingular spectral measure in the functions $\{Q(d/dt)\theta_1(t), \ldots, Q(d/dt)\theta_N(t)\}$ and that the condition of* (4.37) *is satisfied. Then the correlation matrix $\{\sigma_{kj}^n\}$ of pseudo-best estimates will satisfy the following asymptotic formula† (compare with* (4.25)):

$$\lim_{n\to\infty} D_n^0\{\sigma_{kj}^n\}D_n^0$$
$$= \frac{1}{2\pi}\left[\int \frac{1}{g(\lambda)} H^0(d\lambda)\right]^{-1} \left[\int h(\lambda)\frac{1}{g(\lambda)} H^0(d\lambda)\right] \left[\int \frac{1}{g(\lambda)} H^0(d\lambda)\right]^{-1}, \quad (4.45)$$

where $H^0(d\lambda)$ is the spectral measure of the functions $\theta_1(t), \ldots, \theta_n(t)$ and $h(\lambda) = f(\lambda)/g(\lambda)$.

We note that for a spectral density of the form $f(\lambda) = 1/2\pi|Q(i\lambda)|$ this formula yields the asymptotic expression for the correlation matrix $\{s_{kj}^n\}$ of best unbiased estimates:

$$\lim_{n\to\infty} D_n^0\{s_{kj}^n\}D_n^0 = 2\pi\left[\int \frac{1}{f(\lambda)} H^0(d\lambda)\right]^{-1} \quad (4.46)$$

(compare with (4.29)).

We shall find out next when the inequality (4.26) established above in the case of discrete time t holds true under the condition $f(\lambda) \asymp 1$. An analog of such a condition is given by

$$f(\lambda) \asymp (1 + \lambda^2)^{-m}, \quad (4.47)$$

where m is some positive integer.

Let us consider the pseudo-best estimates constructed from the pseudo-spectral density $g(\lambda) = 1/2\pi|Q(i\lambda)|^2$ where $Q(z) = (1 + iz)^m$. Let us also assume that there exists a spectral measure $H(d\lambda)$ in the functions $y_k(t) = Q(d/dt)\theta_k(t)$, $k = 1, \ldots, N$. Then, similarly to what was done before for deducing (4.26), we shall have that for the correlation matrix $\{\sigma_{kj}^n\}$ of any linear unbiased estimates $\tilde{\alpha}_1^n, \ldots, \tilde{\alpha}_N^n$ the inequality

$$\{\sigma_{kj}^n\} \geqslant \left\{\int \frac{g(\lambda)}{f(\lambda)} \psi_k^n(\lambda)\overline{\psi_k^n(\lambda)}G(d\lambda)\right\}^{-1}$$

is satisfied (noting that under the condition of (4.47) the function $g(\lambda)/f(\lambda)$ is bounded, more precisely, $g(\lambda)/f(\lambda) \asymp 1$). Using the common relation (4.4?)

* In fact, it suffices to assume that the function $h(\lambda) = f(\lambda)/g(\lambda)$ is H-admissible.

† It should be mentioned that (4.45) has been obtained and extended to positive uniformly continuous pseudo-spectral densities $g(\lambda)$ for which the limits $\lim_{\lambda\to\pm\infty} g(\lambda)\lambda^{2m}$ exist by A. S. Kholevo: "On estimation of regression coefficients," *Teoria Verojatn. i Primen.* **14** (1969), 79–104 (English translation); "On the asymptotic effectiveness of pseudobest estimates," *Teoria Verojatn. i Primen.* **16** (1971), 516–527 (English translation).

we can deduce that in the case of the continuous spectral density $f(\lambda)$

$$\lim_{n\to\infty} D_n\{\sigma_{kj}^n\}D_n \geqslant \lim_{n\to\infty}\left\{\int \frac{g(\lambda)}{f(\lambda)} \frac{\psi_k^n(\lambda)\overline{\psi_j^n(\lambda)}}{\|\tilde{y}_k^n\|_{G^0}\|\tilde{y}_j^n\|_{G^0}} G(d\lambda)\right\}^{-1} = \left[\int \frac{g(\lambda)}{f(\lambda)} H(d\lambda)\right]^{-1}.$$

Going from $H(d\lambda)$ over to $H^0(d\lambda)$ (see (4.44)) and noting that $g(\lambda) = 1/2\pi|Q(i\lambda)|^2$, we obtain an equality of the type (4.26), that is,

$$\lim_{n\to\infty} D_n^0\{\sigma_{kj}^n\}D_n^0 \geqslant 2\pi\left[\int \frac{1}{f(\lambda)} H^0(d\lambda)\right]^{-1}.$$

We assume now that the spectral density $f(\lambda)$ satisfies instead of (4.47) the condition

$$f(\lambda) \geqslant k(1+\lambda^2)^{-m} \tag{4.48}$$

(where k is some positive constant). We introduce an auxiliary spectral density

$$f_r(\lambda) = \min\{r(1+\lambda^2)^{-m}, f(\lambda)\}.$$

It is seen that $f_r(\lambda)$ satisfies (4.47). If $\{\sigma_{kj}^n\}_r$ is the correlation matrix of the linear unbiased estimates $\tilde{\alpha}_1^n, \ldots, \tilde{\alpha}_N^n$ (calculated with respect to the spectral density $f_r(\lambda)$), then according to the inequality we have proved we shall obtain

$$\lim_{n\to\infty} D_n^0\{\sigma_{kj}^n\}_r D_n^0 \geqslant 2\pi\left[\int \frac{1}{f_r(\lambda)} H^0(d\lambda)\right]^{-1}.$$

Since $f(\lambda) \geqslant f_r(\lambda)$, obviously

$$\{\sigma_{kj}^n\} \geqslant \{\sigma_{kj}^n\}_r$$

and

$$\lim_{n\to\infty} D_n^0\{\sigma_{kj}^n\}D_n^0 \geqslant 2\pi\left[\int \frac{1}{f_r(\lambda)} H^0(d\lambda)\right]^{-1}$$

for any $r > 0$. But as $r \to \infty$ the monotone decreasing sequence of the functions $1/f_r(\lambda)$ has $1/f(\lambda)$ as the limit so that

$$\lim_{r\to\infty} \int \frac{1}{f_r(\lambda)} H^0(d\lambda) = \int \frac{1}{f(\lambda)} H^0(d\lambda)$$

and

$$\lim_{n\to\infty} D_n^0\{\sigma_{kj}^n\}D_n^0 \geqslant 2\pi\left[\int \frac{1}{f(\lambda)} H^0(d\lambda)\right]^{-1} \tag{4.49}$$

(where for the bounded spectral density $f(\lambda)$ the matrix $\int(1/f(\lambda))H^0(d\lambda)$ will be nonsingular if the matrix $H^0 = \int H^0(d\lambda)$ is nonsingular). Thus we have arrived at the following.

Theorem 12. *Let the spectral density $f(\lambda)$ be continuous and satisfy* (4.48), *let there exist a nonsingular spectral measure in the functions $\{Q(d/dt)\theta_1(t), \ldots, Q(d/dt)\theta_N(t)\}$ where $Q(d/dt) = (1 + (d/dt))^m$, and let* (4.37) *be satisfied. Then* *for the correlation matrix* $\{\sigma_{kj}^n\}$ *of any linear unbiased estimates the asymptotic inequality given by* (4.49) *is satisfied, where* $H^0(d\lambda)$ *is a nonsingular spectral measure of the primary functions* $\{\theta_1(t), \ldots, \theta_N(t)\}$.

Similar to the case of discrete time (see Theorem 8) under the condition that (4.45) and (4.49) are satisfied, a sufficient condition for the pseudo-best estimates constructed from the pseudo-spectral density $g(\lambda)$ to be asymptotically effective is that the matrix function $(f(\lambda)/g(\lambda))E$ be constant almost everywhere with respect to the spectral matrix measure $H^0(d\lambda)$. This is also necessary under the additional condition of (4.46).

Example. Let $N = 2$ and let

$$\theta_1(t) = P_1(t) \int e^{i\lambda t} m_1(d\lambda), \qquad \theta_2(t) = P_2(t) \int e^{i\lambda t} m_2(d\lambda),$$

where $P_1(t)$ and $P_2(t)$ are polynomials in t of degrees n_1 and n_2, respectively, with coefficients of highest degree equal to 1, and where $m_1(d\lambda)$ and $m_2(d\lambda)$ are finite measures.

We denote by Λ_1 and Λ_2 the set of points λ for which $m_1(\lambda) \neq 0$ and $m_2(\lambda) \neq 0$, respectively. It can easily be computed that the spectral measure $H^0(d\lambda) = \{H_{kj}^0(d\lambda)\}$ of the functions $\{\theta_1(t), \theta_2(t)\}$ is purely discrete and concentrated at particular points $\lambda \in \Lambda_1 \cup \Lambda_2$, more precisely,*

$$H^0(\lambda) = d^{-1} \begin{pmatrix} \dfrac{1}{2n_1 + 1} |m_1(\lambda)|^2 & \dfrac{1}{n_1 + n_2 + 1} m_1(\lambda) m_2(\lambda) \\ \dfrac{1}{n_1 + n_2 + 1} \overline{m_1(\lambda)} m_2(\lambda) & \dfrac{1}{2n_2 + 1} |m_2(\lambda)|^2 \end{pmatrix} d^{-1},$$

where d denotes a diagonal matrix of the form

$$d = \begin{pmatrix} \sqrt{\dfrac{1}{2n_1 + 1} \sum |m_1(\lambda)|^2} & 0 \\ 0 & \sqrt{\dfrac{1}{2n_2 + 1} \sum |m_2(\lambda)|^2} \end{pmatrix}.$$

In the case where the sets Λ_1 and Λ_2 do not intersect the condition that $(f(\lambda)/g(\lambda))E$ is constant almost everywhere with respect to $H^0(d\lambda)$ is equivalent to the fact that the relation $f(\lambda)/g(\lambda)$ is constant on either of the spectral elements Λ_1 and Λ_2. This condition is obviously satisfied if there are single points $\lambda_1 \in \Lambda_1$ and $\lambda_2 \in \Lambda_2$ only. In the case when the spectrum $\Lambda_1 \cup \Lambda_2$ consists of the finite number of points $\lambda_1, \ldots, \lambda_m$ the condition for asymptotic effectiveness indicated above will be satisfied if a pseudo-spectral density $g(\lambda)$ of the type (4.35) is such that $g(\lambda) = f(\lambda)$ for $\lambda = \lambda_1, \ldots, \lambda_m$.

References

1. Akhiezer, N. I. *Lectures on Approximation Theory.* Moscow, 1965; New York, 1956 (first ed.).
2. Akhiezer, N. I., and Glazman, I. M. *The Theory of Linear Operators in Hilbert Space.* Moscow, 1966; New York (without year).
3. Bari, N. K. *Series and Approximation.* Moscow, 1961; Providence, 1962.
4. Bari, N. K. "Biorthogonal systems and bases in a Hilbert space," *Utchenye Zapiski MGU Ser. Matem.* **4** (1961), 69–107.
5. Cramer, H., and Leadbetter, M. *Stationary and Related Stochastic Processes.* New York and London, 1967.
6. Doob, J. L. *Stochastic Processes.* New York, 1953.
7. Gelfand, I. M., and Vilenkin, N. Ya. *Generalized Functions, IV. Applications of Harmonic Analysis: Equipped Hilbert Spaces.* Moscow, 1961; London, 1964.
8. Gelfand, I. M., and Shilov, G. Ye. *Generalized Functions and Their Applications.* Moscow, 1958.
9. Gelfond, A. O. *Calculation of Finite Differences.* Moscow and Leningrad, 1952.
10. Gikhman, I. I., and Skorokhod, A. V. *Introduction to the Theory of Random Process.* Moscow, 1965; Baltimore, 1969.
11. Goluzin, G. M. *The Geometric Theory of Functions of a Complex Variable.* Moscow, 1957; Providence, 1969.
12. Grenander, U., and Rosenblatt, M. *Statistical Analyses of Stationary Time Series.* New York, 1956.
13. Grenander, U., and Szego, G. *Toeplitz Forms and Their Applications.* Los Angeles, 1958.
14. Hille, E. *Functional Analysis and Semi-Groups.* New York, 1948.
15. Hoffman, K. *Banach Spaces of Analytic Functions.* London, 1962.
16. Ibragimov, I. A., and Linnik, Yu. V. *Independent and Stationary Associated Variables.* Moscow, 1965; Groningen, 1971.
17. Krylov, V. I. "On functions regular in a half-plane," *Matem. Sbornik* **4** (1938), 9–30.

18. Lehmann, E. L. *Testing Statistical Hypotheses.* New York, 1959.
19. Levin, B. Ya. *Distribution of Zeros of Entire Functions.* Moscow, 1956; Providence, 1972.
20. Markushevitch, A. I. *The Theory of Analytic Functions.* Moscow, 1950.
21. Pinsker, M. S. *Information and Information Stability of Random Variables and Processes.* Moscow, 1960; San Francisco, 1964.
22. Pontrjagin, L. S. *Ordinary Differential Equations.* Moscow, 1961; Reading, 1962.
23. Privalov, I. I. *Boundary Properties of Analytic Functions.* Moscow and Leningard, 1950.
24. Rozanov, Yu. A. *Stationary Random Processes.* Moscow, 1963; San Francisco, 1967.
25. Rozanov, Yu. A. *Gaussian Infinite-Dimensional Distributions.* Moscow, 1968.
26. Szegö, G. *Orthogonal Polynomials.* Providence, 1939.
27. Timan, A. F. *The Theory of Approximation of Functions of a Real Variable.* Moscow, 1960; London, 1963.
28. Zygmund, A. *Trigonometric Series.* 2nd ed., Cambridge, 1959.

Applications of Mathematics

Vol. 1
Deterministic and Stochastic Optimal Control
By **W.H. Fleming** and **R.W. Rishel**
1975. ix, 222p. 4 illus. cloth

Vol. 2
Methods of Numerical Mathematics
By **G.I. Marchuk**
1975. xii, 316p. 10 illus. cloth

Vol. 3
Applied Functional Analysis
By **A.V. Balakrishnan**
1976. x, 309p. cloth

Vol. 4
Stochastic Processes in Queueing Theory
By **A.A. Borovkov**
1976. xi, 280p. 14 illus. cloth

Vol. 5
Statistics of Random Processes I
General Theory
By **R.S. Liptser** and **A.N. Shiryayev**
1977. x, 394p. cloth

Vol. 6
Statistics of Random Processes II
Applications
By **R.S. Liptser** and **A.N. Shiryayev**
1978. x, 339p. cloth

Vol. 7
Game Theory
Lectures for Economists and Systems Scientists
By **N.N. Vorob'ev**
1977. xi, 178p. 60 illus. cloth

Vol. 8
Optimal Stopping Rules
By **A.N. Shiryayev**
1978. x, 217p. 7 illus. cloth

Vol. 9
Gaussian Random Processes
By **I.A. Ibragimov** and **Y.A. Rosanov**
1978. x, 275p. 2 illus. cloth